Materials Science
&
Engineering

Materials Science
&
Engineering

Giles F. Carter
Professor of Chemistry
Eastern Michigan University

and

Donald E. Paul
President
D.E. Paul Associates

**The Materials
Information Society**

Library of Congress Catalog Card No.: 90-83686
ISBN: 0-87170-399-8
SAN: 204-7586

Editorial and production coordination by
Kathleen Mills Editorial & Production Services

PRINTED IN THE UNITED STATES OF AMERICA

To Dorothy and to Allan, David, Brian, and Terri

G.F.C.

To Christian, Lise, David, and Jean Lorelle

D.E.P.

Preface

Materials science is a discipline that integrates concepts. Materials engineering involves designing based upon function or property requirements, not upon decisions, for instance, to design an object out of metal just because it has been done that way in the past. Materials science and engineering encourage using an open, unprejudiced mind when solving problems.

This book has been written primarily (1) as a text for scientists and engineers who need to learn about materials science on their own, i.e., outside a classroom, (2) as a text for an introductory course in materials science, and (3) as a handy reference book for scientists and engineers.

Most chemists do not take a course in materials science. This is unfortunate, because a knowledge of materials science is useful to most chemists performing research and/or development. In fact, it is difficult to imagine the usage of chemistry where materials science would *not* be useful. Most chemistry curricula, however, do not require materials science. Perhaps the wisdom of incorporating materials science in chemistry curricula will some day be realized. In the meantime, however, most chemists, whether B.S., M.S., or Ph.D., should find it advantageous to self-teach themselves about materials science. Because of the clear explanations and the many worked examples, this book should be helpful to persons who want to learn for themselves.

Happily, most engineering students are required to take a course in materials science. This is fortunate, because it encourages engineers to think broadly about the possibilities of using a variety or combination of materials.

Most students find current texts in materials science at a level that is somewhat too high for them. This discourages honest attempts to learn from the text before attending lectures on a given subject. The present text has been written with the student in mind, rather than for professors, who sometimes find it hard to believe that students cannot learn materials science because the text is too difficult to understand. Most derivations have been omitted; advanced students should look to specialized texts to supply detailed derivations. Many examples of problems are included to help students understand how to work a variety of problems. Students are encouraged to read the worked examples carefully and not to skip over these as being obvious. Also, students should always read the text first before attending a lecture on a given topic.

When an engineer or scientist has read and studied this text, it should be a handy reference for general information and data. Familiarity with the book should make it easy to locate needed information. If detailed information or data are needed, then one should refer to the ASM International series of metals and materials handbooks.

One of the authors of this book is well versed in polymers. The treatment of polymers is somewhat expanded here and presumably more authoritative than that found in most materials science texts.

The authors appreciate the many contributions made by persons associated with ASM International: editors (especially Kathleen Mills and Mary Thomas Haddad), reviewers, draftsmen, printers, and all others.

The authors, whose friendship has lasted since graduate school at the University of California at Berkeley many years ago (and through the writing of this book), wish you, the reader, every success in your study and the subsequent application of principles of materials science and engineering in your career.

Ypsilanti, Michigan
August 17, 1990

Contents

3. Properties of Metals,   37

4. Imperfections in Metals & Their Effects on Properties,   59

5. Phases & Phase Diagrams,   81

Appendix, 337

Index,   339

1
Introduction

1.1 Classification of Materials

Solids, liquids, and gases are the three states of matter. In materials science solids are defined as matter having *crystallinity:* atoms, ions, or molecules spaced at regular, repeating distances and angles from each other in three dimensions. Liquids are a condensed state of matter in which the particles are touching one another, as in solids. However, the distinguishing feature of liquids is that they are *amorphous*—without crystallinity. Gases, of course, comprise matter in which the particles are separated from one another by comparatively large distances.

In real life there are usually gray areas where materials do not always fit simple definitions. For instance, there are liquid crystals in which some regularities occur over long distances, but the regularities are not in three dimensions. Likewise in supercritical states of matter, at high temperatures and pressures, matter is essentially gaseous but in a condensed state—similar to both gases and liquids. Nevertheless, it is convenient to refer to solids, liquids, and gases according to the definitions above.

These definitions have some interesting consequences: glass, for instance, is defined as a liquid. Actually it is a rigid, supercooled liquid. Several common plastics, such as PVC (poly-vinylchloride), are also liquids, or rigid, supercooled liquids.

In materials science one is almost always concerned only with condensed states of matter that are rigid or somewhat rigid. Liquids having moderate or high fluidity are of some interest, but usually materials science is concerned with solids and rigid liquids.

There are many other classifications of materials, including inorganic or organic, metallic or nonmetallic, biological or inanimate, or even (remember the old childhood game?) animal, vegetable, or mineral. However, the classification used by most materials scientists is the following: metals, polymers, or ceramics.

From our experiences from childhood we recognize metals by their characteristic luster, high thermal and electrical conductivities, and often by their ductility. Polymers comprise materials made from large molecules, usually organic by nature. The plastics are the most important commercial polymers, but polymers include such materials as skin, hair, wood, and so on. Ceramics, according to materials scientists, are a much broader class of materials than usually implied by the term ceramics. Indeed, ceramics include not only pottery, dishes, and other fired clay products, but also graphite, diamond, sodium chloride and other salts and minerals, glass, bricks, concrete, and cement. In science and technology certain words have special meanings that are different from common usage, and the term ceramics is a good example.

1.2   What Is Meant by "Materials Science"?

Materials science began to evolve 30 to 40 years ago when materials became increasingly competitive with one another: plastics competed with metals, and composites of ceramics and plastics grew in importance. In each of the three main classes of materials—metals, ceramics, and polymers—new materials were rapidly developed with an entire spectrum of properties. Furthermore, the fundamental theory for one class of materials usually overlapped considerably the theory for the other classes. Materials science is interdisciplinary because of its dependence on several sciences: chemistry, physics, and metallurgy primarily.

Materials science may be defined as the relationship of properties of an object to its chemical composition and structure. If one knows what atoms are present in a given material and how these atoms are structured, then often the properties of the material may be understood either qualitatively or even calculated quantitatively.

Chemical composition is the weight percent of each element present in a material. The dependence of properties on chemical composition is rather obvious. If one has pure iron, certainly the properties would be expected to be far different from pure carbon. However, the structure, or the "architecture," of the material also affects the properties critically: compare the difference in properties of graphite and diamond, which are both pure carbon. Only the structures are different, and later we shall discuss the structures in detail to understand the properties of both these valuable materials.

Structure, which may be defined as the way in which atoms are arranged in materials, is much more complex than chemical composition. Structure exists on several different size scales. For instance there is *nuclear structure*, which involves the composition of the nucleus in terms of neutrons and protons; the structure of these particles results in important differences in nuclear properties of matter. Nuclear structure will not be dealt with in this text, except briefly. Next in terms of scale is *atomic structure*. This is important in determining the chemical properties of materials. Atomic structure primarily involves the electron structure in atoms. This area will be discussed in some detail, but more extensive information may be found in chemistry texts.

Next in scale is *crystal structure*, in which the three-dimensional, repeating array of atoms or molecules is important. On a still larger scale we have microstructure, which is used extensively by metallurgists and ceramists. *Microstructure* is the structure visible using the light or electron microscope. It involves the size of individual crystals, or grains as they are frequently called, the shape of grains, the presence of precipitates, the distribution of precipitates, and so on. Crystal structure and microstructure are of major importance in materials science.

The last and largest on the scale of size of structure is *macrostructure*, which is defined as that structure visible to the unaided eye. Macrostructure has many of the same features as microstructure, but the scale is different. For instance, one can easily see the rocks and gravel in a cross-section of concrete. Sometimes individual crystals may be seen in materials: the spangles on galvanized steel, which are individual zinc crystals, or the grains on an old brass doorknob, where the knob has been polished and etched by extensive use. The grains are large crystals of brass in the doorknob.

1.3   Why Study Materials Science?

Although some students in universities enroll in courses in materials science only because it is required, one must realize that required courses are not decided upon lightly. A great deal of study goes into the decision of what courses are required. When they are a part of a given curriculum, then there are usually sound reasons for the choice.

Materials science is important to persons dealing with the science and technology of matter. Because the subject is interdisciplinary, involving the disciplines of chemistry, physics, metallurgy, etc., it provides the interested student with an excellent perspective. Metals are not studied to the exclusion of polymers and ceramics, or vice versa. Areas of similarity of various materials are studied, as are areas of dissimilarity. For instance, the way metals plastically deform is compared and contrasted with the way plastics deform. The underlying fundamentals of mechanical properties are the same for the three classes of materials.

Composites, which involve more than one class of materials, are of increasing importance. Glass- or graphite-reinforced plastics, metal-glass combinations, ceramic-metal insulators, transistors, and many other composites are best understood from the perspective of materials science.

1.4.   Who Should Study Materials Science?

Materials science is required for most engineering students because of the understanding presented by an interdisciplinary approach to materials in general. Although most curricula in physics and chemistry do not require materials science, they really should. Materials science provides a solid background involving principles of physics and chemistry applied to the theory of materials and to the practice of processing materials. It is hoped and urged that curricula in physics and chemistry require, or at the very least strongly recommend, that physics and chemistry students take at least one course in materials science.

Materials science is valuable to many persons who are involved in a major way with materials. Those who did not take a course in materials science, but who fall into the above categories, should study materials science at their earliest convenience. Therefore, materials science is an important area of continuing education.

Materials science is an important subject for teachers because it summarizes a great deal of knowledge and gives perspective in understanding why materials behave the way they do. A knowledge of materials is useful at every level of teaching, from grade school to graduate school. Because materials are so interesting, and frequently exhibit bizarre behavior, they are excellent subject matter for demonstrations. Frequently no more than a rubber band, a paper clip, a magnet, or a piece of a plastic garbage bag is needed for a demonstration.

Materials engineering is the application of the principles of materials science in the design of objects or in the processing of materials. Materials engineering deals with the materials themselves and their properties.

1.5 Bonding in Materials

All bonding in materials is due to attractions between positive and negative charges. Because protons, situated in atomic nuclei, have positive charges and electrons have negative charges, bonding is caused ultimately by the attractions between electrons and protons in nuclei. The primary, or strong, types of chemical bonds are covalent, ionic, and metallic bonds. However, there is no clear distinction among these types, as we shall see.

There are two properties that distinguish the types of chemical bonds: the electronegativity difference between two atoms in a chemical bond and the mobility of the electrons in the bond. *Electronegativity* is the electron drawing power of an atom in chemical bonds. Atoms having low electronegativities do not have a strong tendency to draw electrons toward themselves in chemical bonds with other elements; examples are the Group IA metals—lithium, sodium, potassium, etc. In fact, these atoms usually form bonds that are ionic, and the Group IA ion is positively charged due to its low electron-drawing power. *Mobility* of electrons from one bond to another may be measured by determining the electrical conductivity of a material. If the electrons have very low mobility, the material is an insulator; i.e., it has very low electrical conductivity. Examples are most ceramics and polymers. Materials having relatively high electrical mobility are, of course, the metals as well as some ceramics (e.g., graphite) and a few polymers.

Electronegativities may be calculated by referring to bonding in various materials. Table 1.1 contains the electronegativities of the elements. Note that the metals in general have low electronegativities, and that the lowest values are found in the elements at the bottom left of the periodic table. The highest values are at the upper right of the table. Note that even the inert gases have high electronegativities! This may be rather surprising because one may equate high electronegativity with strong bonding; obviously, high electronegativities have nothing to do with the strength of bonds in general. There are a few compounds of the heavier inert gases, and electronegativity values may be obtained by calculation from these compounds. However, the electronegativities of the lighter inert gases are estimates.

When the electronegativity of one element is subtracted from the electronegativity of the other element, the positive difference is called the *electronegativity difference*. Note that if an element is bonded to several other atoms, such as carbon in CH_4, the electronegativity difference is found based on only two atoms, namely the two atoms that are bonded to each other.

If one element in a chemical bond has a relatively high electronegativity and the other element has a low electronegativity, then the electronegativity difference is relatively high. One element draws the electrons in the bond strongly toward itself, while the other element does not have a strong drawing power for the atoms. The result is a bond that is *ionic* in character: the valence electron or electrons have been largely transferred from one atom to the other, resulting in the formation of two ions, one positively charged and one with a negative charge. Generally speaking, ionic solids have low conductivities because their electrons are immobile. Accord-

Table 1.1   Electronegativities of the elements

IA	IIA	IIIB	IVB	VB	VIB	VIIB	—VIIIB—			IB	IIB	IIIA	IVA	VA	VIA	VIIA	VIIIA
H																	He
2.1																	2.7
Li	Be											B	C	N	O	F	Ne
1.0	1.5											2.0	2.5	3.0	3.5	4.0	4.4
Na	Mg											Al	Si	P	S	Cl	Ar
0.9	1.2											1.5	1.8	2.1	2.5	3.0	3.5
K	Ca	Sc	Ti	V	Cr	Mn	Fe	Co	Ni	Cu	Zn	Ga	Ge	As	Se	Br	Kr
0.8	1.0	1.3	1.5	1.6	1.6	1.5	1.8	1.8	1.8	1.9	1.6	1.6	1.8	2.0	2.4	2.8	3.0
Rb	Sr	Y	Zr	Nb	Mo	Tc	Ru	Rh	Pd	Ag	Cd	In	Sn	Sb	Te	I	Xe
0.8	1.0	1.2	1.4	1.6	1.8	1.9	2.2	2.2	2.2	1.9	1.7	1.7	1.8	1.9	2.1	2.5	2.6
Cs	Ba	La(a)	Hf	Ta	W	Re	Os	Ir	Pt	Au	Hg	Tl	Pb	Bi	Po	At	Rn
0.7	0.9	1.1	1.3	1.5	1.7	1.9	2.2	2.2	2.2	2.4	1.9	1.8	1.8	1.9	2.0	2.2	2.4
Fr	Ra	Ac	Th	Pa	U	Np											
0.7	0.9	1.1	1.3	1.5	1.7	1.3											

(a) Lanthanides: 1.1 to 1.2

ingly, ionic materials are located in the upper left of the diagram in Fig. 1.1, which is a plot of electronegativity difference versus electron mobility.

If two identical atoms chemically bond with each other, the attraction is from the nucleus of one atom for the electrons of the other and vice versa. This positive-negative attraction holds the two atoms together, and the valence electrons are shared equally between the two atoms. The electronegativity difference between two identical atoms bonded to each other must be zero. Even though the electrons are shared equally *on the average* between the two atoms, the electrons at any given instant may be closer to one atom than to the other. If the electronegativity difference is zero or relatively low, and if the electron mobility is low, then the bonds are said to be *covalent*. Covalent bonds are located in the lower left portion of the diagram in Fig. 1.1.

If the electronegativity difference is small in bonds and if the electron mobility is high, the bonds are said to be *metallic*. Obviously the metals have metallic bonds by definition: their electrons have high mobility (they can move from one bond to another throughout the material with the expenditure of only a small amount of energy). In metals, however, the atoms are held together by the same forces as in covalently bonded atoms: the same valence electrons are attracted by two nuclei at the same time. Metallic bonding has been described as positive ions in a sea of valence electrons. Actually, there are not enough electrons to ''go around,'' and the electrons can easily move from one atom to another to another. Metallic bonding is found in the lower right portion of Fig. 1.1.

One ''pure'' type of chemical bonding is pure covalent, in which the electrons are equally shared between two identical atoms. The electronegativity difference is zero, and often the electron mobility is very low (e.g., solid hydrogen, chlorine, nitrogen, etc.). ''Pure'' metallic bonding may be said to occur in superconducting materials in which there is *no* resistivity. ''Pure'' ionic bonding in solids probably does not exist because the valence electrons are not *always* in the vicinity of the atom with the higher electronegativity.

It is certain that many chemical bonds fall somewhere in between the three extremities of pure covalent, metallic, or ionic. For instance, a compound, such as aluminum chloride, has

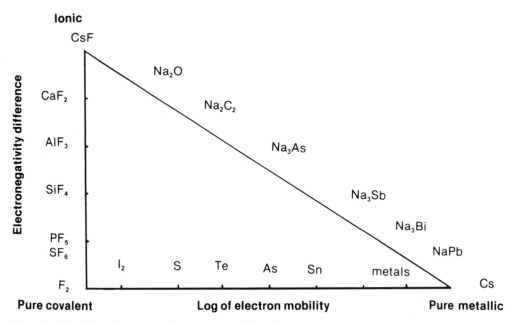

Fig. 1.1   Gradation between extreme types of bonding

characteristics of both covalent and ionic materials, and its electronegativity difference of 1.5 places it about midway between the most ionic compound known, CsF (or perhaps francium fluoride), and pure covalent compounds, such as chlorine, Cl_2. Of course, the electrons in compounds do not care how we label their bonding, whether ionic, covalent, or mixed ionic-covalent. The fact is that the valence electrons in $AlCl_3$ are usually in the vicinity of the chlorine atoms, but sometimes they are in the vicinity of the aluminum atoms.

The same sort of hybrid bonding may occur in metallic materials. Some of these are intermetallic compounds, such as $CaAl_2$, a stoichiometric compound which has moderately mobile electrons and a small electronegativity difference. Actually, there may be some ionic character as well as some covalent and the predominant metallic character of the bonds in $CaAl_2$. The gradation between covalent and ionic bonds is a continuous one, ranging from pure covalent compounds, to compounds that have very weak polar bonds, to covalent compounds with strong polar bonds, to mixed ionic-covalent bonding (very strong polarity), to compounds having strongly ionic bonds.

Likewise, there are gradations between pure covalent bonds and metallic bonds. Semiconductors, such as silicon and germanium, defect structures, metalloids, and low-conductivity metals lie in between.

1.6   Secondary Bonding in Materials

Although the primary bonds described above, namely covalent, ionic, and metallic, are usually strong bonds, there are other, weaker bonds that exist among atoms. *The importance of these weaker bonds must not be underestimated.* In metals and many ceramic materials, primary chemical bonds exist in a network throughout the material: the object could be considered to be a giant molecule according to the usual definition of a molecule. Thus a wedding ring or the hull of a battleship could be considered to be a single molecule! Likewise, a huge block of granite could also be considered a single molecule. For metals and many ceramics, the concept of molecules is not fruitful. Some materials, such as solid hydrogen or chlorine, are defined as ceramics under the broad definition generally used by materials scientists. These materials are, indeed, molecular solids, and the discussion below applies to these materials as well as to polymers.

Polymeric materials are an entirely different matter: most (but not all) are molecular in nature. Many polymers are comprised of molecules, within which are strong covalent and, sometimes also, ionic bonds. Some polymers, such as the highly cross-linked thermosets (see Chapter 10), are, indeed, single molecules, and therefore the secondary bonding in these materials is relatively unimportant. When polymers are comprised of discrete molecules, then the forces of attraction between atoms of different molecules (or between atoms of the same molecule folded back on itself) may be of critical importance in determining properties.

The secondary bonds include polar bonds, hydrogen bonds, and van der Waal's dispersion forces. *Polar bonds* are attractive forces between an atom having a relatively high electronegativity in one molecule and an atom in another molecule (or the same molecule folded back on itself) in which the second atom has a lower electronegativity. For instance acetals, having many –C–O–C– bonds, have polar bonds between oxygen atoms in the middle of the carbon chain and hydrogen atoms bonded to carbon atoms in another molecule. Polar bonds have a great effect in strengthening polymers by "tacking" together polymer chains from different molecules.

An especially strong polar bond is the *hydrogen bond*, which is a bond between an oxygen (or nitrogen or fluorine) atom of one molecule and a hydrogen atom bonded to oxygen, nitrogen, or fluorine in a second molecule. Oxygen has a fairly high electronegativity, while hydrogen has only a moderate electronegativity. Because the oxygen (or nitrogen or fluorine) atom is small and has a partial negative charge due to its high electronegativity, it is strongly attracted to the more positively charged hydrogen atom of another molecule. Examples include water, where there are two hydrogen bonds per molecule on the average. Hydrogen bonding is known to affect almost all properties of water: its melting and boiling points are much higher than expected of a lightweight molecule (compared with other molecules of similar size), its heat capacity is high, the density of the solid, ice, is lower than that of the liquid (highly unusual), and so on. Hydrogen bonding is of great importance in affecting the properties of materials.

The last of the three types of secondary bonding is least in strength, but, even so, it has great importance. *Van der Waal's dispersion forces* are due to fluctuating dipoles (positive-negative attractions) in *all* materials due to the oscillation of electrons in unison in solids and liquids. Electrons oscillate around atoms: at one instant one side of an atom may be slightly negative due to a temporary excess of valence electrons. The next instant that side of the atom is slightly positive because there is a temporary excess of electrons on the opposite side of the atom. Because of these temporary charges, it is possible for the positive side of one atom to attract the negative side of an adjacent atom. Because the electrons are oscillating in unison, there is *always* an attractive force between two atoms. Furthermore, all atoms in the solid or liquid are oscillating in unison, so *all* atoms attract each other due to these van der Waal's forces of attraction. Even though the forces are comparatively weak, there are many of them in a material, and they serve to bind molecules to each other. Polyethylene, for instance, owes the bonding between polymer chains almost entirely to van der Waal's forces of attraction. Frequently, adhesion depends entirely or largely upon these forces, when bonding metals or ceramics with polymers.

1.7   Price and Performance

In the competitive world of business and industry, objects are designed to meet certain performance specifications. The goal of engineers who design objects or processes is to meet the required specifications at minimum cost. The price for a given performance should be as low as possible. For instance, it makes no sense to design an automobile having a projected lifetime of 12 years or so in which one part, such as the bumper, will have corrosion resistance for 30 years, especially if it costs more to provide the extra corrosion protection. Because there is an immense number of materials—metallic, ceramic, and polymeric—the materials engineer has a challenging job to find the best overall material with respect to price and performance for a given application. The result of applying this principle, however, is an increased standard of living for our society.

1.8   The Fun of Materials Science

To many persons with a lively curiosity about the world and universe, the study of materials science is exciting. Understanding *why* materials behave as they do is highly interesting. Of course, there are some who find study to be drudgery, but there is hope that at some time a spark may land and set their interest on fire. More than once such an event has turned a plodding student into a scholar. The world does not need to be filled with scholars, but if you, the reader, gain an improved understanding of materials through hard study, then society as a whole, including you, will benefit when you apply your knowledge to real-life problems.

The authors hope that your study of materials science will be interesting as well as informative. Just think of the great individual accomplishments that have been made—many lifetimes of dedicated work have gone into the development of the principles in this book. The usefulness to humans is difficult to overestimate. Most of the world has benefited from the advances described in the following chapters. However, remember that there are still many interesting problems to be worked on in materials science. Perhaps *you* will be instrumental in developing a new concept or in applying known concepts to new products or processes.

2

Structure of Metals

The structure of a material is the way in which its atoms are arranged. Chemical composition is the weight percent of each chemical element present in the metal. Both the structure of a material and its chemical composition are of critical importance in determining the properties of the metal. If one metal contains 99.8% iron compared with 99.8% copper in a second, obviously the two metals will have quite different mechanical properties (e.g., tensile strength, yield strength, elongation, toughness, hardness), chemical properties (e.g., solubility in acids, corrosion resistance, oxidation resistance), physical properties (e.g., density, crystal structure), electrical and magnetic properties (e.g., conductivity, magnetism), and optical properties (e.g., appearance of fresh metal surfaces). On the other hand, two specimens of steel may contain 99.6% Fe and 0.4% C, yet have quite different properties because of different structures obtained by different heat treatments.

2.1 Macrostructure of Metals

Macrostructure refers to the way in which atoms are put together on a scale discernible either by the unaided eye or a low-power microscope. For instance, the spangles frequently seen on freshly galvanized steel surfaces (see Fig. 2.1) are single crystals of zinc that can be easily differentiated from one another by the human eye.

2.2 Microstructure of Metals

Microstructure concerns structure at the scale visible with an optical or electron microscope at about 50 to 75,000×. The properties of a given metal depend critically on its microstructure. Because metals crystallize during solidification, each crystallite in a freezing metal continues to grow until it impinges on other crystals. Even then, the small crystals, called *grains*, may grow larger at the expense of their neighbors or in turn be consumed by a growing neighboring grain. At room temperature, the microstructure of most metals is static—it does not change appreciably with time. Hence, it is possible to investigate microstructures of metals using (1) the metallograph, which is a special metallurgical microscope, or (2) the electron microscope, etc. These experimental procedures are described later in this chapter.

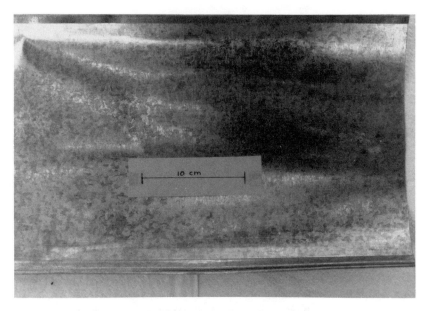

Fig. 2.1   Spangles (zinc crystals) on freshly galvanized steel

Important aspects of microstructure include: (1) grain size and shape, (2) grain orientation, (3) phase or phases present, (4) size and distribution of second phases when present, (5) nature of grain boundaries and precipitates, and (6) subgrain boundaries and dislocations (these are crystal imperfections and are discussed in Chapter 4).

Very large grains are sometimes obtained on slow cooling of ingots or castings. Large grains are frequently objectionable, particularly when the metal must undergo plastic deformation, because large grains cause an uneven appearance after deformation, called *orange peel* (see Fig. 2.2; this is another example of macrostructure). Additionally, large-grained metals are generally not as strong as fine-grained metals. Consequently, control of grain size is important to the manufacturer and user of metals.

Grain orientation affects properties such as magnetism (see Chapter 13). The phase or phases present are very critical to mechanical properties such as yield strength and elongation. In steel, the phases present, the phase size, and the phase distribution can be controlled by heat treatment in addition to mechanical treatment. In some alloys one phase may be distributed along the grain boundaries of the matrix (predominant phase), and the grain boundary precipitate may embrittle the entire metal specimen or make it much less corrosion resistant. When metals are deformed at temperatures far below their melting temperatures, imperfections called ''dislocations'' are proliferated in the metal. These dislocations may be thought of as the linear imperfection at the end of an extra partial plane of atoms. The concentration and distribution of dislocations strongly affect mechanical properties, and dislocations may lead to crack initiation and growth, causing metal failure.

Microstructures will be discussed further in the section on metallography later in this chapter.

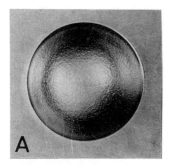
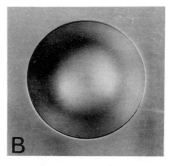

Fig. 2.2   Orange peel: an undesirable surface texture
(A) Rough orange peel surface. (B) Smooth surface

2.3   Crystal Structure of Metals

Crystal structure is the way in which atoms are assembled in a *solid* material. A *crystal* is a three-dimensional, repeating array of atoms. The atoms are spaced at certain distances and angles from one another. These distances and angles may be determined by a method called X-ray diffraction. Because the wavelength of light is much too long to distinguish details of crystal structure, it is necessary to use radiation having short wavelengths, i.e., on the order of the size of atoms.

Fortunately, the most important crystal structures of metals are usually fairly simple. Atoms are packed efficiently in nature, and because the bonds in metals are largely nondirectional, most metals have one of three major crystal structures. If a metal atom were placed at each corner of a cube, and if this unit were repeated over and over in space, the structure would be called *simple cubic* (see Fig. 2.3). However, this structure is very wasteful of space because a relatively low fraction of space in the structure is occupied by atoms. *Occupied space* is the space in which the probability of finding electrons belonging to the atoms is relatively high. The remaining space is referred to as free space, where the probability of finding electrons is very low.

Example 2.3A. Calculate the fraction (percentage) of occupied space in a simple cubic crystal, assuming that the atoms are incompressible spheres.

Solution: The radius of the atom is r. The *interatomic distance* is the distance between the centers of two atoms touching one another. When the atoms are identical, the interatomic distance is 2r. For a simple cubic structure, 2r is the length of the edge of the cube, or a. Each atom is a part of eight identical unit cells (a *unit cell* is the smallest repeating array of atoms representative of the crystal); therefore, each corner atom contributes ⅛ of an atom to the unit cell of interest. Because there are eight corner atoms, each contributing ⅛ atom to our unit cell, there is an average of 8 × ⅛ or 1 atom per unit cell.

% volume of space occupied by atoms

$$= \frac{\text{number of atoms per unit cell} \times \text{volume of one atom}}{\text{volume of unit cell}} \times 100$$

$$= \frac{1 \times (4\pi r^3/3)}{a^3} \times 100$$

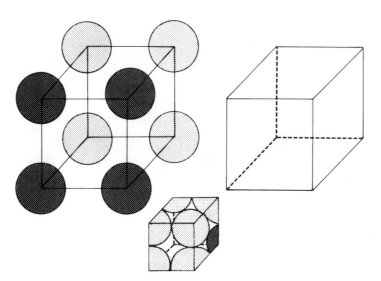

Fig. 2.3   Simple cubic structure

In the simple cubic unit cell, a = 2r by inspection (the length of the edge equals twice the atomic radius). Therefore,

% volume occupied

$$= \frac{4\pi r^3/3}{(2r)^3} \times 100$$

$$= \frac{4\pi r^3}{3 \times 8r^3} \times 100$$

$$= \frac{\pi}{6} \times 100\%, \text{ or } 52.4\%$$

Note: In real crystal structures, atoms are polarizable; that is, their electron clouds may be deformed, and the atoms are somewhat more similar to grapefruit packed in a crate than to billiard balls stacked in a box. However, the simple cubic system is very wasteful of space, so nature chooses a different way of packing metal atoms in space: only one metal, polonium, crystallizes in a simple cubic structure, but this structure actually is complex and contains more than one polonium atom per unit cell.

2.4   Body-Centered Cubic Crystal Structure

One of the three most favored crystal structures for metals is the *body-centered cubic structure* (BCC), illustrated in Fig. 2.4. The BCC unit cell is comprised of eight atoms located

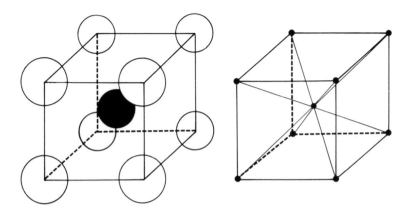

Fig. 2.4   Body-centered cubic structure

at the corners of a cube plus one atom located with its center coinciding with the center of the cube. In the average BCC unit cell, there are eight corner atoms × ⅛ atom + 1 center atom wholly inside the unit cell, for an average of two atoms per unit cell.

Each and every atom in the BCC crystal is surrounded by eight nearest neighbors (touching atoms). The *coordination number* is defined as the number of equivalent nearest (touching) neighbors and hence is eight for the BCC crystal.

The BCC crystal structure is stable for many of the elements listed in Table 2.1, which gives the crystal structures and atomic radii of most metals. The pinnacle of popularity for the BCC structure was probably attained at the 1958 Brussels World's Fair, where the symbol of the event was a monument 102 m (334 ft) high made of a unit cell of iron (see Fig. 2.5); each iron "atom" was a metal sphere 18 m (59 ft) in diameter. This model shows the BCC unit cell mounted along a body diagonal.

Notice that the body diagonal of the BCC unit cell is a *closest-packed direction:* atoms along the diagonal touch each other, and the length of a body diagonal in one unit cell is 4r, where r is the atomic radius. Because "a" is the unit cell dimension (the distance from the center of one corner atom to the center of another corner atom along one edge of the cube), from geometry, $4r = \sqrt{3}a$, because the body diagonal has a length of $\sqrt{3}a$ (according to the Pythagorean Theorem, the length of a face diagonal is the square root of $a^2 + a^2$, or $\sqrt{2}a$; the length of the body diagonal is the square root of $a^2 + 2a^2$, or $\sqrt{3}a$).

Example 2.4A. Calculate the percent of volume occupied by the atoms in a BCC crystal structure.

Solution: There are two atoms in the average BCC unit cell; the volume of the unit cell is a^3, and the volume of one atom is $4\pi r^3/3$. The percent volume occupied = $[(2 \times 4\pi r^3/3)/a^3] \times 100$, but $a = 4r/\sqrt{3}$ from above. Therefore, percent volume occupied = $[2 \times (4\pi r^3/3) \times 100]/(4r/\sqrt{3})^3 = [(8\pi r^3/3)/(64r^3/3\sqrt{3}] \times 100 = (\sqrt{3}\pi/8) \times 100 = 68.0\%$.

Note: Because the electron clouds of atoms in BCC structures are deformed, the real percent of volume occupied by atoms is somewhat greater than 68%. However, it is still likely to be less than the percent volume occupied in the closest-packed structures discussed in Section 2.5.

Table 2.1   Crystal structures and atomic radii of miscellaneous metals

Data taken and adapted from the International Tables for X-Ray Crystallography, Birmingham, England, 1968

Element	Crystal structure	Atomic radius, nm	Unit cell dimensions, nm
Aluminum	FCC	0.14317	0.40496
Antimony	Hex.	· · ·	0.4307, 1.1273(b)
Arsenic	Hex.	· · ·	0.3760, 1.0548
Barium	BCC	0.2176	0.5025
Beryllium	HCP	0.1143, 0.1113(a)	0.22858, 0.35842
Bismuth	Hex.	· · ·	0.4546, 1.1860
Cadmium	HCP	0.14897, 0.16469	0.29793, 0.56181
Calcium	FCC	0.1976	0.5588
Cerium	FCC	0.18247	0.51610
Cesium	BCC	0.265	0.613
Chromium	BCC	0.12491	0.28846
Cobalt	HCP	0.12536, 0.12484	0.25071, 0.40686
Copper	FCC	0.1278	0.36151
Dysprosium	HCP	0.17958, 0.17522	0.35915, 0.56501
Erbium	HCP	0.17780, 0.17335	0.35592, 0.55850
Europium	BCC	0.19844	0.45827
Gadolinium	HCP	0.18168, 0.17858	0.36336, 0.57810
Germanium	FCC(c)	0.20002	0.56575
Gold	FCC	0.14420	0.40786
Hafnium	HCP	0.15942, 0.15608	0.31883, 0.50422
Holmium	HCP	0.17889, 0.17433	0.35778, 0.56178
Iridium	FCC	0.13573	0.38389
Iron	BCC	0.12412	0.28664
Lanthanum	HCP	0.18870, 0.1871	0.37740, 1.217
Lead	FCC	0.17497	0.49489
Lithium	BCC	0.15194	0.35089
Lutetium	HCP	0.17526, 0.17172	0.35052, 0.55494
Magnesium	HCP	0.16044, 0.15982	0.32089, 0.52095
Molybdenum	BCC	0.13626	0.31468
Nickel	FCC	0.12459	0.35238
Niobium	BCC	0.1429	0.3301
Osmium	HCP	0.13670, 0.13376	0.27341, 0.43197
Palladium	FCC	0.13754	0.38902
Platinum	FCC	0.13870	0.39231
Potassium	BCC	0.2310	0.5334
Rhenium	HCP	0.1380, 0.1370	0.2760, 0.4458
Rhodium	FCC	0.1345	0.3804
Rubidium	BCC	0.244	0.563
Ruthenium	HCP	0.13520, 0.13248	0.27041, 0.42814
Scandium	HCP	0.1655, 0.1628	0.331, 0.527
Silicon	FCC(c)	0.19200	0.54307
Silver	FCC	0.14447	0.40862
Sodium	BCC	0.1857	0.4289
Strontium	FCC	0.2152	0.6087
Tantalum	BCC	0.1430	0.3303
Technetium	HCP	0.1368, 0.1352	0.2735, 0.4388
Terbium	HCP	0.18028; 0.17599	0.36055, 0.56966
Thallium	HCP	0.1728, 0.1704	0.3457, 0.5525
Thorium	FCC	0.180	0.509
Titanium	HCP	0.14752, 0.14478	0.29503, 0.46831
Tungsten	BCC	0.1367	0.3158
Vanadium	BCC	0.1316	0.3039
Yttrium	HCP	0.1825, 0.1778	0.365, 0.573
Zinc	HCP	0.13324, 0.14565	0.26649, 0.49470
Zirconium	HCP	0.16156, 0.15894	0.32312, 0.51477

Note: (1) For BCC, r = $(\sqrt{3}/4)$a. For FCC, r = $(\sqrt{2}/4)$a. For HCP, r_a = a/2 and r_{ac} = $\frac{1}{4}\sqrt{4a^2/3 + c^2}$. (2) Manganese has a complex cubic structure. Tin is tetragonal; mercury is hexagonal. Uranium and gallium are orthorhombic. Plutonium and polonium are monoclinic.

(a) The first value is r_a, the radius of atoms in the basal plane. The second value is r_{ac}, the radius of atoms which are touching: one atom is in the basal plane (A-plane), and the second atom is in the next plane (B-plane), parallel with the basal plane. (b) First value is a, second value is c. (c) The diamond structure is a special case of FCC.

Fig. 2.5   The Atomium monument built for the 1958 Brussels World Fair

Example 2.4B. The radius of atoms in BCC iron is 0.1241 nm; calculate the unit cell dimension, a.

Solution: The length of the body diagonal is $4r = \sqrt{3}a$. Therefore, $a = 4r/\sqrt{3} = (4 \times 0.1241 \text{ nm})/\sqrt{3} = 0.2866$ nm.

Note: Table 2.1 presents the structures, unit cell dimensions, and atomic radii for miscellaneous metal atoms. The accepted value for a in iron is 0.2866 nm.

Example 2.4C. Calculate the density of BCC iron given the unit cell dimension a = 0.2866 nm.

Solution: Density = no. of g/no. of cm^3. Determine the density based on the weight and volume of one unit cell. Weight of one unit cell of iron = 2 atoms of Fe $\times$ (55.85 g/mol)/6.022 $\times 10^{23}$ atoms/mol = 1.855×10^{-22} g. The volume of one unit cell = $a^3 = (0.2866 \times 10^{-7}$ cm$)^3 = 2.354 \times 10^{-23}$ cm^3. Therefore, density = g/cm^3 = 1.855×10^{-22} g/2.354×10^{-24} cm^3 = 7.88 g/cm^3 (remember that 1 nm = 10^{-9} m = 10^{-7} cm).

Note: The handbook value for the density of iron is 7.86 g/cm^3. Densities calculated from X-ray data are usually slightly higher than experimentally determined densities because of imperfections in real crystals.

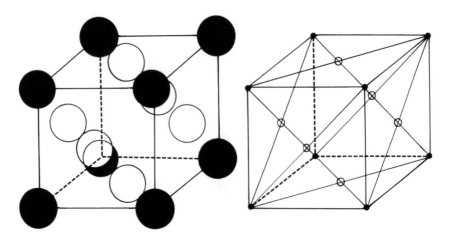

Fig. 2.6   Face-centered cubic structure

2.5   Face-Centered Cubic Crystal Structure

One of the two types of structures having the closest packing of spheres is the *face-centered cubic structure* (FCC), pictured in Fig. 2.6. The closest packing of spheres in two dimensions is the arrangement one obtains with billiard balls in a rack: the center ball is surrounded by six nearest neighbors located at the corners of a hexagon. Note that there are *no* closest-packed layers of atoms in the BCC structure.

If additional layers of closest-packed atoms in the billiard ball arrangement are stacked on each other so that free space is minimized, there are two common arrangements. One of these consists of layers with atoms in odd-numbered layers directly over one another, and also the even-numbered layers have atoms in lines perpendicular to the planes of closest-packed atoms (see Fig. 2.7). This arrangement, labeled ABABAB to indicate that A-layer atoms are lined up with other A atoms and B-layer atoms with B atoms, is the hexagonal closest-packed structure. The other ordered arrangement of layers of closest-packed spheres is ABCABCABC, in which A-layer atoms are lined up with other A atoms perpendicular to the planes of closest-packed atoms, and so forth. This latter structure, illustrated in Fig. 2.6, is the FCC crystal structure. The closest-packed layers are oriented perpendicular to the body diagonal. In fact, there are four different sets of closest-packed layers, each of which is perpendicular to a body diagonal.

It is important to understand the geometry of the FCC unit cell. First, there is a coordination number of 12 (number of nearest neighbors). Second, there is an average of four atoms per unit cell: eight corner atoms each contributing ⅛ atom to one unit cell and six atoms placed in the centers of the six faces of the cube. Each face-centered atom contributes ½ atom to any one unit cell. Therefore, the average unit cell contains 8 × ⅛ corner atoms + 6 × ½ face-centered atoms, or a total of four atoms for each unit cell.

The closest-packed directions, in which atoms touch each other in a line, are the face diagonals, and there are six different face diagonals for each unit cell. The closest-packed layers are perpendicular to the four body diagonals, so there are four different closest-packed planes per unit cell. These closest-packed planes are called the octahedral planes because they

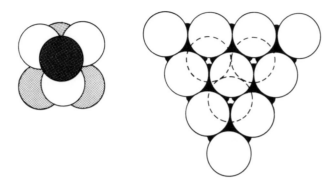

Fig. 2.7   Closest-packed layers in the FCC structure

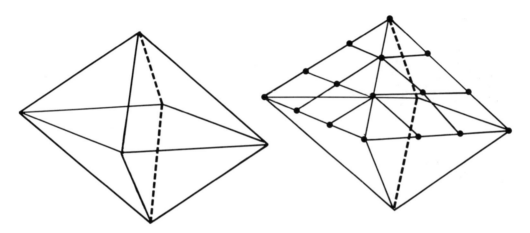

Fig. 2.8   Octahedral (closest-packed) layers in the FCC structure

outline an octahedron when combined with four other parallel closest-packed planes (see Fig. 2.8).

Table 2.1 lists the metals having FCC structures, and some of the temperature ranges for the stabilities of these structures appear in Table 2.2. Because the closest-packed directions lie along the face diagonals, and a face diagonal comprises four atomic radii, then $4r = \sqrt{2}a$, where a is the FCC unit cell dimension.

Example 2.5A. Calculate the atomic radius of FCC nickel given the unit cell dimension $a = 0.3524$ nm.

Solution: The face diagonal has a length of $4r = \sqrt{2}a$. Therefore, $r = \sqrt{2}a/4 = (0.3524$ nm $\times \sqrt{2})/4 = 0.1246$ nm. This answer agrees with the value in Table 2.1.

Example 2.5B. Calculate the percent volume occupied by atoms in FCC structures assuming that the atoms are hard spheres.

Table 2.2 Polymorphism in metals

Element	Structure	Temperature range, °C
Beryllium	HCP	<1250
	BCC	>1250
Calcium	FCC	<450
	BCC	>450
Cerium	HCP	−150 to −10
	FCC	−10 to 725
	BCC	>725
Cobalt	HCP	<425
	FCC	Quenched from >450
Dysprosium	HCP	<1381
	BCC	>1381
Gadolinium	HCP	<1264
	BCC	>1264
Hafnium	HCP	<1950
	BCC	>1950
Holmium	HCP	<906
	BCC	>906
	?	
Iron	BCC	<910
	FCC	910 to 1400
	BCC	>1400
Lanthanum	HCP	−271 to 310
	FCC	310 to 868
	BCC	>868
Lithium	HCP	<−202
	BCC	>−202
Lutetium	HCP	<1400
	?	>1400
Manganese	Four cubic phases	
Mercury	BC tetragonal	<−194
	Rhombohedral	>−194
Neodymium	HCP	<868
	BCC	>868
Neptunium	Orthorhombic	<280
	Tetragonal	280 to 577
	BCC	577 to 637
Plutonium	Several phases	
Polonium	Simple cubic	
	Rhombohedral	>76
Praseodymium	HCP	<798
	BCC	>798
Samarium	Rhombohedral	<917
	BCC	>917
Scandium	HCP	1000 to 1300
	BCC	>1335
Sodium	HCP	<−237
	BCC	>−237
Strontium	FCC	<540
	BCC	>540
Terbium	HCP	<1317
	BCC	>1317
Thallium	HCP	<230
	BCC	>230
Titanium	HCP	<885
	BCC	>885
Uranium	Orthorhombic	<662
	Tetragonal	662 to 774
	BCC	774 to 1132
Ytterbium	FCC	R.T. to 798
	BCC	>798
Yttrium	HCP	R.T. to 1460
	BCC	>1460
Zirconium	HCP	<865
	BCC	>865

Summary:
In 14 metals HCP transforms to BCC as temperature increases.
In 3 metals HCP transforms to FCC as temperature increases.
In 6 metals FCC transforms to BCC as temperature increases.
In 0 metals FCC transforms to HCP as temperature increases.
In 0 metals BCC transforms to HCP as temperature increases.
In 1 metal BCC transforms to FCC as temperature increases.

Solution: Percent volume occupied for one FCC unit cell is equal to [(4 atoms/unit cell) $\times$ $(4\pi r^3/3)/a^3$] $\times$ 100. Because a = $4r/\sqrt{2}$, the percent volume occupied by atoms is equal to $[(4 \times 4\pi r^3/3)/(4r/\sqrt{2})^3] \times 100 = [(16\pi r^3/3)/(64r^3/2\sqrt{2})] \times 100 = (\pi/3\sqrt{2}) \times 100 = 74.1\%$.

Note: The actual volume occupied by atoms in the FCC structure is undoubtedly greater than 74.1% due to the deformation of electron clouds in atoms bonded to one another. The occupied volume is appreciably greater than that of BCC structures, namely 68.0%. The fraction of space occupied by atoms is dependent only on the type of crystal structure and is independent of r, the atomic radius.

The *hexagonal closest-packed crystal structure* (HCP) is also important: zinc, magnesium, and titanium, as well as several other metals, have this structure. Nevertheless, we shall not study this structure in detail. The percent volume occupied by atoms in the HCP structure is 74.1%, exactly the same as for FCC crystals. Each atom has 12 nearest neighbors, and a = 2r. The ratio c/a has a value of 1.633 for hard spheres. Note the departure from the hard sphere ratio for most HCP metals in Table 2.3.

2.6 Polymorphism in Metals

Many metals are *polymorphic:* they have more than one crystal structure. As the temperature is increased from near 0 °K to the melting point of a given metal, similar types of phase transformations frequently occur (see Table 2.2). In fact, it is clear from Table 2.2 that the HCP structure is the favored low-temperature crystal structure of metals. This does not mean that all metals have an HCP structure if the temperature is low enough, but at least 30 metals do (see Table 2.3). Furthermore, the BCC structure is obviously the structure most favored at high temperatures. In fact, only one metal changes from BCC to FCC as temperature *increases* (iron!), and no metals change from BCC to HCP upon heating. Iron is a distinguished exception, although it does change back again to BCC at about 1400 °C. The FCC structure appears to be a stable one, because less than one-half of the FCC elements undergo phase transformations as a function of temperature.

Polymorphism in iron is a great advantage because carbon is much more soluble in FCC iron than in BCC iron; the useful properties of steel, such as combined toughness and high strength, are the result of being able to dissolve carbon in FCC steel at high temperatures and then heat treat the steel in such a way as to control the size and distribution of carbide precipitates in BCC steel at room temperature. This control over the carbide precipitates gives steel its tremendous breadth of attainable properties. It is fortunate that the only exception to the rule that BCC structures do not transform with increasing temperature is iron, and this exception is of tremendous practical importance.

2.7 Miller Indices of Directions in Cubic Crystals

Frequently it is useful to specify given points, directions, or planes in crystals. A coordinate system is usually chosen in which the origin coincides with the center of an atom (any atom will do). Then a given point is designated as having coordinates—for example, of one unit cell

Table 2.3   c/a ratios for HCP metals

Element	Crystal structure	c/a	c/a 1.633
Americium	HCP(ABAC . . .)(a)	3.229	0.989(a)
Beryllium	HCP	1.568	0.960
Cadmium	HCP	1.886	1.155
Cerium	HCP(ABAC . . .)(−150 to −10 °C)	3.24	0.992
Cobalt	HCP	1.623	0.994
Dysprosium	HCP	1.574	0.964
Erbium	HCP	1.571	0.962
Gadolinium	HCP	1.591	0.974
Hafnium	HCP	1.581	0.968
Holmium	HCP	1.571	0.962
Lanthanum	HCP(ABAC . . .)	3.218	0.985
Lithium	HCP(< −202 °C)	1.637	1.002
Lutetium	HCP	1.583	0.969
Magnesium	HCP	1.624	0.994
Neodymium	HCP(ABAC . . .)	3.226	0.988
Osmium	HCP	1.579	0.967
Praseodymium	HCP(ABAC . . .)	3.223	0.987
Rhenium	HCP	1.615	0.989
Ruthenium	HCP	1.582	0.969
Scandium	HCP	1.592	0.975
Sodium	HCP(<−237 °C)	1.634	1.001
Strontium	HCP(235 to 540 °C)	1.635	1.001
Technetium	HCP	1.604	0.982
Terbium	HCP	1.583	0.969
Thallium	HCP	1.598	0.979
Thulium	HCP	1.527	0.963
Titanium	HCP	1.586	0.971
Yttrium	HCP	1.572	0.963
Zinc	HCP	1.856	1.137
Zirconium	HCP	1.593	0.976

(a) The ABAC HCP crystals have a hard sphere c/a of 3.266.

dimension along the x-axis, and two unit cell dimensions along the y-axis, and one unit cell dimension along the z-axis. The coordinates of this point would be 1,2,1 (see Fig. 2.9). Any other desired point may be defined in terms of the number of unit cell dimensions (including fractions or negative numbers) along each of the three axes.

To specify a given direction, it is necessary to define two separate points. By agreement, one of the two points in defining a direction is always the origin, 0,0,0. To define a given direction in a crystal, it is simply necessary to give the three coordinates of the second point. Any direction may be given by listing three coordinates of one point only and drawing a line from the origin through the indicated point.

Example 2.7A. Draw a figure illustrating the three axes of a cubic crystal, the unit cell dimension a, and the direction [123]. In Fig. 2.10, what are the Miller indices for the direction of line B?

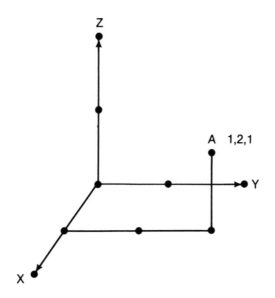

Fig. 2.9   Coordinates of points

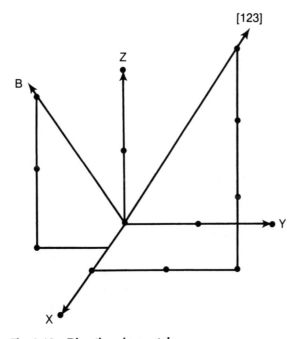

Fig. 2.10   Directions in crystals

Solution: Figure 2.10 shows the x, y, and z axes, the origin, and unit cell dimensions, a, along the axes. To draw [123], proceed one unit cell dimension along the x-axis, then a distance of 2a along y, and finally 3a along z. Construct a point here, and draw a line between the origin

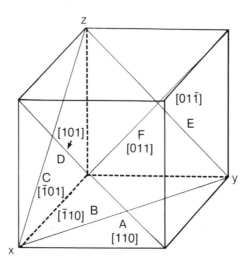

Fig. 2.11   Miller indices for various directions

and this point. To find the direction of line B, determine the coordinates of point B in terms of multiples of a; any fractions must be cleared by multiplying by the proper integer. Thus, the coordinates are ½, −1, and 2, and the direction is [½,−1,2]. However, by convention, no fractions are accepted, so one must multiply all the indices by two; also, no commas are used, and the minus sign is placed above the appropriate index: [1$\bar{2}$4].

Note: Square brackets are used to designate a direction.

If the crystal orientation remains fixed (the crystal is not rotated by changing the directions of the three axes), then any direction in the crystal is parallel to any other direction having the same indices: any atom or position in the crystal could be chosen as the origin. The direction would be the same for a given set of Miller indices. Reversing the signs of the indices by multiplying all three of the indices by −1 simply reverses the direction. Frequently it is convenient to shift the origin to find the Miller indices of a given direction.

Example 2.7B. Give the indices of the six closest-packed directions in an FCC unit cell.
Solution: These directions are the face diagonals shown in Fig. 2.11. Direction A is [110], B is [$\bar{1}$10], C is [$\bar{1}$01], D is [101], E is [0$\bar{1}$1], and F is [011].

Note: These directions are permutations of two indices having a value of +1 or −1 and the third having a value of 0. Direction [110] is the opposite direction of [$\bar{1}\bar{1}$0] and therefore is not another, independent close-packed direction. It is necessary to shift the origin from one lattice point to another equivalent lattice point in order to derive all of the above directions. Shifting the origin is legitimate as long as the crystal orientation remains unchanged.

Example 2.7C. What are the closest-packed directions in BCC crystals?
Solution: Atoms touch one another along the four body diagonals of the cube. One body diagonal is [111]; permutations of +1 and −1 for the various indices will give the four body diagonals: [111], [$\bar{1}$11], [1$\bar{1}$1], and [11$\bar{1}$].

Note: [$\bar{1}\bar{1}\bar{1}$] is not a separate direction; it is the direction opposite to [111].

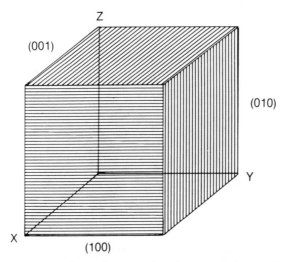

Fig. 2.12 Miller indices for the planes outlining a cube

2.8 Miller Indices of Planes in Cubic Crystals

A plane is defined by three nonlinear points. In order to describe a plane, three Miller indices are given in parentheses: (hkl), where h, k, and l are positive or negative integers. Each Miller index indicates the *reciprocal* of the intercept of the given plane with either the x, y, or z axis. Thus, the index, h, is the reciprocal of the intercept in multiples of a made by the plane on the x-axis; k denotes the reciprocal of the y-intercept of the plane and the y-axis, and l the reciprocal of the z-intercept. Obviously the plane cannot pass through the origin; if it does, then it is necessary to shift the origin to another point, but the axes must maintain the same directions. Once again, only integers are permitted—fractions must be cleared by multiplication by a suitable integer. The indices are not separated by commas, and negative indices have a negative sign over the index.

Example 2.8A. What are the Miller indices for the planes that outline a cubic unit cell?

Solution: According to Fig. 2.12, the Miller indices for the plane intersecting the x-axis at 1a, the y-axis at ∞ × a (the plane is parallel to the y-axis; according to a theorem in geometry, a line that is parallel to a plane intersects the plane at infinity), and the z-axis at ∞ × a are (100). The reciprocals of the intercepts of the plane in units of a along the three axes are $1/1$, $1/\infty$, $1/\infty$, or 1, 0, 0. The plane perpendicular to the y-axis at 1a is (010), and the plane perpendicular to the z-axis, and intersecting z at 1a, is (001). It is most important to note that each of these planes is in reality a *family* of parallel planes spaced one unit cell dimension apart and extending throughout the crystal. Thus, (100) denotes not only the plane intersecting the x-axis at 1a and parallel to the y and z axes, but also refers to a whole family of planes throughout the crystal. Therefore, the families of (100), (010), and (001) planes intersect to form a unit cell, or, actually, many unit cells.

Example 2.8B. Explain the difference between (200) planes and (100) planes.

Solution: The (200) plane intersects the three axes at ½, ∞, ∞. Thus, it intersects the x-axis at a/2. The next parallel plane to the first (200) plane intersects the x-axis at (½ + ½)a, or 1a.

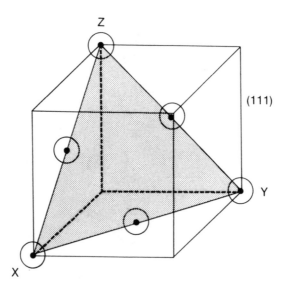

Fig. 2.13   (111) plane in an FCC crystal

The (200) planes are a family of parallel planes having an *interplanar spacing* (perpendicular distance between two neighboring parallel planes) of a/2, whereas the (100) planes are a family of planes having an interplanar spacing of a. There are twice as many (200) planes as (100) planes, and all (100) planes are included in the (200) family, but only one-half of the (200) planes are in the (100) family.

Example 2.8C. What are the Miller indices for the plane having intercepts of 2, 3, −1 on the x, y, and z axes?

Solution: The reciprocals of the intercepts are ½, ⅓, and −1, or (½ ⅓ $\bar{1}$), but fractions must be eliminated, so the plane is 6 × (½ ⅓ $\bar{1}$), or (32$\bar{6}$).

Example 2.8D. What are the intercepts of the next plane in the (32$\bar{6}$) family parallel to the plane having intercepts of 2, 3, and −1?

Solution: Take the reciprocals of the Miller indices: ⅓, ½, −⅙. Add these values to the intercepts of the original plane: 2 + ⅓, 3 + ½, −1 + −⅙. The intercepts of the next parallel plane in the (32$\bar{6}$) family are ⁷⁄₃,⁷⁄₂,−⁷⁄₆.

Example 2.8E. What are the Miller indices of the closest-packed planes in FCC crystals?

Solution: Figure 2.13 shows one of the four closest-packed planes in an FCC crystal. The indices of the plane are (111). The other three closest-packed planes are permutations of indices having a value of +1 or −1: ($\bar{1}$11), (1$\bar{1}$1), and (11$\bar{1}$).

Note: These four planes and the planes parallel to them intersect to form an octahedron in the FCC crystal, and the closest-packed planes are called the octahedral planes. Note also that in an FCC crystal structure each atom has four atoms surrounding it at the corners of a square in each of three planes. These are the (100), (010), and (001) planes. Because cubic crystals have the same properties along the x, y, and z axes, they are said to be *isotropic*.

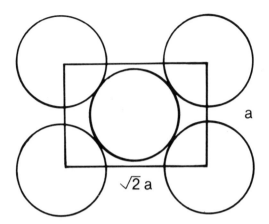

Fig. 2.14   Density of atoms in the BCC (110) plane

In cubic crystals the direction [hkl] is always perpendicular to the plane (hkl) provided that the values of h, k, and l are the same in both cases. Hence, [100] is perpendicular to (100), [110] is perpendicular to (110), [123] is perpendicular to (123), and so on.

Example 2.8F. Calculate the density of atoms in the (110) plane of BCC iron (density of atoms in a plane means the number of atom centers occurring in 1 cm² of the plane).

Solution: From Table 2.1, a = 0.2866 nm for BCC iron. Figure 2.14 is a sketch of the (110) plane in iron. Inside the given rectangle, measuring 0.2866 nm by $\sqrt{2}$ × 0.2866 nm, there are 4 × ¼ + 1, or 2, iron atoms. The density of iron atoms in the (110) plane = 2 atoms/(0.2866 × $\sqrt{2}$ × 0.2866 nm²) = 17.22 atoms/nm² or 17.22 atoms/nm² × (10^7 nm/1 cm)² = 1.72 × 10^{15} atoms/cm².

2.9   Methods for Determining Structure

For many years, metallurgists, chemists, and physicists have used certain instruments and techniques to determine crystal structures and microstructures of metals. Several of these techniques are also used to study ceramics and polymers. These techniques include X-ray diffraction for the study of crystal structures and metallography, the microscopic examination of structure, such as phase distribution and grain size and shape, by means of the metallograph and electron microscope. Variations of these instruments and techniques include the electron microprobe, the scanning electron microscope, and X-ray fluorescence. Some of the most important methods and techniques are discussed in the following sections.

2.10   X-Ray Diffraction and The Bragg Equation

The wavelengths of visible light are far too long to interact with atoms in such a way that crystal structures may be directly seen or deduced. The resolving power of light is insufficient to permit us to see atoms or electron clouds. However, X-rays are photons having wavelengths of the order of the dimensions of atoms, namely about 0.1 nm (1 nm = 10^{-9} m or 10 Angstrom

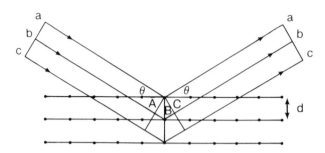

Fig. 2.15   X-ray diffraction from parallel planes of atoms in a crystal

units, Å). X-rays interact with electrons such that the electrons vibrate with the frequency of the X-rays. The electrons reradiate, or "scatter," X-rays with no change in frequency and in random directions. When the reradiated X-rays reinforce each other, rather than cancelling to form heat, they are diffracted.

When atoms are arranged in an ordered, three-dimensional array in space, such as a crystal, the scattered X-rays undergo interference. In most directions, destructive interference occurs with a cancellation of X-rays. In certain directions, however, constructive interference may occur in which the scattered X-rays have their wavelengths in phase. When a beam of X-rays strikes one plane of atoms, constructive interference may occur when the angle of incidence equals the angle of reflection. This is much the way light reflects from a mirror. However, when there are many layers of atoms, the X-rays, unlike light, penetrate through the atoms, and eventually they are scattered or absorbed. The scattered X-rays emerge from the specimen only at very specific angles, depending on the wavelength of the X-rays and the spacing between the layers of atoms. This specific "reflection" is called diffraction of X-rays.

Example 2.10A. Show that X-rays are diffracted from several layers of atoms in accordance with Bragg's equation, $n\lambda = 2d \sin \theta$.

Solution: Construct a drawing (Fig. 2.15) representing atoms in several parallel planes. Ray "a" will be reinforced by ray "b" when the difference in their path length is a multiple of their wavelength. According to Fig. 2.15, ray "b" travels through a path longer than ray "a" by one wavelength, and ray "c" travels a distance greater than ray "a" by two wavelengths. This relationship holds when $AB + BC = 1$ (or 2 or 3, etc.) $\times \lambda$. Because $\sin \theta = AB/d$, AB is equal to $d \sin \theta$, as is BC, where d is the perpendicular distance between planes of atoms (d is the interplanar spacing), and θ is either the angle of incidence or the angle of diffraction. This diffraction will occur when the extra pathway, $AB + BC$, equals 1, 2, 3, or $n \times \lambda$, the wavelength. Therefore, $AB + BC = 2d \sin \theta = n\lambda$. Both λ and d are currently expressed in nanometers (nm); however, previously it was customary to use Angstrom units, where 10 Å = 1 nm. In the Bragg equation, n, which is always an integer, is the "order" of the diffraction. When n = 1, we call this the first order, which leads to the most intense diffraction. n = 1 when the diffracted X-ray path length is one extra wavelength for the second plane of atoms compared with the first plane of atoms. In most problems involving the Bragg equation, n = 1.

Note: The Bragg equation, $n\lambda = 2d \sin \theta$, is one of the most fundamental and important equations in materials science; it describes the diffraction of X-rays from metals, as well as from all other crystals, and enables crystal structures to be deduced.

Example 2.10B. Copper $K\alpha_1$ X-rays have a wavelength of 0.15406 nm. What is the interplanar spacing, d, for the (200) planes in iron when the Bragg angle, θ, is 32.52°?

Solution: Using the Bragg equation, $n\lambda = 2d \sin \theta$, assuming n = 1, or first-order diffraction (if no statement is made to the contrary, assume that n = 1), 0.15406 nm = 2d sin 32.52°, or 0.15406 nm = 2d $\times$ 0.5375. Therefore, d = 0.15406 nm/(2 $\times$ 0.5375) = 0.1433 nm.

Note: d for (200) planes is 0.1433 nm, and this interplanar spacing is exactly half the interplanar spacing for (100) planes, which is the unit cell dimension, a: $d_{(200)}$ = 0.1433 nm, and $d_{(100)}$ = 0.2866 nm = a.

Example 2.10C. Calculate the Bragg angle of diffraction, θ, for the (110) planes in iron using copper $K\alpha_1$ X-rays, λ = 0.15406 nm; the interplanar spacing, d, is equal to 0.2027 nm.

Solution: Using the Bragg equation, $n\lambda = 2d \sin \theta$, and assuming that n = 1, λ = 0.15406 nm, d = 0.2027 nm, and therefore 0.15406 nm = 2 $\times$ 0.2027 nm $\times$ sin θ; therefore, sin θ = 0.15406 nm/(2 $\times$ 0.2027 nm) = 0.3800. Using an electronic calculator, you will find that the angle with a sine of 0.3800 is 22.33°.

Example 2.10D. Calculate the wavelength of $K\alpha_1$ X-rays of a common metal used in X-ray tubes as a target material when diffraction occurs from the (220) planes in iron having an interplanar spacing of 0.1013 nm; the Bragg angle, θ, is 20.49°. From the calculated wavelength, identify the material by looking in Table 2.4, which gives the wavelengths of important X-rays of many common metals.

Solution: Using the Bragg equation, $n\lambda = 2d \sin \theta$, and assuming that n = 1, since d = 0.1013 nm, θ = 20.49°, and sin θ = 0.3500, then 1 $\times$ λ = 2 $\times$ 0.1013 nm $\times$ 0.3500, or λ = 0.07091 nm. This is very close to the wavelength of $K\alpha_1$ X-rays of molybdenum ($K\alpha_1$ = 0.07093 nm, as given in Table 2.4).

2.11 Interplanar Spacings for Planes in Cubic Systems

In Example 2.10B, we found that the interplanar spacing, $d_{(200)}$, of the (200) planes in iron is just one-half the unit cell dimension, a. This follows directly from inspection of (200) planes in a cubic unit cell. The (200) family of planes intersect the x-axis at 0, a/2, 1a, 3a/2, 2a, 5a/2, 3a, and so on, or at intervals of a/2. Therefore, the interplanar spacing $d_{(200)}$ = a/2. Inspection of a cubic unit cell shows the following: $d_{(100)}$ = a, $d_{(110)}$ = a/$\sqrt{2}$, $d_{(200)}$ = a/2, $d_{(111)}$ = a/$\sqrt{3}$, $d_{(300)}$ = a/3, and so on. A general equation has been derived from geometry to relate d and a in cubic crystals:

$$d_{(hkl)} = \frac{a}{\sqrt{h^2 + k^2 + l^2}}$$

Table 2.4   X-ray wavelengths of various metals

Element	Atomic no.	Designation	Wavelength, Å	Energy, keV	Designation	Wavelength, Å	Energy, keV
Lithium	3	$K\alpha$	228	0.0543	$\cdots$	$\cdots$	$\cdots$
Beryllium	4	$K\alpha$	114	0.1085	$\cdots$	$\cdots$	$\cdots$
Sodium	11	$K\alpha$	11.9101	1.0410	$K\beta$	11.575	1.0711
Magnesium	12	$K\alpha$	9.8900	1.25360	$K\beta$	9.521	1.3022
Aluminum	13	$K\alpha_2$	8.34173	1.48627	$K\alpha_1$	8.33934	1.48670
		$K\beta$	7.960	1.5574			
Silicon	14	$K\alpha_2$	7.12791	1.73938	$K\alpha_1$	7.12542	1.73998
		$K\beta$	6.753	1.8359			
Potassium	19	$K\alpha_2$	3.7445	3.3111	$K\alpha_1$	3.7414	3.3138
		$K\beta_1$	3.4539	3.5896			
Calcium	20	$K\alpha_2$	3.36166	3.68809	$K\alpha_1$	3.35839	3.69168
		$K\beta_1$	3.0897	4.0127			
Titanium	22	$K\alpha_2$	2.75216	4.50486	$K\alpha_1$	2.74851	4.51084
		$K\beta_1$	2.51391	4.93181			
Vanadium	23	$K\alpha_2$	2.50738	4.94464	$K\alpha_1$	2.50356	4.95220
		$K\beta_1$	2.28440	5.42729			
Chromium	24	$K\alpha_2$	2.293606	5.40551	$K\alpha_1$	2.28970	5.41472
		$K\beta_1$	2.08487	5.94671			
Manganese	25	$K\alpha_2$	2.10578	5.88765	$K\alpha_1$	2.101820	5.89875
		$K\beta_1$	1.91021	6.49045			
Iron	26	$K\alpha_2$	1.939980	6.39084	$K\alpha_1$	1.936042	6.40384
		$K\beta_1$	1.75661	7.05798			
Cobalt	27	$K\alpha_2$	1.792850	6.91530	$K\alpha_1$	1.788965	6.93032
		$K\beta_1$	1.62079	7.64943			
Nickel	28	$K\alpha_2$	1.661747	7.46089	$K\alpha_1$	1.657910	7.47815
		$K\beta_1$	1.500135	8.26466			
Copper	29	$K\alpha_2$	1.544390	8.02783	$K\alpha_1$	1.540562	8.04778
		$K\beta_1$	1.3926	8.9029			
Zinc	30	$K\alpha_2$	1.43900	8.61578	$K\alpha_1$	1.435155	8.63886
		$K\beta_1$	1.29525	9.5720			
Zirconium	40	$K\alpha_2$	0.79015	15.6909	$K\alpha_1$	0.78593	15.7751
		$K\beta_1$	0.70173	17.6678	$L\alpha_1$	6.0705	2.04236
Niobium	41	$K\alpha_2$	0.75044	16.5210	$K\alpha_1$	0.74620	16.6151
		$K\beta_1$	0.66576	18.6225	$L\alpha_1$	5.7243	2.16589
Molybdenum	42	$K\alpha_2$	0.713590	17.3743	$K\alpha_1$	0.709300	17.47934
		$K\beta_1$	0.632288	19.6083	$L\alpha_1$	5.40655	2.29316
Silver	47	$K\alpha_2$	0.563798	21.9903	$K\alpha_1$	0.5594075	22.16292
		$K\beta_1$	0.497069	24.9424	$L\alpha_1$	4.15443	2.98431
Tin	50	$K\alpha_2$	0.495053	25.0440	$K\alpha_1$	0.490599	25.2713
		$K\beta_1$	0.435236	28.4860	$L\alpha_1$	3.59994	3.44398
Antimony	51	$K\alpha_2$	0.474827	26.1108	$K\alpha_1$	0.470354	26.3591
		$K\beta_1$	0.417085	29.7256	$L\alpha_1$	3.43941	3.60472
Barium	56	$K\alpha_2$	0.389668	31.8171	$K\alpha_1$	0.385111	32.1936
		$K\beta_1$	0.340811	36.3782	$L\alpha_1$	2.77595	4.46626
Tungsten	74	$K\alpha_2$	0.213828	57.9817	$K\alpha_1$	0.209010	59.31824
		$K\beta_1$	0.184374	67.2443	$L\alpha_1$	1.47639	8.3976
Platinum	78	$K\alpha_2$	0.190381	65.122	$K\alpha_1$	0.185511	66.832
		$K\beta_1$	0.163675	75.748	$L\alpha_1$	1.31304	9.4423
Lead	82	$K\alpha_2$	0.170294	72.8042	$K\alpha_1$	0.165376	74.9694
		$K\beta_1$	0.145970	84.936	$L\alpha_1$	1.17501	10.5515
Bismuth	83	$K\alpha_2$	0.165717	74.8148	$K\alpha_1$	0.160789	77.1079
		$K\beta_1$	0.141948	87.343	$L\alpha_1$	1.14386	10.8388

Example 2.11A. Calculate the interplanar spacing, $d_{(111)}$, of the (111) planes in nickel, with a = 0.3524 nm.

Solution: $d_{(111)} = a/\sqrt{h^2+k^2+l^2} = 0.3524$ nm/ $\sqrt{1^2+1^2+1^2} = 0.3524$ nm/$\sqrt{3}$ = 0.2035 nm.

Example 2.11B. What family of planes in nickel has an interplanar spacing of 0.1246 nm?

Solution: Because $d_{(hkl)} = a/\sqrt{h^2+k^2+l^2}$, 0.1246 nm = 0.3524 nm/$\sqrt{h^2+k^2+l^2}$. Therefore, $\sqrt{h^2+k^2+l^2}$ = 0.3524/0.1246 = 2.828, or $(h^2 + k^2 + l^2) = 2.828^2 = 7.998$. Because h, k, and l are integers, then $(h^2 + k^2 + l^2)$ must also be an integer, and 7.998 may be rounded to 8. The value of (hkl) for which $h^2 + k^2 + l^2 = 8$ is (220). This is derived by trial and error substitution of values of small integers for h, k, and l. In conclusion, the (220) planes in nickel have an interplanar spacing of 0.1246 nm.

Example 2.11C. What is the unit cell dimension of BCC chromium given that the (220) planes have an interplanar spacing of 0.1020 nm?

Solution: Because $d_{(220)} = a/\sqrt{h^2+k^2+l^2}$, 0.1020 = a/$\sqrt{2^2+2^2+0^2}$, or a = 0.1020 $\sqrt{2^2+2^2+0^2} = 0.1020\sqrt{8} = 0.2885$ nm. This value may be checked in Table 2.1. The combination of the above equation with the Bragg equation enables many important and practical problems to be solved.

Example 2.11D. Calculate the Bragg angle of diffraction from the (111) planes in nickel using copper $K\alpha_1$ X-rays.

Solution: Use both equations: $n\lambda = 2d \sin \theta$ and $d_{(hkl)} = a/\sqrt{h^2+k^2+l^2}$. First, find λ for Cu $K\alpha_1$ X-rays in Table 2.4: $\lambda = 0.15406$ nm. Next find the unit cell dimension of nickel from Table 2.1: a = 0.3524 nm. Third, calculate $d_{(111)}$ for nickel: $d_{(111)} = 0.3524$ nm/ $\sqrt{1^2+1^2+1^2} = 0.3524$ nm/$\sqrt{3}$, or $d_{(111)} = 0.2035$ nm. Fourth, substitute this value into the Bragg equation, $n\lambda = 2d \sin \theta$: 1×0.15406 nm = 2×0.2035 nm $\times \sin \theta$, or $\sin \theta = 0.15406/(2 \times 0.2035) = 0.3785$, and $\theta = 22.24°$.

2.12   Identification of Unknown Phases Using X-Ray Diffraction

The X-ray diffractometer, shown in Fig. 2.16, is a valuable instrument for determining simple crystal structures. The use of this instrument results in x-ray diffractometer scans, as shown in Fig. 2.17. Different planes of atoms, having various Miller indices, give rise to diffraction peaks when the conditions of the Bragg equation are fulfilled.

Even though the X-ray diffractometer is a valuable instrument for determining simple crystal structures, it is of even more value for identifying unknown crystalline materials. If an unknown metal is comprised of one or two pure phases (no solid solutions are present), then the diffractometer scan may be used to identify the crystal phase or phases present. This is carried out by measuring all the interplanar spacings, d, and then looking up the d-spacings of the most intense peaks in a reference book. This technique is used widely in industry not only for identifying unknown metallic phases, but also for identifying ceramic phases.

Fig. 2.16   X-ray diffractometer

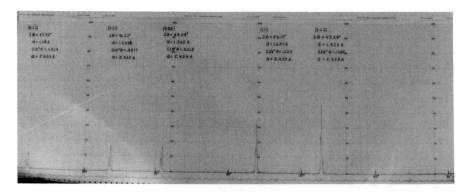

Fig. 2.17   X-ray diffractometer scan of elemental silicon

2.13   Additional Applications of X-Ray Diffraction

Solidification, cold rolling, or cold drawing, sometimes followed by recrystallization (annealing at an elevated temperature after plastic deformation), frequently results in *preferred orientation,* in which crystals tend to be aligned along certain directions. The intensities of various peaks are either enhanced or reduced due to preferred orientation, and the exact nature of the preferred orientation may be determined by certain techniques.

When metals are plastically deformed, such as by cold rolling, the crystals are elongated in the direction of rolling and compressed in a direction perpendicular to the surface of the sheet. The interplanar spacings of planes of a metal that are more or less parallel with the surface will be slightly compressed, resulting in a slight decrease in d for these planes. Planes that are more or less perpendicular to the rolling direction, however, will have a slight increase in d, the interplanar spacing. By measuring this spread using X-ray diffraction, one may calculate the stress level that is retained in the plastically deformed metal (the diffraction peaks are broadened by cold work).

Peak broadening in X-ray diffraction may also be caused by the presence of very small crystals. The number of parallel planes may be too small in very small crystals for diffraction to occur at precise angles. The peak broadening may then be used to calculate the approximate size of the crystals present.

2.14   X-Ray Fluorescence Analysis

Another useful tool for the materials scientist is X-ray fluorescence analysis, which depends on the same principles as X-ray diffraction. Whereas X-ray diffraction enables one to determine the crystal structure of a material or to identify the phase or phases present, X-ray fluorescence enables one to determine the chemical composition of a specimen, that is, the weight percent of each element present, provided that the element has an atomic number of at least 11. X-ray fluorescence is complementary to X-ray diffraction.

X-ray fluorescence depends on using "white" X-rays, which are X-rays having a continuous spectrum of energies or wavelengths within a given range. The white X-rays are directed onto the specimen. Incoming X-rays eject electrons from the K and L shells of electrons from some of the atoms. Other electrons fall into the vacated electron orbitals in the K and L shells, giving rise to K and L X-rays having specific wavelengths characteristic of the particular element involved (wavelengths and energies are given for miscellaneous metals in Table 2.4). For example, some of the iron atoms present in a copper alloy would have their K electrons knocked out in a given time interval. Electrons fall back into the K shell immediately, and as a result, $K\alpha_1$ and $K\alpha_2$ X-rays are emitted from the iron atoms; these X-rays have precise wavelengths and energies.

$K\alpha_1$ and $K\alpha_2$ X-rays, which are emitted in random directions from the specimen, are collimated by passing through a collimator having several thin metal sheets parallel to one another. The collimator absorbs all X-rays that are not parallel, leaving a beam of parallel X-rays. Iron $K\alpha_1$ and $K\alpha_2$ X-rays, which vary only slightly from each other in wavelength, then strike an analyzing crystal, usually LiF, which is oriented at an angle of 28.78° (Bragg angle θ) to the X-ray beam. Diffraction occurs at this angle, and the X-rays leave the crystal at an angle of 28.78°. The total angle is $2\theta = 57.56°$. A suitable detector is located in the path of the diffracted iron X-rays and counts the X-rays as they enter. The count rate for a given wavelength of X-rays is dependent on the concentration of the element giving rise to these X-rays in the specimen. If the "fluorescing" X-rays are iron $K\alpha_1$ and $K\alpha_2$, then the number of counts per second is dependent on the iron concentration in the specimen: the higher the counts per second of iron X-rays, the greater the concentration of iron in the specimen.

Example 2.14A. Verify that the spectrometer setting (the spectrometer includes the detector, the analyzing crystal, the collimators, and the mechanism for setting the Bragg angle, θ; experimentally, the setting is 2θ) of 2θ for iron $K\alpha_1$ and $K\alpha_2$ X-rays is $57.55°$.

Solution: Iron $K\alpha_1$ X-rays have a wavelength of $\lambda = 0.19360$ nm. The interplanar spacing in the LiF analyzing crystal is $d_{(200)} = 0.2013$ nm. Therefore, $n\lambda = 2d \sin \theta$, or $1 \times 0.19360 = 2 \times 0.2013 \sin \theta$, or $\sin \theta = 0.19360/(2 \times 0.2013) = 0.48087$, and $\theta = 28.74°$. Therefore, $2\theta = 57.49°$. For iron $K\alpha_2$, $\lambda = 0.19400$ nm and $\sin \theta = 0.19400/(2 \times 0.2013) = 0.48187$, or $\theta = 28.81°$ and $2\theta = 57.61°$. The average 2θ for $K\alpha_1$ and $K\alpha_2$ is $57.55°$.

To calculate the weight percent of a given element in a metallic specimen, generally it is necessary to obtain the counts per second for a known material. In general, the composition of the known specimen should be as close as possible to that of the unknown specimen due to interelement effects: each element absorbs X-rays differently, and one element may enhance or diminish the fluorescent X-rays of another. Computer programs are available to adjust automatically for interelement absorption effects.

Example 2.14B. A National Bureau of Standards certified sample containing 0.105% Fe in a brass alloy gave a counting rate of 595 counts per second (cps). The background is 110 cps, and an unknown brass specimen gave 467 cps. Calculate the weight percent iron in the unknown.

Solution: The net counts per second for the known standard is the total counts per second minus the background counts per second, or net cps $= 595—110 = 485$ cps. Because the standard contains 0.105% Fe, the cps for 1.00% Fe are 485 cps/0.105% $= 4619$ cps/1.00% Fe. Because the net cps for the unknown are $467—110 = 357$ cps, the concentration of iron in the unknown is wt% Fe $= 357$ cps/(4619 cps/1.00% Fe) $= 0.077%$ Fe. In some cases, when the composition of the unknown is not near that of the standard, calculations involving two or more standards must be used because the cps/1.00% of the given element is a variable that depends on composition. Formulas and examples of their use may be found in standard texts on X-ray fluorescence.

It should be noted that X-ray fluorescence often depends on X-rays emitted from a thin surface layer of the material, for example, about 5 μm (0.0002 in.) from metals such as copper. The X-rays of elements having high atomic numbers, e.g., silver, will come from much thicker layers of materials because the X-rays are high energy and therefore more penetrating. Matrices of low atomic number materials, such as organic materials, enable X-rays of many elements to come from a much greater depth than 5 μm. To use X-ray fluorescence effectively, the composition of the surface must be representative of the composition of the interior, but this can usually be accomplished by proper surface preparation, such as cleaning and abrading.

2.15 Metallography

The metallograph is a microscope that has been specially designed for investigating the microstructures of metallic and ceramic materials. Because properties strongly depend on microstructure, it is frequently necessary to investigate the microstructure of a material in order to understand its properties and behavior.

Metallography involves the preparation of a metal for visual inspection by means of the metallograph or electron microscope. In metallography, a specimen generally is first sectioned to obtain a convenient-size specimen, and then the specimen is ground with several abrasive papers having successively finer abrasives. Finally the specimen is polished using very fine abrasives suspended in water or oil—billiard or nylon or other cloths are used—before a chemical etching step. This step increases the contrast among various features of the microstructure, such as the grain boundaries or precipitate particles. Many illustrations in this text utilize metallography to emphasize various facts or principles.

2.16   Electron Microscopy

An electron beam has two important advantages over visible light: its short wavelength and its ability to penetrate a specimen. Because the wavelengths of electrons are very short, the limit of resolution should be about 0.01 nm. However, instrumental factors limit the resolution of present microscopes to 0.2 or more nanometers. Electrons are able to penetrate from 1.0 to 100 nm in most materials.

Electron beam microscopes require different designs than optical microscopes because of the difference between electrons and photons. The main column of the electron microscope must be evacuated to prevent absorption of the electrons by air. Collimating, objective, and ocular "lenses" comprise magnetic fields that may be controlled to produce the desired results. The electrons transmitted through the specimen are focused to form an image that may be viewed or photographed. Magnification up to at least 250,000× is possible.

The scanning electron microscope (SEM), with its relatively large depth of focus, is used to observe the surface of a specimen directly. The beam is extremely small and strikes only a tiny fraction of the specimen at an instant. In a manner similar to that of a television screen, the beam is scanned rapidly over the surface of the specimen. The resulting backscattered and secondary electrons form the image on a fluorescent screen.

2.17   The Electron Microprobe

The electron microprobe enables the chemical composition of an extremely small volume of material (a few cubic micrometers) to be determined by focusing a fine beam of electrons into a metallographic specimen. When electrons strike the various elements present in the excited small volume of metal, characteristic X-rays are emitted as in X-ray fluorescence analysis. The greater the percentage of a given element present, the higher the counting rate of its characteristic X-rays. The electron microprobe is useful for determining the compositions of various phases, such as that of small inclusions, and segregation due to precipitation, diffusion, and so forth.

2.18   Electron Diffraction

Electron diffraction is analogous to X-ray diffraction in determining crystal structures because of the wave nature of electrons. Because electrons are strongly absorbed by materials, the diffraction pattern gives information only on the structure of the surface layer. In electron diffraction, electrons from a heated filament are accelerated to a desired velocity (energy and therefore wavelength) by a potential of about 50,000 V. Magnetic lenses adjust the beam, and it then strikes the specimen and produces a diffraction pattern on a photographic plate.

The Bragg angle, θ, is much smaller than in X-ray diffraction because lambda for electrons is much smaller than λ for X-rays. Electron diffraction may be used to identify substances and to obtain information on grain size and orientation and the state of stress. Electron diffraction is frequently combined with electron microscopic examination.

2.19   Neutron Diffraction

Because a nuclear reactor is needed as an intense source of neutrons, the usefulness of this technique is limited. The neutrons coming from a reactor have a variety of wavelengths or energies and are made monochromatic (having a single energy or wavelength) by diffraction from a lead crystal. This reduces the intensity of the neutron beam by a factor of about 1000. In neutron diffraction, the nucleus causes most of the diffraction, while electrons are responsible for X-ray and electron diffraction. Thus, low atomic number elements, such as carbon and hydrogen, are poor scatterers for diffraction of X-rays or electrons. It is difficult to locate hydrogen atoms by X-ray diffraction when heavy elements are present. Additionally, it is difficult to distinguish between isoelectronic (the same number of electrons) ions or atoms nearly isoelectronic with each other. In neutron diffraction, various elements show an irregular variation in neutron-scattering power. Because hydrogen, beryllium, and carbon have higher relative scattering powers, these atoms may be located in structures containing these atoms.

2.20   Problems

1. Calculate the radius of an atom in an FCC metal that has a unit cell dimension of a = 0.4049 nm. What is the metal?

2. The atomic radius of BCC molybdenum is 0.1362 nm; calculate the unit cell dimension (check your answer by referring to Table 2.1).

3. Calculate the weight in grams of one sodium atom; the atomic weight of sodium is 23.0 g.

4. Calculate the volume of a unit cell of BCC niobium; the radius of a niobium atom is 0.1430 nm.

5. Calculate the density of aluminum based on the mass and volume of one unit cell.

6. According to Table 2.2, which of the lanthanide metals are *not* known to be polymorphic? (If the element is not in the table, assume that it is not known definitely whether it is polymorphic.) Would you predict that these metals are probably polymorphic? Why?

7. If you were to investigate crystal structure changes as a function of pressure, which metals would you study first? Last? Why?

8. Would you predict that most actinides are polymorphic? Explain your answer.

9. Draw cubic coordinate systems and sketch the following directions: [110], [120], [30$\bar{1}$], [01$\bar{1}$].

10. Draw cubic coordinate systems and sketch the following planes: (110), ($1\bar{1}1$), (021), (302).

11. How many atoms per centimeter lie in the [110] direction in calcium? In the [111] direction?

12. What is the interplanar spacing for the (310) planes in chromium if copper $K\alpha_1$ X-rays are diffracted at an angle of 57.63°? $\lambda = 0.15406$ nm for Cu $K\alpha_1$.

13. What metal was used as an X-ray tube target in X-ray diffraction when the (400) planes of iron were diffracted at a Bragg angle of 7.44°?

14. What is the Bragg angle θ for diffraction of copper $K\alpha_1$ X-rays having a wavelength of 0.15406 nm from the (211) planes of vanadium having a unit cell constant of 0.304 nm?

15. Niobium cannot be distinguished from tantalum by X-ray diffraction alone. Why not? What additional simple, nonchemical test could be made to distinguish between the two metals (the two elements are very similar chemically)?

16. Using tungsten $K\alpha_1$ X-rays, how many orders of diffraction could be obtained for the (111) planes of calcium metal?

17. In an X-ray diffractometer scan using copper $K\alpha_1$ X-rays, a peak occurred at $2\theta = 65.14°$. What are the interplanar spacing and the indices of the planes giving this peak in FCC aluminum having a = 0.4049 nm?

18. The following data were obtained from a diffractometer scan using copper $K\alpha_1$ X-rays:

Relative peak intensity	2θ, degrees
47	39.96
11	58.04
30	72.98

Determine the crystal structure of this cubic metal, calculate a for each peak, and identify the metal by referring to Table 2.1.

19. Show why X-ray diffraction will not occur from (111) planes of molybdenum.

20. A sheet of copper was used for obtaining an X-ray diffractometer scan. No peak was found for the (111) planes, while the (200) peak was unusually intense. What is the explanation for these results?

21. Assuming that the $L\alpha_1$ peak is the best one for determining antimony by X-ray fluorescence, what angle 2θ would be used to detect antimony using a LiF analyzing crystal (d = 0.2013 nm)?

22. In an X-ray diffractometer scan of a cold-rolled steel, the background was unusually intense and interfered with interpreting the scan. The tube target was copper. Why was the background so high?

23. A steel wire was cold drawn through a die. How would you mount the wire specimens to see all the features of the microstructures?

24. Despite careful grinding and polishing procedures, scratches were still obtained on a soft brass specimen. How could you ensure that no scratches would be present on the specimen?

3

Properties of Metals

The properties of materials determine their usefulness. For a given application, a combination of properties is usually required. There are relatively few applications where cost is *not* a major factor. Consequently, for a given application engineers are interested in materials that provide certain combinations of specified properties or performance characteristics at minimum cost. Design engineers cannot afford to overdesign more than a normal amount as a safety factor: it is senseless to pay more for an automobile, for example, with bumpers that last five times as long as the rest of the car, if bumpers that last as long as the other parts are available at lower cost.

In this chapter not only are the properties of metals discussed, but also some general principles, particularly relating to mechanical properties of all types of materials. The properties discussed in this chapter include *mechanical properties*, such as tensile strength, yield strength, elongation (ductility), toughness and hardness; *physical properties*, such as optical and thermal (Chapter 14), electrical and electronic (Chapter 13), and magnetic properties (section 3.9 in this chapter); and *chemical properties*, such as chemical reactivity, solubility, surface energy, diffusivity (Chapter 4), basicity, and corrosion and oxidation (Chapter 15).

3.1 Elasticity and Young's Modulus

When materials are stressed, one or more of the following will occur: (1) the material will undergo temporary deformation that lasts until the stress is removed, at which time the material returns to its original dimensions, for example, the stretching of a rubber band; this is *elastic deformation;* (2) the material may undergo permanent deformation, called *plastic deformation,* in which the material does *not* return to its original shape when the stress is removed (for example, the bending of a metal coat hanger); (3) the material may fracture into two or more pieces, such as the shattering of a glass windowpane.

Obviously, gases do not have bulk properties of elastic or plastic deformation, but in a given gas, each atom collides elastically with other atoms in the gas. Liquids that have high flow rates (low viscosities) do not normally have useful mechanical properties, whereas extremely high viscosity liquids and crystalline solids do exhibit elasticity and sometimes plasticity as well. Glass is an amorphous material and is considered by materials scientists to be a rigid, supercooled liquid. Over long periods of time, it exhibits the property of flow to a small extent, and it may crystallize to form a solid. On our time scale, glass appears to be a "solid" because it is elastic (glass fibers bend easily), it will fracture, and it has minimal flow at room temperature. Several common polymers, such as polyvinylchloride (PVC), are amorphous and are rigid, supercooled liquids or are flexible, high-viscosity liquids (pseudo-solids). The distinguishing characteristic of a solid is that it is crystalline. Liquids are amorphous, condensed states of matter.

Metals are almost always crystalline when cooled far beneath the melting point, although it is possible by extremely rapid cooling to obtain amorphous metallic ''glass.'' All crystalline metals are elastic. Elasticity is due to the deformability of electron clouds: if a material is compressed, then the electron clouds are compressed slightly, and the volume decreases. When a material is stretched, then the chemical bonds, involving electron clouds, are elongated, and the material stretches. Many metals deform plastically, and all metals fracture when stressed beyond certain limits.

Stress is force per unit area, such as newtons/square meter (pascals, Pa) or pounds/square inch (psi). Because a newton is not a large force and a square meter is a large area, one pascal is a very small stress: 1 Pa = 1.451 × 10^{-4} psi. Stress is classified as follows: (1) *tensile stress* is a force or forces applied in opposite directions, for example, a load hung on a hook; (2) *compressive stress* is a force or forces applied toward a common point, for instance, a weight placed on a brick (hydrostatic stresses are forces applied on all sides of a material, such as a material immersed in a liquid that is compressed by a piston); (3) *shear stresses* are those that are due to forces on an object when the forces are slightly offset from one another, for example, the use of scissors in cutting a material. In real life when a material is stressed, it is usually subjected to a combination of the above types of stress; in fact, it is difficult to obtain any single type of stress in its pure form.

When a metal wire is loaded with a weight, it is subjected to a tensile stress. If the stress level is not excessive, the wire deforms elastically by stretching a small amount. The stretching of the wire is proportional to the load provided that the load does not exceed a certain value. When the load is removed, the wire contracts to its original length: it has temporarily deformed elastically, provided that the wire has undergone no permanent deformation. The increase in the length of the wire is proportional to the tensile stress (see Fig. 3.1): E = σ/ε, where E is the *Young's modulus of elasticity* in pascals or usually megapascals (1 MPa = 10^6 Pa), or psi; σ is the tensile stress in pascals (or psi), and ε is the strain in m/m, or cm/cm, or in./in.

Strain is fractional deformation (stretching or elongation): ε = Δl/l, which is unitless. Young's modulus is constant for a given metal at a given temperature. Table 3.1 lists E for a

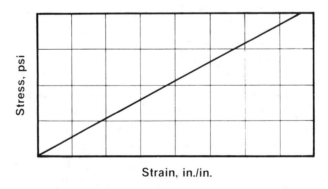

Fig. 3.1   Hooke's law

Table 3.1   Elastic constants for polycrystalline metals at 20 °C

Metal	Young's modulus, E, 10^6 psi	Bulk modulus, K, 10^6 psi	Shear modulus, G, 10^6 psi	Poisson's ratio, ν
Aluminum	10.2	10.9	3.80	0.345
Brass, 30 Zn	14.6	16.2	5.41	0.350
Chromium	40.5	23.2	16.7	0.210
Copper	18.8	20.0	7.01	0.343
Iron (soft)	30.7	24.6	11.8	0.293
(cast)	22.1	15.9	8.7	0.27
Lead	2.34	6.64	0.811	0.44
Magnesium	6.48	5.16	2.51	0.291
Molybdenum	47.1	37.9	18.2	0.293
Nickel (soft)	28.9	25.7	11.0	0.312
(hard)	31.8	27.2	12.2	0.306
Nickel-silver, 55 Cu, 18 Ni, 27 Zn	19.2	19.1	4.97	0.333
Niobium	15.2	24.7	5.44	0.397
Silver	12.0	15.0	4.39	0.367
Steel, mild	30.7	24.5	11.9	0.291
Steel, 0.75 C	30.5	24.5	11.8	0.293
Steel, 0.75 C, hardened	29.2	23.9	11.3	0.296
Steel, tool	30.7	24.0	11.9	0.287
Steel, tool, hardened	29.5	24.0	11.4	0.295
Steel, stainless, 2 Ni, 18 Cr	31.2	24.1	12.2	0.283
Tantalum	26.9	28.5	10.0	0.342
Tin	7.24	8.44	2.67	0.357
Titanium	17.4	15.7	6.61	0.361
Tungsten	59.6	45.1	23.3	0.280
Vanadium	18.5	22.9	6.77	0.365
Zinc	15.2	10.1	6.08	0.249

number of metals. Young's modulus may be thought of as being related to stiffness: the higher the value of E, the stiffer the material. For instance, metals in general are stiffer than plastics, which have Young's moduli ranging from about 689 to 10,300 MPa (0.1 to 1.5 × 10^6 psi), compared with 13,800 to 413,000 MPa (2 to 60 × 10^6 psi) for metals. Notice that tungsten is stiffer than steel, which is stiffer than lead (see the values for E in Table 3.1).

Young's modulus is the slope of the straight line when stress is plotted versus elastic strain (see Fig. 3.1). In general, rubbery materials, called *elastomers*, do not have a strictly linear stress-strain curve, and therefore a single value for E, but most solids and rigid liquids have a straight-line relationship between stress and elastic strain.

Example 3.1A. Aluminum has a Young's modulus of 1.0 × 10^7 psi (68,900 MPa). Calculate the elongation in 100 ft of aluminum wire, 0.1 in. in diameter, loaded with 20 lb.

Solution: E = σ/ϵ and E = 1.0 × 10^7 psi (given). σ = force/area = 20 lb/(π × 0.05^2) = 20/0.0025π = 2546 psi. ϵ = σ/E = 2546 psi/(1.0 × 10^7) = 2.546 × 10^{-4} in./in. Because the aluminum wire is 100 ft long, 100 ft × 12 in./ft = 1200 in. long, the total elongation or total strain is 1200 in. × 2.546 × 10^{-4} in./in., or 0.306 in.

Note: The modern engineer should be able to make calculations using either psi or MPa.

Example 3.1B. Calculate the modulus of elasticity of a steel rod that has a diameter of 0.505 in. and a strain of 1.66 × 10^{-4} in./in. when loaded with 1000 lb.

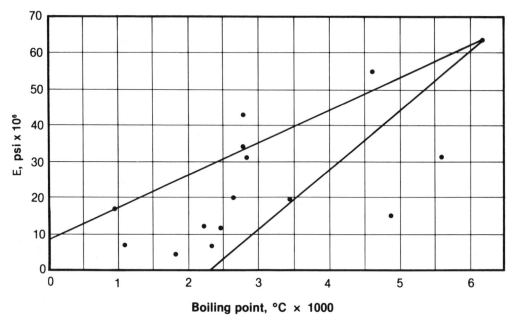

Fig. 3.2   Young's modulus versus boiling point of metals

Solution: Using $E = \sigma/\epsilon$, σ is the tensile stress = force/cross-sectional area, or $\sigma = 1000/[\pi \times (0.505/2)^2] = 1000/0.0638\pi = 1000/0.2003 = 4993$ psi. Therefore, $E = \sigma/\epsilon = 4993/(1.66 \times 10^{-4}) = 3.01 \times 10^7$ psi.

Young's modulus is very roughly proportional to the boiling point of metals (Fig. 3.2). Actually, both properties are related to the strength of the bonds in metals: the stronger the metal bonding, the higher the Young's modulus and the boiling point.

Young's moduli are anisotropic: the value of E depends on crystallographic direction in a single crystal. Table 3.2 presents some of the variations of E parallel to [111] and to [110]. Polycrystalline metals have values of E lying between the maximum and minimum values of E for single crystals. The Young's modulus also decreases with increasing temperature in a metal (for example, see Fig. 3.3) because the strength of bonding decreases with increasing temperature. Note the abrupt change in E in Fig. 3.3 when BCC iron transforms to FCC iron.

Table 3.2   Anisotropic Young's moduli in metals

Metal	Maximum E, [111] direction, 10^6 psi	Minimum E, [100] direction, 10^6 psi	Random crystal orientations, 10^6 psi
Aluminum-FCC	11	9	10
Copper-FCC	28	10	16
Gold-FCC	16	6	12
Iron-BCC	41	19	30
Lead-FCC	4	1	2
Tungsten-BCC	57	57	57

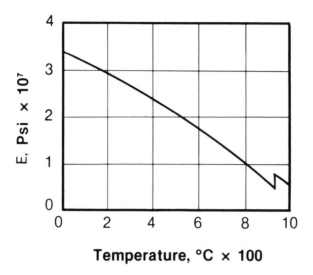

Temperature, °C × 100

Fig. 3.3   Young's modulus versus temperature in iron

3.2   Poisson's Ratio

When a material is subjected to an axial tensile stress, it expands in the direction of the stress and contracts laterally. Figure 3.4 shows the top view of a cube of metal that has been pulled axially along the z-axis. If the material is isotropic, such as polycrystalline metals, ϵ_x and ϵ_y are equal. *Poisson's ratio* is defined as:

$$\nu = \frac{-\epsilon_x}{\epsilon_z}$$

where $-\epsilon_x$ is the lateral contraction (the negative sign indicates contraction), and ϵ_z is the axial elongation. If a material maintains constant volume when it is elongated, it may be shown that $\nu = 0.5$. The value of 0.5 is the theoretical upper limit for Poisson's ratio: polyisoprene, a type of rubber, has $\nu = 0.49$. However, most metals have Poisson's ratios in the range of 0.25 to 0.40 (see Table 3.1).

Example 3.2A. A cubic inch of iron is subjected to a tensile stress of 25,000 psi. What are the dimensions of the elastically deformed iron? Poisson's ratio for iron is 0.291 and E is 3.07×10^7 psi.

Solution: Using $E = \sigma/\epsilon_z$, $E = 3.07 \times 10^7$ psi, and $\sigma = 25,000$ psi, $\epsilon_z = \sigma/E = 25,000$ psi/$(3.07 \times 10^7$ psi$) = 8.14 \times 10^{-4}$ in./in. The dimension in the direction of tensile stress is $1.0000 + 8 \times 10^{-4}$ in. $= 1.0008$ in. Because $\nu = -\epsilon_x/\epsilon_z$, and $\nu = 0.291$, $\epsilon_x = -0.291\epsilon_z$, and $\epsilon_x = -0.291 \times 8.14 \times 10^{-4} = -2.37 \times 10^{-4}$ in./in. The length in the x-dimension is 1.00000 in. $-$ 0.00024 in., or 0.99976 in. This is also the y-dimension. The deformed steel block has dimensions of 1.0008 in. $\times$ 0.99976 in. $\times$ 0.99976 in.

Note: The volume of the deformed steel block is slightly greater than that of the original block: 1.00034 in.^{3} versus 1.00000 in.^{3}.

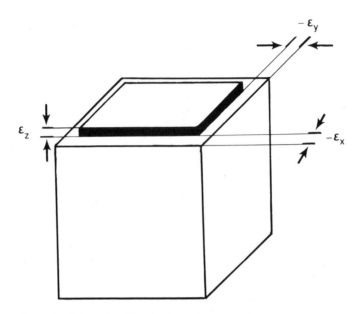

Fig. 3.4   Poisson's ratio and lateral contraction

3.3   Bulk Modulus of Elasticity

Hydrostatic compression causes metals to contract in volume, $\Delta V/V$, and this relative volume is initially proportional to the hydrostatic pressure:

$$\frac{\Delta V}{V} = \beta\sigma_{hyd}$$

where $\Delta V/V$ is the relative volume contraction in m^3/m^3 (or $in.^3/in.^3$). β is a constant having units $(m^3/m^3)Pa$ or $(in.^3/in.^3)psi$. σ_{hyd} is the hydrostatic pressure in pascals or pounds per square inch. β is called the *compressibility* and is constant for a given metal.

The *bulk modulus of elasticity* is the constant that relates the ratio of hydrostatic pressure to the resulting volume change, or:

$$K = \frac{1}{\beta} = \frac{\sigma_{hyd}}{\Delta V/V}$$

Values of bulk moduli, K, are given in Table 3.1. Young's modulus, E, and the bulk modulus, K, are related through Poisson's ratio:

$$E = 3K(1 - 2\nu)$$

Note that if $\nu = 0.50$, representing zero volume expansion on application of tensile stress, then $K = \infty$, or $\beta = 0$, representing no compressibility.

Example 3.3A. When iron is compressed hydrostatically with 20,000 psi, the volume is changed by 0.0816%. Calculate the compressibility and the bulk modulus. Poisson's ratio is 0.291.

Solution: $\Delta V/V$ is 0.0816%, or 8.16×10^{-4}. $\sigma_{hyd} = 20,000$ psi. Because $\Delta V/V = \beta \times \sigma_{hyd}$, then $\beta = (\Delta V/V)/\sigma_{hyd} = (8.16 \times 10^{-4})/20,000 = 4.08 \times 10^{-8}$. $K = 1/\beta = 1/(4.08 \times 10^{-8}) = 2.45 \times 10^7$ psi.

Note: Since $E = 3K(1 - 2\nu)$, then $E = 3 \times 2.45 \times 10^7 (1 - 2 \times 0.291)$, or $E = 7.35 \times 10^7 (1 - 0.582) = 7.35 \times 10^7 \times 0.418 = 3.07 \times 10^7$ psi, which agrees with the value for Young's modulus for iron. This last step is merely a check on the above calculations.

3.4 Plastic Deformation

When a metal is stressed beyond a certain level, it deforms permanently and will not return to its original size and shape when the stress is removed. This is *plastic deformation*. The stress at which plastic deformation begins is called the *yield stress* or elastic limit. Brittle materials, such as glass and ceramics in general, chromium or other brittle metals, and some plastics such as Bakelite (phenol formaldehyde), fracture before appreciable plastic deformation occurs.

Plastic deformation is a useful property in shaping many materials, but once the material is in final form, further plastic deformation is usually undesirable. Some metals, such as iron, have a sharp yield point, while most materials have gradual transitions from elastic to plastic deformation, e.g., aluminum or polyethylene. *Elongation* is the relative permanent change in length of a plastically deformed metal. This is the same as plastic strain:

$$\text{Plastic strain} = \text{elongation} = \frac{l_f - l_o}{l_o}$$

where l_f is the final length and l_o is the original length. Notice that elongation, or strain, is unitless; actually strain has the units of meters/meter or inches/inch, etc. *Ductility* is the ability of a material to be elongated, usually by drawing through a die or stretching. Malleability refers to the ability of a metal to be deformed plastically by hammering. Although technically there is a difference between ductility and malleability (lead is highly malleable but not very ductile), the term ductility is commonly used to mean the ability to undergo plastic deformation. Ductility may also be measured by determining the reduction of area:

$$\text{Reduction of area} = \frac{A_o - A_f}{A_o}$$

where A_o indicates the original cross-sectional area, which is larger than A_f, the final cross-sectional area.

Tensile strength is defined as the maximum strength based on original dimensions: it is measured in pascals (usually megapascals, MPa, or 10^6 Pa) or in pounds per square inch. When a piece of metal is stressed in a tensile testing machine, it stretches elastically until the yield

Table 3.3 Mechanical properties of steel and ferrous alloys

Alloy and condition(a)	Yield strength, psi	Tensile strength, psi	Elon-gation, %	Hardness(b)	Impact strength, Izod, ft-lb
Carbon and Alloy Steels					
1015, annealed	41,250	56,000	37.0	111 HB	84.8
1022, annealed	46,000	65,250	35.0	137 HB	89.0
1030, annealed	49,500	67,250	31.2	126 HB	51.2
1030, normalized	50,000	75,500	32.0	149 HB	69.0
1030, Q&T 204 °C	94,000	123,000	17	495 HB	· · ·
1030, Q&T 316 °C	90,000	116,000	19	401 HB	· · ·
1030, Q&T 427 °C	84,000	106,000	23	302 HB	· · ·
1030, Q&T 538 °C	75,000	97,000	28	255 HB	· · ·
1040, annealed	51,250	75,250	30.2	149 HB	32.7
1050, annealed	53,000	92,250	23.7	187 HB	12.5
1060, annealed	54,000	90,750	22.5	179 HB	8.3
1080, annealed	54,500	89,250	24.7	174 HB	4.5
4140, annealed	60,500	95,000	25.7	197 HB	40.2
4340, annealed	68,500	108,000	22.0	217 HB	37.7
4340, Q&T 204 °C	243,000	272,000	10	520 HB	· · ·
4340, Q&T 316 °C	230,000	250,000	10	486 HB	· · ·
4340, Q&T 427 °C	198,000	213,000	10	430 HB	· · ·
5140, annealed	42,500	83,000	28.6	167 HB	30.0
6150, annealed	59,750	96,750	23.0	197 HB	20.2
8650, annealed	56,000	103,750	22.5	212 HB	21.7
9310, annealed	63,750	119,000	17.3	241 HB	58.0

(continued)

stress is exceeded. Then the metal deforms plastically, usually by *necking down*, which is a local moderate decrease in the cross-sectional area due to plastic deformation. Failure need not occur yet, and indeed in some materials, such as polyethylene, it occurs long after necking down. Practically speaking, tensile strength is a very important property because it is a measure of the maximum load a material may withstand before failure. The *true tensile strength* equals the load divided by the reduced cross-sectional area due to necking down. Hence, the true tensile strength may be much greater than the tensile strength as normally measured, but the concept of true tensile strength is not very important from the practical standpoint: the maximum *load* that a given material will sustain is important rather than the load divided by the "necked-down" cross-sectional area. The *breaking strength* is the stress at failure. Often this is appreciably less than the tensile strength, indicating that necking down and drawing can occur extensively prior to failure, causing the strength to diminish from the maximum, i.e., the tensile strength. Table 3.3 contains values of various mechanical properties for selected steels and ferrous alloys. Tables 3.4 and 3.5 tabulate mechanical properties of nonferrous metallic elements and nonferrous alloys, respectively.

3.5 Stress-Strain Curves

One of the most useful diagrams in materials science, particularly to design engineers, is the *stress-strain curve*, which is a plot of stress in MPa (or psi) versus strain, usually measured in meters/meter or inches/inch along the x-axis. Stress-strain curves summarize quite a lot of useful information on mechanical properties: (1) yield stress, (2) tensile stress, (3) breaking

Table 3.3   Mechanical properties of steel and ferrous alloys (continued)

Alloy and condition(a)	Yield strength, psi	Tensile strength, psi	Elon-gation, %	Hardness(b)	Impact strength, Izod, ft-lb
Stainless Steels					
202	55,000	105,000	55	90 HRB	. . .
302	40,000	90,000	50	85 HRB	. . .
304	42,000	84,000	55	80 HRB	. . .
310	45,000	95,000	45	85 HRB	. . .
316	42,000	84,000	50	79 HRB	. . .
405	40,000	65,000	25	75 HRB	. . .
430	50,000	75,000	25	85 HRB	. . .
440A	60,000	105,000	20	95 HRB	. . .
446	50,000	80,000	20	83 HRB	. . .

Metal	Modulus of elasticity psi × 10^6	Brinell hardness	Tensile strength, psi	Yield strength, psi(c)	Elonga-tion in 2 in., %	Endurance limit, psi
Some Cast Ferritic Metals						
Gray cast iron, class 20	12	160	22,000	. . .	. . .	10,000
Gray cast iron, class 40	17	220	44,000	. . .	. . .	. . .
Gray cast iron, class 60	20	260	63,000	. . .	. . .	. . .
Malleable cast, iron, ferritic	25	135	50,000	32,500	10	28,000
Malleable cast iron, pearlitic	26	185	65,000	45,000	10	28,000
Nodular cast iron, 80-55-06	24	220	100,000	67,000	6	~30,000
White cast iron	. . .	400	35,000	. . .	. . .	. . .
Cast low alloy steel	30	143	74,000	44,000	28	35,000
Cast carbon steel	30	131	63,000	35,000	30	30,000
Cast stainless steel	28	140	77,000	37,000	55	. . .

(a) For compositions, see Tables 6.2 and 6.3. Q&T = quenched and tempered. (b) HB = Brinell hardness; HRB = Rockwell B hardness. (c) 0.2% offset

stress, (4) elongation, (5) Young's modulus of elasticity, and (6) toughness, which has not yet been discussed.

Figure 3.5 shows a stress-strain curve for a brittle material, such as chromium, glass, or Bakelite. Failure occurs before any appreciable plastic deformation. There is no real yield stress—only a breaking or fracture strength, which may also be called the tensile strength. Figure 3.6 shows a stress-strain curve typical of iron; this element has a *sharp yield stress*, which is a definite stress at which plastic deformation suddenly begins. As soon as the yield stress is exceeded, the stress required for plastic deformation decreases appreciably—the minimum value is called the lower yield stress. Soon afterwards, the steel becomes stronger with increasing plastic deformation, and the stress increases with increasing strain until the tensile stress is achieved; this is the maximum of the stress-strain curve. At higher elongations, the stress is somewhat lower, until finally the breaking stress occurs.

Figure 3.7 shows a stress-strain curve typical of a metal such as nickel. Because it has no sharp yield point, the yield stress is defined as the stress at which 0.2% elongation has occurred. In order to locate this point, first locate 0.2% elongation on the x-axis; then from this point draw a line parallel to the elastic portion of the stress-strain curve until the newly

Table 3.4 Mechanical properties of nonferrous metals

Metal	Condition	0.2% yield stress, psi	Ultimate tensile strength, psi	Elonga-tion, %	Hard-ness(a)
Aluminum	...	...	7,600	61	15 HB
Calcium	Soft cast	1,800	7,000	55	17 HB
	Hard rolled	11,000	15,000	7	42 R15-T
Chromium	Annealed	...	28,000	0	110 HV
Cobalt	Soft	25,000	32,000	4–8	124 HB
	Hard	...	89,000	2–8	165 HB
Copper	Annealed OFHC	6,000	28,000	60	42 HB
	30% cold rolled OFHC	38,000	41,000	16	96 HB
Gold	Soft	0	16,000	30	33 HB
	Hard 60%, cold rolled	28,000	30,000	4	58 HB
Molybdenum	Annealed	40,200Y(b)	53,800	40–60	160 HV
	Hard sintered	145,000	155,000	1–3	225 HV
Nickel	Annealed	8,000	41,000	28	85 HB
Niobium	Soft sintered	...	35,200	49	40 HV
	Hard rolled	70,000	75,000	5	140 HV
Palladium	Annealed	4,600	25,000	40	37 HV
	Hard 50% cold drawn	27,000	42,000	1.5	106 HV
Platinum	Annealed	3,000	16,400	30–40	37 HV
	Hard 50% cold rolled	24,000	26,000	3	92 HV
Rhenium	Annealed	...	147,000	28	280 HV
	Hard 30% cold rolled	278,000	288,000	2	820 HV
Rhodium	Annealed	...	60,000	25	100 HB
	Hard rolled	...	272,000	...	260 HB
Silver	Soft	1,000	18,000	50	26 HV
	Hard	...	50,000	4	90 HV
Tantalum	Annealed	23,600	26,400	36	90 HV
	Hard 95% cold rolled	44,000	54,000	5	160 HV
Thorium	Annealed	6,200	15,400	36	38 HV
	Hard rolled	38,400	39,200	6	70 HV
Tungsten	Swaged and recrystallized	25,600	53,200	16	200 HV
	Hard swaged	...	230,000	1–4	490 HV
Vanadium	Annealed	11,800	25,600	38	55 HV
	Hard	38,400	74,400	24	145 HV
Zirconium	Annealed	10,000	28,000	25	60 HB
	Hard 60% cold rolled	75,000	80,400	10	200 HB

(a) HB = Brinell hardness; R15-T = Rockwell 15-T hardness; HV = Vickers hardness. (b) Sharp yield

constructed line intersects the stress-strain curve. The stress at this point is then called the yield stress. Note that at any point on the stress-strain curve the metal contracts elastically when stress is removed, and this contraction follows a line parallel with the elastic part of the stress-strain curve.

Example 3.5A. A specimen of steel wire 0.06 cm in diameter began to deform plasticly at a load of 22.0 lb. Calculate the yield stress.

Solution: The cross-sectional area is πr^2, and r is 0.06 cm/2 = 0.030 cm or 0.0118 in. The area is $\pi \times 0.0118^2 = 4.37 \times 10^{-4}$ in.^{2}. The yield stress is the load/cross-sectional area when plastic deformation starts, or Y.S. = 7.0 lb/4.37 $\times 10^{-4}$ in.^{2} = 16,000 psi.

Table 3.5 Mechanical properties of nonferrous alloys

Metal	Chemical composition	Condition	0.1% yield stress, psi	Ultimate tensile strength, psi	Elonga- tion, %	Brinell hard- ness, kg/mm²
Alclad 2024-T4	4.4 Cu, 0.75 Mn, 1.0 Mg, 0.4 Si, bal Al	Heat treated	32,000	54,000	15	· · ·
Alclad 7075-T6	1 Cu, 5.7 Zn, 2.5 Mg, 0.35 Mn, 0.12 Cr, bal Al	Heat treated	58,000	68,000	10	· · ·
Aluminum bronze	95 Cu, 5 Al	Annealed	19,000	55,000	63	80
		40% cold rolled	72,000	87,000	13	175
Brass	70 Cu, 30 Zn	Annealed	11,000	41,400	67.5	62
		30% cold rolled	49,000	60,000	19.5	132
Cu-Be	97.9 Cu, 1.85 Be, 0.25 Co	· · ·	28,000	62,000	47	110(a)
		Solution treated and aged	138,000	158,000	7	370(a)
		40% cold rolled and aged	176,000	188,000	2	425(a)
Cu-Ni	80 Cu, 20 Ni	Annealed	15,000	43,000	49	85
		40% cold rolled	61,000	66,000	9	148
Hastelloy B	28 Mo, 5 Fe, a little Mn, Si, bal Ni	Cold rolled	55,000	121,000	42	230
Inconel 6000	16 Cr, 6.5 Fe, bal Ni	Annealed	31,000	76,000	45	150
		Cold rolled	98,000	121,000	5	260
Mg-Al-Zn	6.0 Al, 1.0 Zn, bal Mg	Press forged	22,000	38,000	10	· · ·
Phosphor bronze	90 Cu, 10 Sn	· · ·	18,000	36,000	· · ·	90

(a) Vickers hardness

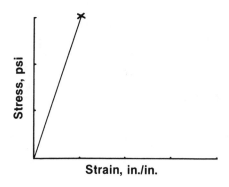

Fig. 3.5 Stress-strain curve for brittle metals

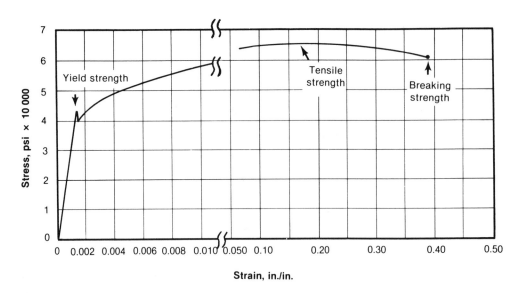

Fig. 3.6   Stress-strain curve for iron (containing 0.15% C)

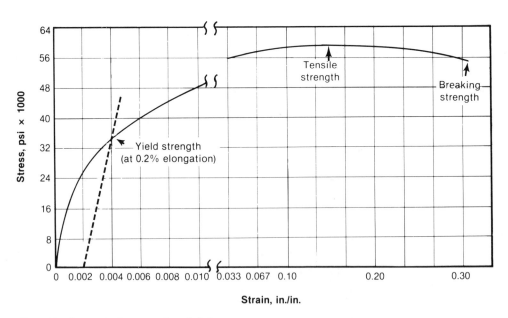

Fig. 3.7   Stress-strain curve for nickel

Example 3.5B. An aluminum rod originally had marks on it exactly 2 in. apart. At 6000 psi, which is the yield strength, the marks were 2.0012 in. apart, and at failure the marks were 3.22 in. apart. Calculate the Young's modulus and the percent elongation.

Solution: $E = \sigma/\epsilon$, and at 6000 psi, $E = 6000$ psi/[(2.0012−2.000)/2.000], or $E = 6000/(0.0012/2) = 6000/0.0006 = 1.0 \times 10^7$ psi. Percent elongation = $[(l_f−l_o)/l_o] \times 100 = [(3.22−2.00)/2.00] \times 100 = 61.0\%$.

Example 3.5C. What is the maximum load that may be applied to a brass rod 0.2 in. in diameter without having plastic deformation occur? The maximum elastic elongation is 0.00125 in./in. and $E = 1.46 \times 10^7$ psi.

Solution: Just below the yield stress, $E = \sigma/\epsilon$, or $\sigma = E\epsilon = 1.46 \times 10^7 \times 0.00125 = 1.82 \times 10^4$ psi. Tensile stress σ = load in pounds/cross-sectional area, or load = $\sigma \times$ cross-sectional area. Thus, load = $18,200 \times \pi r^2 = 18,200 \, \pi(0.2/2)^2 = 18200 \, \pi(0.01)$, or the load = 572 lb.

3.6 Toughness

Toughness is a measure of the amount of energy a material absorbs when it fractures. Usually, one is interested in being able to absorb impact without failure. In general, brittle materials are not tough at all: the ability of glass to absorb energy of impact is relatively small unless the glass has been toughened by a thermal process to place the surface under compression compared with the interior. In order to absorb a fair amount of energy in impact, a material either should be highly elastic or ductile. High strength is also important if a very tough material is desired. Rubber is highly elastic and therefore may absorb a fair amount of energy without failing; however, it does not have a high tensile strength. Armor plate is very strong yet fairly ductile. This combination makes an extremely tough material and, of course, thickness increases the overall toughness because the thicker the material, the more energy can be absorbed on impact without rupture.

Toughness is related to the area under the stress-strain curve: if this area is large, then the material is very tough, and if the area is relatively small, the material is brittle. Brittle materials have low toughness; however, they may be very strong but have poor ductility, or they may have adequate ductility but very low strength, or they may show some combination of these two factors. Figure 3.5 illustrates stress-strain curves of brittle materials, while Fig. 3.6 shows the stress-strain curve of a tough material, iron.

Toughness can be measured by any of a variety of impact tests. A common toughness tester holds a specimen securely and has a weighted arm or pendulum that falls freely from a preset height. The specimen must be broken to make the test, and the test comprises measuring the height of the swing of the weighted arm *after* fracturing the given specimen. In this way, the energy absorbed in impact is measured, and this is a measure of toughness. The specimen often has a notch cut in it; in certain instances, materials are notch sensitive and fail much more easily when notches are present. The American Society for Testing and Materials (ASTM) may be consulted for standard impact tests. The units of toughness are often given as foot-pounds of energy absorbed for a given size specimen. Several toughness values are presented in Table 3.3.

3.7   Hardness

Hardness is the resistance of a material to plastic indentation. This may involve a simple scratch test or indentation by loading a ball, diamond, or other penetrator with a weight and measuring the length, width, or depth of the indentation. The harder the material, the smaller the indentation for a given load. The higher the load, the larger the size of indentation for a given metal. Of course, the indenter material must be much harder than the material tested to prevent deformation of the indenter material. That is why indenters are frequently made of diamond, tungsten carbide, or hardened steel. The most common hardness tests are (1) Rockwell, using a diamond cone or hardened steel sphere, (2) Brinell, using a 10-mm steel or tungsten carbide sphere, (3) Vickers, using a square-based diamond pyramid, and (4) Knoop, using an elongated diamond pyramid. The hardness numbers are usually read directly from a table of values calculated for one parameter, such as the length, width, or depth of the indentation. Hardness values for various metals are given in Tables 3.3, 3.4, and 3.5.

Hardness testing may be either macro or micro; the latter involves using a microscope and either a Vickers or Knoop diamond indenter. Because the indenter is extremely small, the hardness of coatings or of different phases, such as a tiny precipitate particle in a metal, may be measured rather than the overall hardness.

Hardness is related to wear and abrasion resistance: it is usually desirable to have a very hard material in order to obtain good abrasion resistance. Soft materials scratch easily, and scratches generally are undesirable from the viewpoint of appearance. Therefore, a decorative metal should be hard enough to resist scractching.

Another useful concept is that, for a given metal, hardness is roughly proportional to strength (see Table 3.3). Thus, if one steel is harder than a second alloy, then the first one is also stronger. Hardness may therefore be used as a rapid comparative strength test, and frequently specifications call for minimum hardness. In applications such as machine tools, hardness is obviously a valuable property. Because heat treatments are frequently used to harden metals, as shown in Table 3.3 (see also section 4.16 in Chapter 4, on age hardening of aluminum alloys), hardness is used as a quality-control test in production. Hardness also increases with the degree of cold work (refer to Table 3.5).

3.8   General Physical Properties of Metals

Although mechanical properties are a subgroup of physical properties, they have been separated and placed in the foregoing sections because they are so important. Physical properties also include other important groups of properties: electric and electronic (see Chapter 13), optical and thermal (see Chapter 14), magnetic (section 3.9 in this chapter), and general.

General physical properties include crystal structure, coefficient of friction, density, viscosity of liquid, and miscellaneous other properties that do not readily fit into the other categories.

The crystal structures of metals have been discussed in Chapter 2. The *coefficient of friction* between two surfaces is the ratio of the force required to move one over the other to the total force pressing the two surfaces together:

$$\mu = \frac{F}{W}$$

where μ is the coefficient of friction, F is the force required to move one surface over another, and W is the force pressing the surfaces together. Of course, lubricants, which are used to prevent or reduce wear, markedly affect the friction between metals, and normally they are used in moving metal parts. The subject of lubricants is extensive and of great importance in industry.

Several processes may occur in the wear of metals: (1) seizing, in which a particle or *asperity*, which is a projection from one of the surfaces, digs into the other metal and causes local welding; finally, the welded metal asperity is broken; (2) deformation, in which the substrate may be locally deformed plastically followed by tearing loose of the deformed region; (3) corrosion, such as the removal of surface oxides, hydroxides, or sulfides, which may reform rapidly in air or even in oil because truly fresh metal surfaces are highly reactive chemically with nearly any environment; the repeated removal of oxides from the metal surface leads to wear; and (4) heating of the metal surface due to friction; melting rarely occurs, but heating decreases hardness and increases corrosion, both of which lead to wear. Hard surfaces are generally the most wear resistant, but the surface should also be corrosion resistant.

The densities of various metals are given in Tables 3.6 and 3.7. Viscosity is a property of fluids (both liquids and gases) and is a resistance to change of form. It is the reciprocal of fluidity or flow rate. The formal definition of viscosity is:

$$\eta = t \cdot \frac{\tau}{\gamma}$$

where η is the viscosity in the units of poise (dyne $\times$ s/cm^2); τ is the shear stress in dynes per square centimeter; γ is the shear strain in centimeters per centimeter; and t is the time in seconds. Viscosity is very temperature sensitive. Most liquid metals possess low viscosities because it is easy for metal atoms in the liquid to flow past one another due to their small size and nondirectional bonding. Polymers usually have high viscosities because it is difficult for large molecules to slip by one another. Viscosity usually increases with *decreasing* temperature according to:

$$\eta = \eta_0 e^{Q/RT}$$

where η_0 is a constant; Q is an activation energy, measured in calories per mole, for viscous shear of atoms as they pass one another; R is the molar Boltzmann constant (i.e., "gas law constant") 1.987 cal/°Kmol; T is the temperature in degrees Kelvin; and e is the natural logarithm base, 2.71828.

Table 3.6   Viscosities, surface tensions, and other properties of metals at their melting points

Metal	Melting point, °C	Density, ρ_d, g/cm^3	Electrical resistivity, ρ, 10^{-6} $\Omega \cdot$ cm	Thermal conductivity, k, cal $\cdot$ cm/s°K $\cdot$ cm^2	Viscosity, η, cP	Surface tension, ergs/cm^2
Aluminum	660.1	2.39	20.0	0.22	4.5	860
Antimony	630.5	6.49	113.5	0.052	1.30	383
Beryllium	1284	$\cdots$	(5)	$\cdots$	$\cdots$	1100
Bismuth	271	10.06	128.1	0.0262	1.68	393
Cadmium	320.9	8.02	33.7	0.105	1.4	666
Chromium	1875	6.46	36.6	0.06	0.684	1590
Copper	1083	7.96	21.1	0.118	3.36	1285
Gallium	29.8	6.20	2.8	0.08	2.04	735
Gold	1063	17.32	2.3	$\cdots$	$\cdots$	754
Indium	156.4	7.03	33.1	0.1	1.69	559
Iridium	2443	20.0	5.3	$\cdots$	$\cdots$	2250
Iron	1537	7.15	9.71	$\cdots$	2.2	1675
Lead	327.4	10.68	95.0	0.039	2.634	470
Lithium	180	0.516	24.0	0.11	0.60	398
Magnesium	650	1.585	27.4	0.333	1.24	556
Mercury	Values at 20 °C	13.55	98.4	0.020	1.554	465
Molybdenum	2620	9.34	5.17	$\cdots$	$\cdots$	2250
Nickel	1453	7.90	6.84	$\cdots$	$\cdots$	1756
Palladium	1552	10.7	10.8	$\cdots$	$\cdots$	1500
Potassium	63.6	0.825	13.2	$\cdots$	0.534	101
Platinum	1769	19.7	9.81	$\cdots$	$\cdots$	1740
Rhodium	1960	10.65	4.3	$\cdots$	$\cdots$	2000
Rubidium	38.8	1.475	11.3	0.1	0.673	76
Silver	960.8	9.33	17.2	$\cdots$	3.9	930
Sodium	97.8	0.929	9.6	0.205	0.726	191
Thallium	303	11.29	73.1	0.06	$\cdots$	490
Tantalum	2980	15.0	13.5	$\cdots$	$\cdots$	2150
Tin	231.9	6.97	48.0	0.08	1.97	579
Titanium	1670	4.13	55	$\cdots$	$\cdots$	1510
Tungsten	3380	17.6	5.5	$\cdots$	$\cdots$	2310
Zinc	419.5	6.64	37.4	0.144	3.93	824

3.9   Magnetic Properties

In a vacuum, a magnetic field, H, produces a magnetic flux density of B, and the proportionality constant is μ_o. When a material is placed in a magnetic field, the magnetic flux density is:

$$B = \mu_o \mu_r H$$

where μ_r is the relative permeability. For a vacuum, $\mu_r = 1$; for *paramagnetic* materials, μ_r is slightly greater than 1, and the materials are weakly magnetic. Paramagnetic metals often have one or more unpaired electrons in the isolated atom. In *diamagnetic* metals, which are slightly repelled by magnetic fields, μ_r is a little less than 1. Several metals have extremely high permeabilities—these are the *ferromagnetic* metals, i.e., the metals that are moderately or strongly magnetic. Iron, cobalt, and nickel are the only elemental metals that are ferromagnetic

Table 3.7   Some thermal and electrical properties of metals

Metal	Melting point, °C	Boiling point, °C	Density, g/cm³	Thermal conductivity, cal·cm/s°K·cm² at 25 °C	Specific heat cal/g°C at 20 °C	Coefficient of thermal expansion, cm/cm°C × 10⁻⁶ (0 to 100 °C)	Resistivity, 10⁻⁶ Ω·cm at 20 °C
Aluminum	660.4	2494	2.699	0.57	0.215	23.5	2.69
Antimony	630.5	1587	6.684	0.042	0.0495	8 to 11	42
Arsenic	⋯	613(subl)	5.727	0.120	0.0785	4.7	33.3
Barium	725	1638	3.51	0.044	0.046	~18	50
Beryllium	1278	2770	1.848	0.40	0.436	12	4.0
Bismuth	271.3	1560	9.80	0.019	0.0296	13.4	116
Cadmium	321.0	765	8.642	0.22	0.0555	31	7.4
Calcium	839	1484	1.54	0.3	0.156	22	4.1
Cerium	798	3433	6.771	0.03	0.0459	8	75.3
Cesium	28.4	670	1.878	0.086	0.057	97	21
Chromium	1875	2680	7.20	0.165	0.107	6.5	12.9
Cobalt	1495	2900	8.9	0.165	0.107	12.5	6.24
Copper	1083	2567	8.92	0.94	0.0922	17.0	1.67
Dysprosium	1409	2567	8.540	0.023	0.0414	9	92.6
Erbium	1522	2868	9.045	0.023	0.0410	9	86.0
Europium	822	1529	5.253	0.033	0.0421	26	91.0
Gadolinium	1311	3273	7.898	0.02	0.055	4	131.0
Gallium	29.8	2403	5.904	⋯	0.089	18.3	17.4
Germanium	937.4	2830	5.32	0.14	0.077	5.75	46,000,000
Gold	1064	2857	19.3	0.70	0.0308	14.1	2.3
Hafnium	2227	4602	13.31	0.05	0.035	6.0	30.6
Indium	156.6	2073	7.30	0.196	0.056	24.8	9.0
Iridium	2447	4500	22.42	0.14	0.0312	6.8	5.3
Iron	1538	2870	7.86	0.17	0.107	12.1	9.71
Lanthanum	920	3464	6.166	0.035	0.0479	5	5.70
Lead	327.5	1750	11.34	0.082	0.031	29.0	20.6
Lithium	180.5	1342	0.534	0.17	0.84	56	9.35
Magnesium	648.8	1107	1.74	0.40	0.244	26.0	3.9
Manganese	1244	2095	7.20	0.0187	0.114	23	160
Mercury	−38.9	357	13.59	0.022	0.0332	61	95.8
Molybdenum	2617	5560	10.2	0.34	0.0598	5.1	5.7
Neodymium	1010	3074	7.003	0.14	0.04	6	64.3
Neptunium	640	3902	20.45	0.015	⋯	⋯	⋯
Nickel	1453	2732	8.90	0.21	0.106	13.3	6.84
Niobium	2468	4927	8.57	0.13	0.064	7.2	14.5
Osmium	2700	5500	22.48	0.14	0.0311	4.57	9.5
Palladium	1552	3480	11.97	0.17	0.0584	11.0	10.8
Platinum	1772	3827	21.45	0.17	0.0317	9.0	10.6
Plutonium	641	3232	19.84	0.08	⋯	54	141
Polonium	254	962	9.4	⋯	0.03	⋯	⋯
Potassium	63.6	756.5	0.86	0.22	0.180	83	6.86
Praeseodymium	931	3520	6.772	0.13	0.046	4	68.0
Radium	700	1140	~5	⋯	0.0288	⋯	⋯
Rhenium	3180	5900	20.53	0.17	0.0329	6.6	19.1
Rhodium	1966	3727	12.4	0.20	0.0582	8.5	4.7
Rubidium	38.9	688	1.532	0.139	0.0860	90	12.5
Ruthenium	2310	4080	12.30	0.280	0.0569	9.6	7.3
Samarium	1072	1794	7.537	0.032	0.0469	⋯	105.0
Scandium	1539	2832	2.989	0.038	0.117	⋯	50.9
Selenium	217	685	4.81	0.005	0.0767	37	12.0
Silicon	1410	3280	2.34	0.2	0.168	7.6	$2.3E+11$
Silver	961.9	2163	10.5	1.00	0.564	19.1	1.6
Sodium	97.8	882.9	0.97	0.32	0.293	71	4.6

(continued)

Table 3.7   Some thermal and electrical properties of metals (continued)

Metal	Melting point, °C	Boiling point, °C	Density, g/cm³	Thermal conductivity, cal·cm/s°K·cm² at 25 °C	Specific heat cal/g°C at 20 °C	Coefficient of thermal expansion, cm/cm°C × 10⁻⁶ (0 to 100 °C)	Resistivity, 10⁻⁶ Ω·cm at 20 °C
Strontium	769	1099	2.6	0.085	0.0719	. . .	22.76
Tantalum	2996	5425	16.6	0.130	0.0334	6.5	13.5
Tellurium	449.5	990	6.25	0.014	0.0481	1.7	400,000
Terbium	1360	3230	8.234	0.027	0.0436	7	114.5
Thallium	303.5	1423	11.85	0.094	0.0307	30	16.6
Thorium	1750	4790	11.7	0.09	0.0276	11.2	18.62
Tin	232.0	2770	7.28	0.155	0.0543	23.5	12.8
Titanium	1660	3287	4.5	0.041	0.125	8.9	55
Tungsten	3410	5700	19.35	0.394	0.0320	4.5	5.5
Uranium	1132	3818	19.05	0.07	0.0277	13	29
Vanadium	1890	3380	5.96	0.07	0.116	8.3	26
Ytterbium	824	1193	6.972	0.083	0.0346	. . .	25.1
Yttrium	1523	3337	4.457	0.041	0.0713	. . .	59.6
Zinc	419.6	907	7.14	0.265	0.0925	31	5.92
Zirconium	1852	3700	6.49	0.05	0.0666	5.9	44.6

at room temperature, but gadolinium and some other rare earth metals are ferromagnetic at very low temperatures. The temperature at which a metal loses its ferromagnetism is called the *Curie temperature*.

In a ferromagnetic material, the electron spins of unpaired electrons align to produce a *magnetic domain*, which is a small region of metal in which the electron spins are aligned. Each domain, which is not confined to one crystal, acts like a small magnet. In soft iron, the domains are not permanently oriented. Rather, they orient themselves randomly so that there is no net magnetic field. However, in the presence of a magnetic field, the domains have their electron spins all lined up, and the iron becomes strongly magnetic (one can magnetize several pins in a row with a magnet—each pin becomes magnetized and lines up many of the domains in the next pin). In a permanent magnet, the domains remain lined up in the absence of an external magnetic field, and the metal is moderately to strongly magnetic.

If a ferromagnetic material that has never been magnetized is placed in a magnetic field and H is increased, the corresponding magnetic flux density, B, is increased nonlinearly as shown by line OC in Fig. 3.8. At point C, the induced magnetic flux density is at a maximum, B_s: this is the *inductance* of the material. All domains are lined up in one direction at B_s; below B_s only a varying number of domains are aligned, depending on the value of B. If at point C the magnetic field is gradually removed, the induced flux density, B, decreases, as shown by line CB_r in Fig. 3.8. When H is reduced to zero, B equals B_r, which is called the remanent induction or *remanence*. If the magnetic field is now reversed and gradually increased in strength, B decreases with H, as shown by the line B_rH_c. At $-H_c$, all the domains have again been randomized, and $-H_c$ is called the coercive field or *coercive force*. If $-H$ is increased still further (in a negative sense, moving to the left), B becomes negative until point F is reached. At this point, all the domains are aligned in the direction opposite to the alignment at point C.

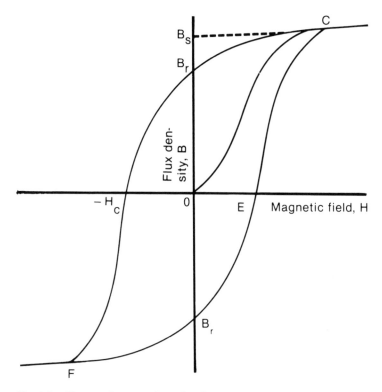

Fig. 3.8   Hysteresis curve for a hard magnet

If the magnetic field is now reversed and increased, B varies with H according to the line FEC. By varying H suitably to cause a complete circuit, curve CH_cFEC, one has traversed one cycle of the *hysteresis loop* (it is called hysteresis because B always lags behind changes in H). The energy consumed in one cycle is equal to the area inside the loop.

A *permanent magnet* (hard magnet) is one that has a relatively large hysteresis loop with relatively high values for the remanence, B_r, and the coercive force, $-H_c$, as illustrated in Fig. 3.8. In a permanent magnet, when the external magnetic field is removed, there is little decrease in B, the induced magnetic field. Furthermore, this value of B_r remains constant for a long period in the absence of an external magnetic field.

Soft magnets or magnetically soft materials, such as silicon iron used in transformers, have only a small energy loss per cycle of the hysteresis loop (see Fig. 3.9). Values of B_r, the remanence, and $-H_c$, the coercive force, are both low. For transformer steel, the lower the energy loss per cycle, the more economical the transformer. Anything that hinders the movement of domain walls adds to the hardness of the magnet (permanent character). Impurities or precipitate particles interfere with the motion of domain walls. Other crystal imperfections, such as dislocations resulting from plastic deformation, also hinder domain-wall movement. Magnetic properties of metals are anisotropic: they vary with crystallographic direction in a single crystal.

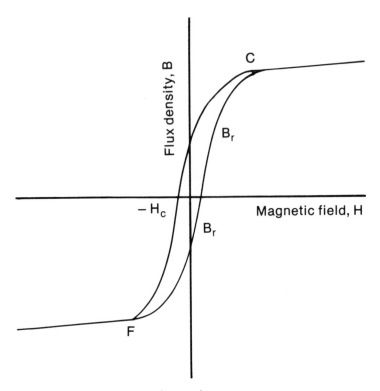

Fig. 3.9   Hysteresis curve for a soft magnet

3.10   Chemical Properties

Chemical properties in general are the properties that involve energy or energy changes and, frequently, changes in electron structure. These include thermodynamic properties such as internal energy, enthalpy, entropy, free energy, and heat capacity, as well as the usual chemical properties, such as chemical reactivity, oxidation resistance in various environments, corrosion resistance in various environments, solubility, surface energy, diffusivity, and basicity.

A discussion of thermodynamic properties is beyond the scope of this text; the reader is encouraged to consult a chemistry text for details. Chemical reactivity is the ease with which metals (or other materials) react with the environment or with various chemicals. Two aspects of chemical reactivity are (1) *oxidation resistance*, which is the slowness of the rate of reaction between metals and oxygen or oxygen-containing compounds, particularly at elevated temperatures; and (2) *corrosion resistance*, which is the slowness of the rate of reaction of metals to form compounds through loss of electrons in electrochemical reactions with other elements or compounds.

Solubility is the ability to form a homogeneous, or single-phase, mixture of elements. In metals, solid as well as liquid solubility is important. Solid solutions are discussed in Chapter 4, while Chapter 5 concerns phases and phase diagrams.

Vapor pressure is the equilibrium gas pressure of a solid or liquid phase at a given temperature. The normal boiling point of a material is the temperature at which the equilibrium vapor pressure equals 760 torr (mm of mercury).

Surface energy or surface tension is the work required to increase the surface area by 1 cm^2. Because surface atoms are attracted only toward the center of a liquid or solid, in order to increase the surface, work must be done to overcome the inward pull. Solid and liquid metals have high surface energies compared with organic materials. Ceramics have surface energies approaching those of metals.

Diffusivity is the rate constant for the relative motion of atoms by processes other than convection or stirring. When solid metals are heated to relatively high temperatures, foreign atoms diffuse, or move through the host lattice.

Most metals are bases: they react with hydrogen ions (protons), and they are capable of supplying one or more valence electrons in a variety of chemical reactions. The greater the tendency to lose electrons in reactions, the stronger the basic character of metals. Hence alkali metals, Group IA, are strong bases, as are Group IIA metals.

Not only are most metals basic, but many metal oxides are bases. Group IA metal oxides, such as Na_2O, are very strong bases. They react with water to form hydroxides, which are also strong bases. Group IIA bases are not as strong as IA bases, and the metals farther to the right in the periodic table form bases that are increasingly weak. Because nearly all metals react with oxygen and moisture in the environment, clean metallic surfaces in reality are covered with extremely thin oxide and hydrated oxide layers.

3.11 Problems

1. A soft nickel wire 0.05 in. in diameter and 12 ft long is loaded with a 10 lb weight. Calculate the elongation of the loaded wire.

2. The Young's modulus of tungsten is 5.96×10^7 psi and the bulk modulus is 4.51×10^7 psi. Calculate Poisson's ratio.

3. Young's modulus for molybdenum is 4.71×10^7 psi and Poisson's ratio is 0.293. What is the compressibility of molybdenum?

4. Calculate the volume change in chromium having $\nu = 0.210$ and $E = 4.05 \times 10^7$ psi when a hydrostatic pressure of 10,000 psi is applied.

5. What is the reduction of area for a steel rod that initially had a diameter of 0.20 cm and a final diameter of 0.14 cm?

6. Plot the yield strengths, tensile strengths, and elongations of annealed steel as a function of carbon concentration; obtain the necessary data from Table 3.3.

7. Plot the yield strengths, tensile strengths, and elongations of 1030 steel tempered at various temperatures (see Table 3.3).

8. Plot Brinell hardness values versus tensile strengths for 10 different ferrous alloys. What conclusions can you draw from the graph?

9. Plot yield stress as a function of atomic number for elements 19 through 33. What statements can you make concerning your graph?

10. Plot the boiling points of elements 37 through 52 versus atomic number. Describe the results.

11. Based on the boiling points of manganese, molybdenum, ruthenium, and rhenium, what temperature do you predict as the boiling point of technetium?

12. What trends in surface tension of liquid metals at their melting points can you observe? Can you correlate surface tension with any other physical property?

13. Multiply the specific heat of at least 10 metals by their atomic weights. What conclusion can you draw? This relationship is well known and will be discussed in Chapter 4.

14. Plot the densities of the lanthanide metals versus atomic number. Try to account for any variations in the general trend.

15. Using Table 3.7, calculate the resistivity of cobalt at 100 °C. Compare this value with the resistivity at 20 °C.

4

Imperfections in Metals & Their Effects on Properties

It is essentially impossible to obtain a perfect material. Imperfections include vibrations and rotations due to thermal energy, point defects (vacancies, interstitials, substitutional solutes), linear defects (edge and screw dislocations), boundary or surface defects (grain boundaries and phase boundaries, including surfaces), and three-dimensional defects (amorphous or liquid state). Imperfections in metal crystals are of extraordinary importance: (1) without imperfections, metals probably would not possess ductility, which is needed to form useful products economically. Civilization itself was likely sparked by the discovery of metal ductility, which enables production of more effective tools and weapons, as well as creation of art objects such as jewelry and sculpture; (2) metals frequently are strengthened by the incorporation of other elements in solution or as second phases; (3) diffusion processes occur because of the presence of vacancies or interstitials in metals; (4) melting of metals may be considered to be a means of obtaining a vast number of imperfections; (5) grain boundaries drastically affect the mechanical properties of metals; (6) properties such as resistivity/conductivity also depend on the imperfections in metals. The overall importance of imperfections in materials science and engineering cannot be overestimated.

4.1 Thermal Disorder

At absolute zero, atomic motion still exists, but it is at an irreducible minimum: this is the zero point vibrational energy. As the temperature increases, there is a gradual increase in the amplitude (i.e., "width") of vibrations of atoms in solids. The thermal energy per mole of a monatomic gas is $3RT/2$. In a simple solid, the mean potential energy (P.E.) is equal to the mean kinetic energy (K.E.), and in a simple solid, $E = P.E. + K.E. \approx 3RT/2 + 3RT/2 \approx 3RT$. R is the molar Boltzmann constant (gas law constant): $R = 1.987$ cal/mol°K (or $R = 8.314$ J/mol°K), and $E \approx 3RT \approx 6$ cal/mol°K. The value of approximately 6 cal/mol°K is the heat capacity at constant volume and is essentially the same for solids as C_p, the heat capacity at constant pressure: this is the law of Dulong and Petit. C_p is approximately six times the number of atoms per molecule or formula unit.

Increased temperature causes the atoms of a solid metal to vibrate with greater amplitudes. Because the vibration is nonlinear (atoms cannot easily be compressed), more room is required for the increased amplitudes of vibration, and the average distance between *all* atoms in the solid increases with increasing temperature. The increase in distance between atoms causes metals to expand upon heating. This expansion is the *thermal coefficient of expansion*, which is measured in cm/cm (i.e., the coefficient is unitless). In cubic metals, thermal expansion is *isotropic*, or the same in all crystallographically equivalent directions (i.e., the same along the x, y, and z axes, etc.). However, in noncubic crystals, such as hexagonal closest-packed metals, the coefficient of thermal expansion parallel to the c-axis is different from the value parallel to the a-axis. Zinc atoms are egg-shaped, with the longer axis parallel to the c-axis (see Table 2.1). Upon heating to higher temperatures, the zinc atoms tend to become more spherical due to different coefficients of thermal expansion along the c and a axes.

4.2   Point Defects

Imperfections that center about one point within a crystal lattice are called *point defects*. In metals, point defects include (1) *vacancies*, which are empty lattice sites (a metal atom is missing where an atom should be), (2) *interstitials*, which are usually small nonmetal atoms placed in between the normal lattice sites, and (3) *substitutional solute atoms*, which are metal atoms different from the solvent atoms (the ones present in the majority) placed at regular lattice sites in the positions normally occupied by solvent atoms.

Point defects affect the crystal more than at just one point: around each point defect is a small volume in which the atoms are strained, or moved slightly from their equilibrium positions. The atoms surrounding a vacancy are drawn in towards the center of the vacant site due to weak bonding with the atoms across the vacancy (see Fig. 4.1). Each vacancy has some strain energy associated with it. In other words, it takes a small amount of work to create a vacancy. An atom in a normal lattice site may move to the surface or, more likely, to the disordered region called the grain boundary. Another mechanism for the creation of vacancies is the climb of edge dislocations, discussed in section 4.7.

Interstitial atoms cause the lattice to be strained, or put into compression around the interstitial: crowding occurs. Substitutional atoms that are large compared with the solvent atoms cause crowding also, whereas relatively small substitutional atoms cause tension, somewhat like a vacancy.

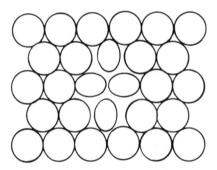

Fig. 4.1   Point defects in metals: vacancies

Point defects are important in the following ways: (1) vacancies are responsible for the diffusion of metal atoms in solids (diffusion is discussed in section 4.17); (2) interstitials affect the strength, hardness, ductility, yield strength, and other mechanical properties; and (3) substitutional solutes affect the strength and hardness of alloys.

4.3   Vacancies

Vacancies are defined as the absence of atoms at regular lattice sites in a metal crystal. Vacancies may be created in metals by movement (diffusion) of an atom from its lattice site (1) into an unoccupied site on the surface of a metal, (2) into a grain boundary, or (3) through edge dislocation climb (to be discussed). Vacancies can move in metals as a result of the movement or diffusion of metal atoms. When a metal atom jumps from its lattice site into a vacancy, then a vacancy occurs where the metal atom used to be. This is equivalent to vacancy movement.

For an atom to jump from a lattice position into a vacancy, an activation energy is necessary to overcome temporarily the net attractive force between the atom moving into the vacancy and the atoms surrounding the moving atom. The activation energy is an energy barrier that must be overcome in order for movement to occur. Depending on the instantaneous thermal energy of a vibrating atom, it may or may not have sufficient energy to surmount the activation energy barrier. At a given temperature, only a small fraction of atoms may have energies greater than the activation energy; however, as the temperature is increased, the fraction of atoms having energies exceeding the energy barrier is considerably greater. Therefore, atom movements increase dramatically with increasing temperature.

At any given temperature in a metal there is an equilibrium concentration of vacancies. The concentration of vacancies is expressed by the following equation:

$$\frac{N_v}{N_o} = e^{-Q_f/RT}$$

where N_v is the number of vacancies; N_o is the number of atoms (N_v/N_o = vacancies/atom); Q_f equals the work required to form 1 mol of vacancies in calories per mole; R is the molar Boltzmann constant, or 1.987 cal/°K · mol; and T is the temperature in degrees Kelvin.

Example 4.3A. The work required to form 1 mol of vacancies in copper was experimentally determined to be 20,000 cal/mol. What is the concentration of vacancies in copper at (a) room temperature, (b) 500 °C, and (c) 1073 °C (melting point = 1083 °C)? How many atoms per vacancy exist at each temperature?

Solution: (a) At 25 °C, or 298 °K, $N_v/N_o = e^{-20,000/1.987 \times 298} = e^{-33.78} = 2.14 \times 10^{-15}$. The number of atoms per vacancy is simply N_o/N_v, or the reciprocal of 2.14×10^{-15}, or 4.67×10^{14} atoms per vacancy. (b) At 500 °C, or 773 °K, $N_v/N_o = e^{-20,000/1.987 \times 773} = e^{-13.02} = 2.22 \times 10^{-6}$. The number of atoms per vacancy is $1/2.22 \times 10^{-6} = 4.51 \times 10^{5}$. (c) At 1073 °C, or 1346 °K, $N_v/N_o = e^{-20,000/1.987 \times 1346} = e^{-7.478} = 5.65 \times 10^{-4}$. The number of atoms per vacancy = $1/5.65 \times 10^{-4} = 1769$ atoms per vacancy.

Note: Since copper is FCC with 4 atoms per unit cell, there is 1 vacancy in 1769/4 or 442 unit cells.

4.4   Interstitials

In metals, *interstitials* are usually small nonmetallic atoms, such as H, B, C, N, or O, that are located in the vacant spaces between metal atoms. Interstitial atoms are usually larger than these free spaces, and therefore they force the solvent or matrix metal atoms apart slightly, introducing strain into the crystal. Even if the interstitial atoms are much smaller than the matrix atoms, strain is introduced somewhat like strain caused by vacancies.

The concentrations of nonmetallic interstitials may be fairly high at elevated temperatures. It is uncommon to find a solvent atom in an interstitial position in a metal crystal because of the very large strain involved—the crystal would be grossly deformed around the interstitial. Consequently, the work to form a mole of interstitial solvent atoms is very high: even at elevated temperatures, the concentration of interstitial solvent atoms is extremely low.

Example 4.4A. Calculate the diameter of the largest interstitial space in FCC iron with a = 0.3654 nm at 732 °C.

Solution: According to Fig. 4.2, the diameter of the interstice (localized, unoccupied space) is "a" minus the diameter of the iron atom. Because $r = \sqrt{2}a/4$, $r = \sqrt{2} \times 0.3654/4$ = 0.1292 nm, or the diameter is 0.2584 nm. The diameter of the interstice is a − 0.2584 nm = 0.3654 − 0.2584 = 0.1070 nm.

Note: At the same temperature a similar calculation shows the largest interstice in BCC iron is only 0.0732 nm, which is considerably smaller. It is surprising that BCC crystals, which have more free space than FCC crystals, have smaller (but more) volumes of free space. Because of

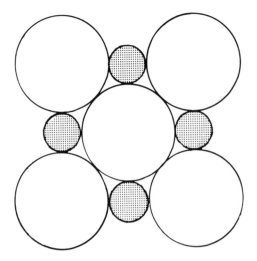

Fig. 4.2   Diameter of the largest interstice in FCC iron
The gray sphere is the largest interstice. This interstice is much larger than the one in BCC iron.

this fact, the solubility of interstitials is much greater in FCC crystals than in BCC crystals if other factors are the same. For example, there is much less strain involved when a carbon atom dissolves in FCC iron than in BCC iron, and the solubility of carbon is greater in FCC by a factor of 10 to 100!

The solubility of interstitials in a metal follows an equation similar to the one describing the concentration of vacancies: wt.% I $= Ae^{-Q_I/RT}$, where wt.% I is the maximum solubility of an interstitial, A is a constant, Q_I is the work required to introduce one mole of interstitial atoms into the metal (it is a constant), R is the molar Boltzmann constant, 1.987 cal/mol · °K, and T is the temperature in degrees Kelvin.

Example 4.4B. Calculate the equilibrium solubility of carbon in BCC iron at room temperature. The value for A is 250 and Q_I is 18,570 cal/mol.

Solution: T $= 25$ °C, or 298 °K. Wt.% C $= 250e^{-18,570/1.987 \times 298} = 250e^{-31.36} = 250 \times 2.40 \times 10^{-14} = 6.0 \times 10^{-12}$% C.

Note: It is most unlikely that all the carbon can precipitate from iron as it is cooled. In order to precipitate, the carbon would have to diffuse and nucleate to form a tiny particle of iron carbide. Diffusion is extremely slow at room temperature, and it is slow at least up to red heat ($\sim$600 °C).

4.5 Substitutional Solid Solutions

In many metals, solid solution is accomplished by the substitution of a solute atom for a solvent atom in one of the normal lattice sites for the solvent. For instance, a nickel atom may be substituted for a copper atom in FCC copper. The substitution occurs randomly. When substitutional solution occurs, the lattice is distorted or strained in the vicinity of each solute atom because of a difference in size and bonding between solvent and solute atoms.

Extensive solid solubility in metals occurs at room temperature only when the strain is relatively small: the radius ratio of the solute and solvent atoms should be within the limits of 1.00 ± 0.15. This is the first rule devised by Hume-Rothery. The other Hume-Rothery rules for extensive solid solubility are (2) the solvent and solute atoms should be similar in electronic structure (i.e., they are located in roughly the same column or vicinity in the periodic table), and (3) the atoms should have about the same chemical reactivity. For complete solid solubility, (4) the two metals must have the same crystal structure.

Because substitutional solid solubility depends on similar principles to interstitial solid solubility, the equation describing solid solubility is similar:

$$\text{Wt.\% S} = Ae^{-Q_s/RT}$$

This equation indicates that solid solubility increases sharply with increasing temperature in most cases.

4.6 Solution Hardening

Solid solutions are harder and stronger than pure elements. It is common practice to alloy metals in order to increase their strengths. Sometimes alloying will result in the formation of more than one phase, but that subject will be discussed later. For instance, gold wedding rings are alloyed, not to cheapen the metal, but to strengthen the gold. Silver is alloyed with copper to increase its strength, as in sterling silver. In some instances, alloying results not only in superior properties, but also lower costs. For instance, zinc costs less than copper, and brass, an alloy of copper and zinc, is stronger than pure copper; another advantage of brass is its attractive gold color.

The extent of solution hardening depends on the concentration of the solute atom as well as on the mismatch in radii of the solvent and solute. Another factor is the ease of deformability of the atom: zinc is elastically softer (i.e., the bulk modulus, K, is lower) than nickel, so nickel hardens copper more than does zinc.

4.7   Edge Dislocations

When calculations are made for the stress required to plastically deform a metal by moving one plane of atoms past another, the calculated stresses are 100 to 1000 times higher than experimental measurements! In fact, calculations show that metals would require 1 to 5 million psi for plastic deformation compared with 5 to 50,000 psi normally found. Single crystals may deform at still lower stresses. Obviously this theory is wrong: planes of atoms do *not* slip by one another by simultaneous relative motion of all atoms involved. Instead, a far different mechanism of deformation is involved: only a relatively few atoms move past one another at a given instant, and the means by which they move involves linear defects, namely, edge or screw dislocations.

An *edge dislocation* is a linear (one-dimensional) mismatch of atoms at the edge of an extra partial plane of atoms. Imagine a perfect crystal having parallel planes of atoms similar to a deck of cards. If an extra half plane of atoms is inserted into the crystal, similar to an extra half of a card, then the crystal is no longer perfect: a line of mismatched atoms occurs at the edge of the extra half plane. This is the edge dislocation (see Fig. 4.3). The extra partial plane of atoms does not have to be a straight line: if a playing card is cut into a circle and if the circle is placed

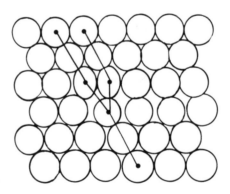

Fig. 4.3   Edge dislocation

in the deck of cards, then the mismatch around the perimeter of the circle corresponds to an edge dislocation. In general, an edge dislocation can take the shape of any closed loop within a crystal.

An edge dislocation can only end at the surface of a crystal. Edge dislocations exist in all metal crystals except perhaps in fine metal whiskers, usually at concentrations of roughly 10^8 cm/cm^3 of annealed metal. Figure 4.4 shows some dislocations visible through the electron microscope. Dislocations are visible in a transmission electron microscope or may be seen as a series of etch pits on the surface where the dislocations end. In a mild etchant, the minute spot where a dislocation intersects the surface occasionally is etched away faster than the surrounding material, leaving an etch pit.

Edge dislocations are involved in the plastic deformation of metals. When metals are stressed, new dislocations can form around old dislocations that are pinned at two points, such as at grain boundaries, surfaces, etc. Under the influence of stress, the dislocations can move quite rapidly. When they move one after another in the same direction, plastic deformation occurs. Part (a) of Fig. 4.5 represents an unstressed crystal. In (b) when a stress is applied, an edge dislocation forms at the edge of the crystal. In (c), we see that the edge dislocation has moved one unit cell dimension. In (d), the dislocation has moved over an additional unit cell

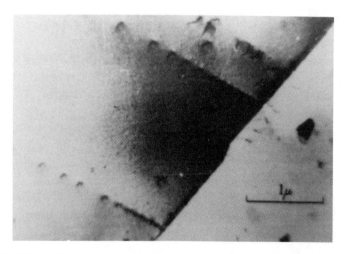

Fig. 4.4   Dislocations along a slip plane near a grain boundary in pure copper

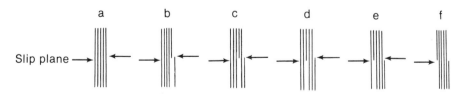

Fig. 4.5   Motion of an edge dislocation during plastic deformation
Arrows represent shear stresses above and below slip plane. Lines represent planes of atoms. Interior half-line is a dislocation; exterior half-line is a step in the crystal caused by shear.

dimension, and it continues in (e). In (f), the edge dislocation has finally moved out of the crystal. The net result of the motion of the dislocation from one side of the crystal to the other is a shear of the top half of the crystal relative to the bottom half by a distance of one unit cell dimension.

Creation and motion of additional edge dislocations across the crystal result in further relative shear or slippage of planes of atoms past one another. Now we can understand why plastic deformation occurs at relatively low stresses: only a relatively few atoms are being strained (moved) from their equilibrium position at a given time. Therefore, the stress required for plastic deformation is only a small fraction of the stress required to slip one entire plane of atoms past the adjacent plane of atoms. The dislocation is a line that moves in a plane that separates the crystal into two parts that move in opposite directions relative to each other. The *slip plane* is the plane in which the dislocation moves and across which relative shear of the metal occurs. Note that the edge dislocation *lies* in the slip plane, and the edge dislocation *moves* in the slip plane in a direction perpendicular to itself—it moves in the direction of shear.

The *Burgers vector* is the smallest unit distance of slip in the direction of shear due to the movement of one dislocation. A vector has both magnitude and direction. The magnitude of the Burgers vector, **b**, is the distance between two atoms, usually touching one another, in the direction of slip (i.e., usually **b** = 2r). The direction of the Burgers vector is the direction of slip. Because the Burgers vector and its edge dislocation are perpendicular to one another, they define the slip plane. The edge dislocation moves in the direction of the Burgers vector, and slip occurs in this direction.

Edge dislocation climb is the motion of an edge dislocation straight up or down out of its slip plane. Energetically this is difficult because it results in the creation of a row of vacancies if the edge dislocation climbs up one row of atoms. For it to move down, the edge dislocation must annihilate a row of vacancies, and this takes much time for the vacancies to come to the correct locations.

4.8   Screw Dislocations

A *screw dislocation* is a linear defect (one-dimensional defect) in which one part of a crystal has been partially sheared from another part by a scissorslike action. Figure 4.6 shows a screw dislocation in which the top half of the crystal has been displaced by one Burgers vector from the front bottom half of the crystal. The slip plane is the sheet of paper provided that the x's represent atoms above the plane of the paper. Shearing occurs from left to right above the slip plane and right to left below the slip plane. As slip proceeds, the screw dislocation moves perpendicular to itself, as shown in Fig. 4.7. When the screw dislocation has moved from the front of the crystal to the back, slip has occurred by one unit cell dimension in the direction of shear. In summary, a screw dislocation (1) lies *in* its slip plane, (2) is *parallel* with its Burgers vector, and (3) *moves* perpendicular to itself.

An unusual and very important property of a screw dislocation is that it is not confined to move in its original slip plane because it is parallel with its Burgers vector. Rather, it can move into any new slip plane that intersects the first slip plane in the direction of the Burgers vector. Hence, a screw dislocation can easily move out of its slip plane and into a new slip plane. Even

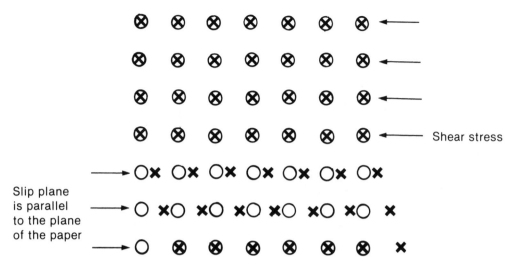

Fig. 4.6   Screw dislocation
Circles represent atoms in plane of paper and behind this plane. Crosses represent atoms above the plane of the paper.

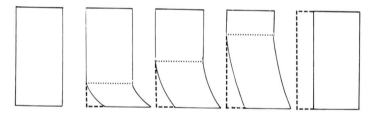

Fig. 4.7   Slip due to motion of a screw dislocation
The slip plane is parallel with the surface of the page. The dotted lines represent the screw dislocation, and the dashed lines represent where slip has occurred.

if this occurs, the crystal is still being sheared in the same direction, the direction of the Burgers vector.

Closed dislocation loops frequently occur in crystals. A real dislocation can be a mixture of screw- and edge-type components, as shown in Fig. 4.8. Additionally, a dislocation may have parts that are pure edge or pure screw type. The important point to realize is that real dislocations are usually mixtures of edge and screw components and are therefore reasonably complicated.

4.9   Plastic Deformation: Slip Planes and Slip Directions

The great importance of dislocations in metals is that they enable plastic deformation to occur at comparatively low stresses, thereby enabling metals to be processed into useful shapes. It is difficult to imagine what life would be like in the absence of dislocations—all metals, although

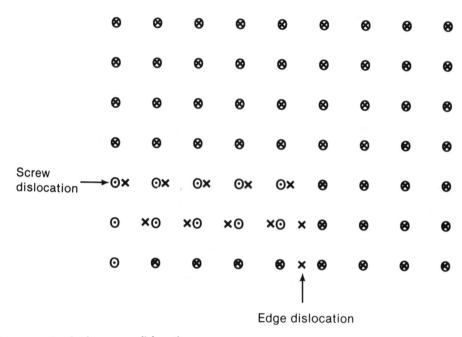

Fig. 4.8 Mixed edge-screw dislocation
Circles represent atoms in plane of paper and behind this plane. Crosses represent atoms above the plane of paper.

strong, would probably be brittle (similar to ceramics) and perhaps formable only through casting or machining.

The movement of dislocations causes shear to occur in the direction of the Burgers vector. Shear occurs across a slip plane or a series of slip planes (a slip surface actually is comprised of many parallel slip planes). In order to deal with the plastic deformation of metals, it is useful to know the slip direction and slip plane, which together comprise the *slip system*. The direction of slip is almost always a closest-packed direction: in FCC metals, slip usually occurs in a face diagonal direction, indicated by <110>*; in BCC, a body diagonal, or <111> direction, is the usual direction of slip. The slip direction is almost always a closest-packed direction because then the Burgers vector, **b**, is a minimum (**b** = 2r): the energy associated with a dislocation depends on the square of b; therefore, the smaller the Burgers vector, the lower the energy of a dislocation (and nature likes to minimize energies).

Slip planes are usually closest-packed planes; these planes have the widest d spacings (interplanar spacings), and shear occurs most easily between these planes. Dislocations move more easily along planes with large d-spacings, and the lattice distortion is less for planes of wide spacing. In BCC crystals, there are no closest-packed planes; slip occurs, however, in the most widely spaced planes, {110}*, and the closest-packed directions are <111>.

---

*The symbol <110> means all directions involving permutations of two ''ones'' and a zero, including negative values: [110], [$\bar{1}$10], [101], [$\bar{1}$01], [011], and [0$\bar{1}$1]. Note that [$\bar{1}$10], is the same direction as [1$\bar{1}$0]. The symbol { } means all planes involving permutations of the Miller indices.

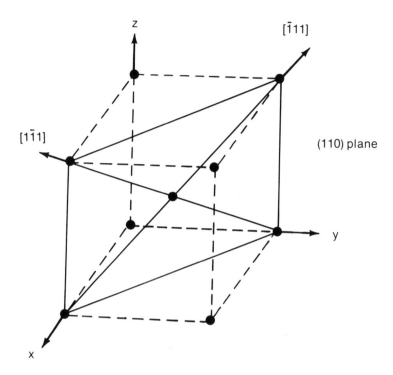

Fig. 4.9   Some directions of slip in BCC crystals

Example 4.9A. One of the slip planes in BCC iron is (110). Determine all the probable slip directions (i.e., closest-packed directions) in this plane.

Solution: Directions of slip must lie in the slip plane. First draw the (110) plane (see Fig. 4.9). The directions of slip are the closest-packed directions, <111>, or body diagonals. There are two body diagonals in the (110) slip plane: $[1\bar{1}1]$ and $[\bar{1}11]$. Therefore, there are two slip systems inolving the (110) plane: $(110)[1\bar{1}1]$ and $(110)[\bar{1}11]$.

Note: In FCC crystals there are 12 slip systems, and these systems are well distributed in space so that one slip system is almost always favorably oriented for slip to occur. Consequently, FCC metals, such as copper, silver, gold, lead, nickel, and aluminum, are nearly always very ductile. Hexagonal closest-packed metals, however, have few slip systems, and these are usually limited to one plane of the crystal; accordingly, HCP metals usually have somewhat poor ductility.

4.10   Cross Slip

Cross slip may occur in a metal in more than one slip plane when two or more slip planes intersect in a common direction of slip. Cross slip is a steplike slip such as shown in Fig. 4.10. A good analogy to cross slip is the motion of a drawer in a desk. The drawer moves in one direction, but the sides and bottom are different planes intersecting in the direction of motion.

During cross slip, dislocations move from one slip plane to another or in more than one slip plane at a time. Cross slip is primarily due to the motion of screw dislocations. Because screw

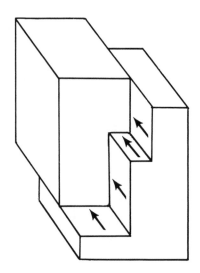

Fig. 4.10   Cross slip
Burgers vectors are drawn in each of the four slip planes. The block on the left has been sheared in the direction of the Burgers vectors.

dislocations and their Burgers vector lie in the same direction and move in the slip plane perpendicular to the Burgers vector, they can easily move from one slip plane to another. *Slip lines or bands* are visible lines seen in chemically etched metals that have been plastically deformed (see Fig. 4.11). Observation of slip lines with the electron microscope reveals that each line is in reality a band of closely spaced slip planes.

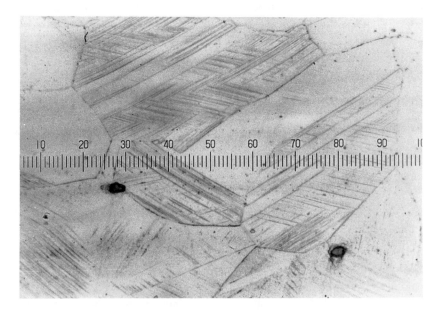

Fig. 4.11   Slip lines or bands on plastically deformed brass
The distance between 40 and 50 on the scale is 25 μm.

4.11   Strain Hardening

Strain hardening is the strengthening of metals by plastic deformation at room temperature or slightly above. It involves the creation of many dislocations due to the plastic deformation. Once dislocations have formed, it becomes difficult for certain other dislocations to cut through these dislocations. A dislocation tangle results. Anything that makes it more difficult for dislocations to move results in a stronger metal. Hence, the creation of dislocations results in the stabilization of other dislocations, with the result that the metal becomes stronger. Strain hardening works only to a certain point: if excessive plastic deformation occurs, then the dislocation tangle starts a crack which may cause the metal to fail.

Cold working is the plastic deformation of metals at room temperature or slightly above (the temperature must be below the temperature at which edge and screw dislocations disappear due to rearrangement of atoms). Cold working causes strain hardening in metals due to the formation of dislocation tangles as explained above.

4.12   Recovery

The cold working of metals affects a number of properties: strength, hardness, and electrical resistance increase while ductility decreases. X-ray diffraction patterns of deformed single crystals show elongated spots, resulting from diffraction from bent and strained portions of the crystal. Cold-worked metals also show line broadening in X-ray diffractometer scans.

Recovery is accomplished by heating a cold-worked metal to a few hundred degrees above the cold working temperature. *Recovery* comprises the return or partial return of properties to their original values prior to the cold working and is due to slight heating. For instance, recovery usually results in a decrease in resistivity and hardness. However, the grain size does not change during recovery, even though the cold working may have caused severe deformation of grains. Recovery is caused by the motion of some dislocations, which results in the annihilation of some dislocations (due to canceling each other out) or to the accumulation of dislocations in certain planes, called subgrain boundaries.

The principal application of recovery heat treatments is in stress relief of metals to prevent stress-corrosion cracking or to minimize distortion caused by residual stresses. Some stress relief is usually possible without decreasing strength and hardness appreciably.

4.13   Recrystallization

Recrystallization is the formation of new and smaller grains in a metal by heating for a certain period following cold working. In metals, grain size control is important because fine grains increase the strength of a metal compared with coarse grains, and fine-grained metals may be plastically deformed with a minimum of orange peel. *Orange peel* is a roughening of the surface due to uneven plastic deformation of relatively large grains. It is usually avoided or minimized because it is generally considered objectionable. Metals are also recrystallized to avoid cracking during further plastic deformation. Cold working produces many dislocations that cause strain hardening; further plastic deformation results in cracking. Thus, annealing is frequently used to enable severe plastic deformation of metal parts.

Recrystallization occurs by a nucleation and growth process: nuclei of recrystallized grains form, and some of these grow to form new crystals. When the old, strained crystals have all been consumed in forming new grains, recrystallization is complete. Recrystallization begins slowly at a given temperature because it takes time to form nuclei. Once many nuclei have formed, then the recrystallization rate at the given temperature increases rapidly as growth of the nuclei occurs. However, the new grains soon begin bumping into each other, and as the amount of unrecrystallized material decreases, the overall rate of recrystallization decreases.

Although recrystallization is a function of both time and temperature, it is convenient to define the recrystallization temperature as the minimum temperature at which a cold-worked metal will essentially recrystallize in one hour. The rate of recrystallization depends on the amount of cold work performed on a metal. In general, the greater the amount of cold work, the faster the recrystallization (i.e., the lower the recrystallization temperature as defined above). The grain size of a recrystallized metal depends on the amount of cold work applied, but not on the temperature of recrystallization: the greater the amount of cold work, the smaller the grains in the recrystallized metal. A minimum amount of cold work is usually required for recrystallization to occur at any temperature. Because recrystallized grains of metals having only a minimum of cold work are large, small amounts of cold work are avoided, and an elongation of at least 10% is normally used commercially for metals undergoing recrystallization.

Extremely pure metals have rapid rates of recrystallization. Only 0.01% of a solute atom may increase the recrystallization temperature by over 100°. The increase in the recrystallization temperature is strongly dependent on the nature of the solute atom. Solute atoms, particularly those differing appreciably in size from the solvent or those which are quite different chemically, interact with grain boundaries and dislocations. The overall grain boundary and dislocation energies are lowered by the presence of solute atoms. These solute atoms can stabilize grain boundaries and prevent them from moving so easily, thereby impeding recrystallization. The net effect is an increased recrystallization temperature.

The grain size of the cold-worked metal also affects recrystallization. Decreasing the grain size of the cold-worked metal increases the number of potential sites for nucleation. Therefore, the smaller the grains before a given amount of cold work and recrystallization, the smaller the recrystallized grains will be. A general rule-of-thumb states that the recrystallization temperature of a metal is approximately 40% of its melting temperature expressed in degrees Kelvin. Recrystallization is highly important to industry.

4.14 Grain Growth

Grain boundaries are disordered regions between crystals. It is not surprising that atoms may move from a grain boundary to an adjoining crystal, or from the surface of a crystal into a grain boundary. If these processes occur repeatedly in the same direction, then grain growth occurs: the grain boundary will move, and one crystal grows at the expense of the second because the number of atoms per unit area of the grain boundary remains essentially constant. Furthermore, there is a driving force causing grain boundary motion, particularly at elevated temperatures: the grain boundary energy is similar to a surface energy—it tends to minimize itself where possible by decreasing the grain boundary area. This means that at elevated temperatures grains tend to grow; some grow larger while others are eliminated.

The energy of the crystal is decreased when the grain boundary moves *toward* its center of curvature. In this way, fewer atoms enter the grain boundary than the number that leave, and the grain boundary area decreases (the thickness remains constant). The tendency towards motion of the grain boundary depends on the size of the radius of curvature: grain boundaries that are sharply bent tend to straighten out; small grains adjacent to large ones tend to disappear quickly because of the relatively large radius of curvature of small grains. In general, grain growth will continue to occur at elevated temperatures until grains extend through a wire or sheet of metal. Then the grain boundaries are nearly planar and have very little tendency to move.

At elevated temperatures, the motion of grain boundaries results in grain growth, which causes certain grains to grow larger and other grains to disappear. Grain growth occurs far more slowly than the movement of boundaries in recrystallization. Frequently grain growth is slowed by the presence of second-phase particles such as iron carbide in steel. Grain growth is also inhibited by other nonmetallic inclusions, such as oxides, sulfides, or silicates, which frequently remain after the refining process. These particles stabilize grain boundaries by lowering their energy, and therefore it is difficult for grain boundaries to move past inclusions. Consequently, inclusions frequently occur in grain boundaries. Pores also inhibit grain growth.

Solutes (dissolved atoms) can also cause grains to grow more slowly through stabilization of grain boundaries. The solute atoms lower the grain boundary energy and therefore prevent the grain boundary from moving as rapidly as it would in the absence of solute atoms. The effect of solutes in retarding grain growth depends on the solute involved: elements that distort the lattice structure the most impede grain growth the most. Second-phase inclusions may also cause grain growth to cease at a certain grain size. When the boundary has trapped a certain number of inclusions, the surface-tension force cannot overcome the restraining force of the inclusions.

4.15   High-Temperature Strengthening of Metals and Inhibition of Grain Growth

Metals can be strengthened at elevated temperatures by dispersing a second phase of fine metal oxide particles throughout the metal. An example is thoria dispersed (TD) nickel, in which the high-temperature strength of nickel is increased by dispersing fine particles of thorium dioxide, ThO_2, in the nickel. A second example is sintered aluminum powder (SAP), in which Al_2O_3 is dispersed in aluminum. This also enables retention of good strength until relatively high temperatures are reached in aluminum. A by-product of high-temperature strengthening of metals with dispersed oxides is that grain growth is also inhibited by the presence of the fine oxide particles.

In some high-temperature applications, it is desirable to minimize grain boundary area because grain boundaries are relatively weak at elevated temperatures, which is the opposite of their strengthening metals at lower temperatures. Consequently, in some metal parts, such as vanes in jet engines, the metal is made from single crystals, effectively eliminating grain boundaries as a source of weakness.

4.16   Age Hardening

Age hardening is the solid-phase strengthening of metals by means of a heat treatment that results in the formation of clusters of solute atoms in a solvent metal. Frequently, age hardening is incorrectly referred to as "precipitation hardening." Age hardening is strengthening an alloy by (1) heat treating at an elevated temperature to dissolve the precipitate or second-phase present, (2) quenching to room temperature to obtain a supersaturated solid solution, (3) heating to an intermediate temperature for a specific time to cause *clusters* of atoms (close-knit groups of like atoms having the same crystal structure as the solvent metal) to form, but avoiding the formation of a discrete precipitate, and (4) cooling to room temperature. To obtain age hardening, several factors must be satisfied: (1) the stable alloy must comprise at least two phases at room temperature, (2) the second phase must dissolve at some elevated temperature, (3) quenching to room temperature must result in a supersaturated solution (that is, insufficient time is available for precipitation of the second phase), (4) by heating for a specific time and temperature, it should be possible to obtain the proper sort of cluster of solute atoms, and (5) the metal should not overage at room temperature. *Overaging* is losing strength and hardness of an age-hardened metal due to the growth of clusters of solute atoms to form discrete precipitate particles.

Age hardening results in *very* significant increases in the strength of several alloys (see Table 4.1) and is of considerable importance in aluminum alloys and in "precipitation hardened" stainless steels. The mechanism of strengthening alloys through age hardening involves impeding the motion of dislocations. Dislocations are apparently stabilized most effectively when the cluster of solute atoms is *coherent* with the crystal structure of the solvent, i.e., the solute atoms have collected into a cluster but still have the same crystal structure of the solvent phase. This causes a great deal of strain because of the mismatch in size between the solvent and solute atoms. Clusters strongly stabilize dislocations, because dislocations tend to reduce the strain a great deal, similar to the reduction in strain energy of a single solute atom by a

Table 4.1   Age-hardened alloys

Alloy composition, wt. %	Yield strength, psi	Tensile strength, psi	Elongation, %	Treatment
95.5Al-4.5Cu	15,000	35,000	40	Solution treated and quenched
	45,000	60,000	20	Age hardened
	10,000	25,000	20	Overaged
	10,000	25,000	15	Annealed
98Cu-2Be	· · ·	72,000	· · ·	Solution treated and cooled rapidly
	· · ·	175,000	· · ·	Age hardened
	· · ·	107,000	· · ·	Cold worked only (37%)
	· · ·	200,000	0	Age hardened, then cold worked
	· · ·	195,000	0	Cold worked, then age hardened
90Ti-6Al-4V	134,000	144,000	14	Annealed
	160,000	170,000	10	Age hardened
75Ti-3Al-8V-	121,000	128,000	15	Annealed
6Cr-4Mo-4Zr	200,000	210,000	7	Age hardened
94Mg-5.5Zn-0.45Zr	29,500	41,500	5.5	As fabricated
	34,500	44,500	5	Age hardened

dislocation. When dislocations are anchored or trapped by coherent solute clusters, the alloy is considerably strengthened and hardened. The surrounding strained volume is appreciably greater than the volume of the solute cluster itself: clusters are nearly intolerable in metals because of their high strain energy. They can either grow at moderately elevated temperatures to form a precipitate, in which case the strain energy decreases markedly, or, second best, they can exist at lower temperatures and lock dislocations, thereby also reducing the strain energy but by a smaller amount.

If growth of the clusters occurs, precipitation will occur, and the strength decreases sharply: overaging has occurred. Overaging depends both on temperature and time—obviously it is advisable to avoid age-hardened alloys that overage at room temperature or at operating temperatures. Thus, age-hardened aluminum alloys are limited in use to temperatures below about 300 °C and therefore cannot be used in high-speed aircraft.

When age hardening is combined with cold working, such as cold rolling or cold drawing, very high strengths may be achieved (see Table 4.1).

4.17   Diffusion

Diffusion is the net relative movement of atoms in a solid, liquid, or gas. However, motion due to atomic vibrations, convection, or stirring is not considered to be diffusion. Diffusion has great theoretical importance: the theory of diffusion is capable of clarifying the processes occurring in phase transformations. For example, when iron in the FCC crystal structure, stable above 727 °C, transforms to a mixture of BCC iron and Fe_3C, diffusion of carbon occurs. There are many practical applications of diffusion, such as carburizing steel (hardening the surface by diffusion of carbon) or the formation of oxidation- or corrosion-resistant coatings on metals. Diffusion is of importance in the migration of moisture or oxygen through polymer films or in the penetration of dye into some textile fibers.

Diffusion coatings on metals may be obtained by heating a metal in certain gaseous mixtures, by heating electroplated metals, or by heating metals in liquid metal or salt baths. Diffusion occurs by atoms jumping into vacancies. Because an activation energy is necessary for an atom to jump into a vacancy, the higher the temperature, the greater the thermal energy of atoms and, consequently, the faster the rate of diffusion. Diffusivity, which is the rate constant for the process of diffusion (i.e., the higher the diffusivity, the faster the rate of diffusion), increases exponentially with temperature according to the following:

$$D_A = D_0 e^{-Q/RT}$$

where D_A is the diffusivity of atom A in cm^2/s, D_0 is a constant for a given diffusion couple, e is the natural logarithm base (e = 2.71828), Q is the activation energy in calories per mole of diffusing atoms, R is the molar Boltzmann constant (R = 1.987 cal/mol°K), and T is the temperature in degrees Kelvin. The above equation may be expressed as:

$$\ln D_A = \ln D_0 - Q/RT$$

where "ln" designates natural logarithms. An alternate form for the equation is:

$$\log D_A = \log D_o - \frac{Q}{2.303RT}$$

where "log" designates logarithms to the base 10. With an electronic calculator it is usually easier to use natural logarithms.

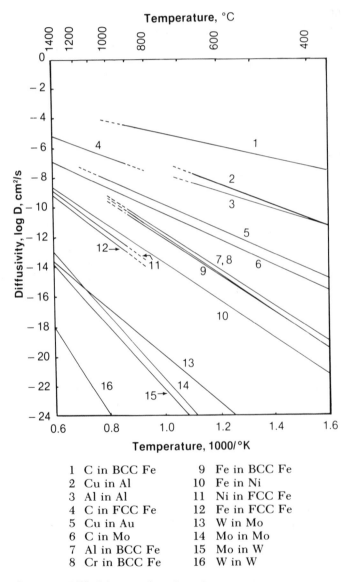

Fig. 4.12   Diffusivity as a function of temperature
Dashed lines indicate that the phase is unstable at these temperatures.

Most diffusion data are presented in terms of D_o and Q for a given temperature range. This enables D_A to be calculated at any temperature within the specified range.

Example 4.17A. Calculate the diffusivity, D, of nickel in FCC iron at 1100 °C, where $D_o = 0.5$ and $Q = 66,000$ cal/mol.

Solution: $D = D_o e^{-Q/RT} = 0.5 e^{-66,000/1.987 \times 1373} = 0.5 e^{-24.19} = 0.5 \times 3.115 \times 10^{-11} = 1.56 \times 10^{-11}$ cm^2/s.

Experimentally, D may be measured as a function of temperature. It is then possible to plot ln D on the y-axis and 1/T on the x-axis to obtain a straight line. One may calculate Q from the slope and D_o from the y-intercept. Figure 4.12 contains a graph showing plots of log D in cm^2/s versus the reciprocal temperature in 1000/°K.

Diffusivity increases exponentially with temperature. Diffusivity depends also on factors such as composition, crystal structure of the solvent metal, and the strength of bonding, which is reflected by the melting or boiling point of solvent metal. More open (greater free space) crystal structures cause diffusivities to be relatively high. High melting metals with strong bonding have low diffusivities compared with other metals. Diffusivity along grain boundaries is more rapid than across grains, again because of more free space in the grain boundaries.

Carburizing is the diffusion of carbon into the surface of a metal, usually steel, for the purpose of hardening the surface for greater wear resistance. The carburizing of steel is called *case hardening*. Case hardening may be accomplished by heating steel in a gas containing carbon (such as methane) or by treatment in a fused salt bath such as sodium cyanide. *Carbonitriding* involves the formation of a diffusion coating ("case") of both carbon and nitrogen. *Chromizing* is coating a metal with a diffusion coating containing chromium for corrosion resistance, wear resistance, or for improved oxidation resistance at elevated temperatures. Siliconizing and aluminizing are processes to coat a metal with either silicon or aluminum, primarily for oxidation resistance. *Cladding* is a method for coating a metal with another metal or alloy in which the metals are bonded together metallurgically with a diffusion coating.

4.18 Problems

1. Estimate the specific heat of sodium metal using the Dulong-Petit law. Compare your answer with the experimental value given in Table 3.7.

2. The specific heat of uranium is 0.0277 cal/g°C. Calculate the molar specific heat and compare this to the value predicted by Dulong and Petit.

3. The work required to form a mole of vacancies in aluminum is 17,500 cal/mol. Calculate the fraction of vacant sites in aluminum at 20 °K and at 10 °C below the melting point of aluminum. What are the numbers of atoms per vacancy at both of these temperatures?

4. Calculate the work required to form 1 mol of vacancies in solid silver if the fraction of vacant sites is 3.70×10^{-5} at the melting point of silver.

5. The fraction of vacant sites is usually about 1 in 2000 atoms just below the melting point of a metal. Calculate the approximate work to form a mole of vacancies in solid tungsten at its melting point.

6. Calculate the approximate strain energy in ergs and calories for 10^{12} cm of screw dislocations per cubic centimeter of lead. The shear modulus, G, of lead is 8.11×10^5 psi, and the atomic radius of lead is 0.175 nm. $E \cong lGb^2$, where E is energy, l is the length of the dislocation in cm, and b is the length of the Burgers vector in cm.

7. What are four of the 12 slip systems employing {112} planes and <111> directions in BCC metals?

8. Compare the length of the Burgers vector for slip in iron in the [111] direction with the Burgers vector for nickel, which slips in the [110] direction.

9. A *cross-slip system* involves two planes intersecting in the direction of slip, such as (101)[111](110) in BCC crystals. Write three other cross-slip systems for BCC systems (hint: make careful sketches; the direction passes through two points that are common to the two different planes).

10. Write three cross-slip systems that exist in FCC crystals. How many different cross-slip systems exist in FCC crystals?

11. It takes 2400 min at 557 °C and 50 min at 637 °C to recrystallize zirconium that has been reduced in area by 13%. Calculate the values of A and Q in the equation for recrystallization: $1/\tau = Ae^{-Q/RT}$.

12. It takes 3000 min at 497 °C and 60 min at 567 °C to recrystallize zirconium that has been reduced in area by 51%. Calculate the values of A and Q in the equation for recrystallization: $1/\tau = Ae^{-Q/RT}$.

13. The solubility of carbon in BCC vanadium is about 1.2% at 1653 °C and 0.3% at 1400 °C. Calculate the values of A and Q in the equation, wt.% $C = Ae^{-Q/RT}$. Calculate the concentration of carbon in vanadium at 723 °C and compare this with the concentration of carbon in BCC iron at 723 °C.

14. The solubility of carbon in FCC iron is 0.77% at 727 °C, 0.94% at 800 °C, 1.22% at 900 °C, 1.92% at 1100 °C, and 2.11% at 1148 °C. Plot ln wt.% C versus 1/T°K. Obtain the slope of the curve and calculate Q from the following: $Q = -R \times$ slope. Then calculate A using your value of Q in the equation, $1.55 = Ae^{-Q/R \times 1273}$.

15. The solubility of carbon in tantalum is about 0.2% at 2180 °C and 0.1% at 1500 °C. Calculate the constants A and Q in the equation, wt.%$C = Ae^{-Q/RT}$.

16. Which should be more soluble in iron at 700 °C: boron or carbon? Why?

17. The solubility of tin in copper is 11.0% at 350 °C and 1.3% at 200 °C. What is the extrapolated solubility of tin in copper at room temperature?

18. Choose three pairs of metals that would show little or no solubility at room temperature. Using the Hume-Rothery rules, explain why each pair does not have much solid solubility.

19. Which of the following pairs of metals have very high, moderate, or very low solubilities: (a) Nb and Ti; (b) Ce amd Mg; (c) Cr and Os; (d) Cu and Fe? Answers may be checked by referring to a book of phase diagrams, such as Volume 8 (8th edition) of ASM's *Metals Handbook*.

20. At 427 °C, cobalt transforms to the FCC structure. For which of the following binary alloys would you predict complete solubility at 430 °C: (a) Co and Mo; (b) Co and Pd; (c) Co and Ni; (d) Co and Re? Explain your reasons.

21. A shipment of aluminum alloy bars was received in the age-hardened condition. The manufacturer wanted to fabricate them first and then age harden them. What could be done to solve this problem other than returning the shipment or melting the bars?

22. Would you expect silicon or chromium to harden steel more at the 2% level? Why?

23. Which hardens copper more effectively, zinc or beryllium? Explain.

24. Explain why a cluster of copper atoms strengthens aluminum much more than does a precipitate at the same weight percent of copper in aluminum.

25. For a hypothetical interstitial metal solute, $Q = 50,000$ cal/mol and $A = 200$. Calculate the solubility of this interstitial in the metal solvent at 1000 and 2000 °C. Does this demonstrate why metal interstitial atoms occur at very low concentrations?

26. The diffusivity of cobalt in iron is $D = 6.38e^{-61,400/RT}$. Calculate D at room temperature, 500, 1000, and 1500 °C.

27. The diffusivity of gold in nickel is $D = 0.02e^{-55,000/RT}$. At what temperature does $D = 10^{-10}$ cm^2/s?

28. The diffusivity of aluminum in silicon is $D = 4.8e^{-77,400/RT}$. If silicon did not melt, what is the maximum value of D at very high temperatures? If in general metals did not melt, write an equation for the diffusivity at infinite temperature.

29. For the diffusion of nickel in copper, $D = 1.93e^{-55,600/RT}$. Plot ln D versus $1/T°K$.

5

Phases & Phase Diagrams

A *phase* is a homogeneous part of a system. Although this definition may seem a little less than meaningful at first, with a little practice the reader will easily be able to identify the phases present in a system. First, most pure materials exist in three phases: solid, liquid, and gas. At the melting point, two phases are in equilibrium: solid and liquid. At the boiling point, the phases in equilibrium are liquid and gas. Variables for a system comprising a single pure material are simply temperature and pressure. A beaker full of sand is one phase even though the particles of sand are physically separated from one another (the gaseous phase between sand particles could be counted as a second phase in this system). Sand placed in water comprises two phases: solid sand and liquid water. However, salt placed in water is one phase if the salt dissolves completely; if it does not, and solid salt remains in the saturated salt solution, there are two phases: $NaCl(s) + H_2O(soln)$. At a unique pressure and temperature fixed by nature, the three phases of a pure substance—solid, liquid, and gas—are in equilibrium with one another. This is called the *triple point*.

The phases present in a metal or ceramic system and the way in which they are distributed (size, shape, and number of precipitate particles) critically affect the properties of materials. Therefore, a knowledge of phases and phase diagrams is a prerequisite for understanding most metal and many ceramic systems and their properties.

5.1 Phases and Components

In metallurgy, experiments frequently involve only solids and liquids; gas may be present as the atmosphere, but it may usually be ignored as an inert material, and gaseous metals usually are not involved in most applications of metallurgy. In any system, there is at most only one gaseous phase because all gases are soluble in one another, assuming that they do not react chemically. Because gaseous metals are relatively unimportant, this chapter will deal almost exclusively with solid and liquid metal systems.

A *component* is a pure element or compound. A single element, such as mercury, is a single component whether it is present as a solid, liquid, or gas, or some combination of these phases. An amalgam of silver and mercury, however, is comprised of two components. The materials scientist or engineer should be able to determine the number of phases and number of components in a given system.

Example 5.1A. In U.S. coins minted before 1964, the alloy was 90%Ag-10%Cu. At high temperatures, the alloy is liquid, and silver and copper are soluble in one another. However, at room temperature the metals are mutually insoluble (actually, only very slightly soluble). How many phases and how many components are present in (a) liquid 90Ag-10Cu and (b) 90Ag-10Cu at room temperature?

Solution: There are two elements present over the entire temperature range. Therefore, there are the same two components in (a) and (b): silver and copper. At the elevated temperature, there is only one phase: a liquid solution; at room temperature, there are two phases: solid silver and solid copper.

Example 5.1B. Carbon steel at room temperature is comprised of iron containing an extremely low concentration of carbon in solution and a precipitate of iron carbide, Fe_3C, a pure compound. Identify the phases and components present.

Solution: The components are Fe and Fe_3C, and the phases are iron solution and Fe_3C solid, which is a stoichiometric compound (i.e., having no solubility of excess iron or carbon).

Note: The two components could also be considered to be iron and carbon.

Binary alloys are defined as two-component systems; ternary, quaternary, etc., systems are relatively complicated, and these will not be dealt with here. The reader should refer to texts written on phases and phase diagrams.

Binary alloys frequently form *intermetallic compounds* that have crystal structures different from either pure element and have no solid solubility; they are stoichiometric compounds. *Intermediate phases* are compounds different from either pure element; many intermediate phases have appreciable solid solubility ranges.

5.2 The Phase Rule

In metal systems, the primary variables are *temperature* and *composition*. The pressure is assumed to be constant at one atmosphere unless specifically stated otherwise. We have already defined phases and components and have seen how these are determined for a given system.

The *phase rule* is a useful equation for analyzing metal or ceramic systems: it relates the number of phases and components to the degrees of freedom, $F = C - P + 1$, where F = degrees of freedom, C = number of components, and P = number of phases. All three of these are positive integers. F, the *degrees of freedom,* gives the number of variables that may vary independently under given circumstances while maintaining the same phase or phases. For instance, nickel and copper dissolve completely in one another, even in the solid state. Therefore, there are $F = C - P + 1 = 2 - 1 + 1 = 2$ degrees of freedom in the solid state, because the number of components is two (copper, nickel) and there is one phase (solid copper-nickel solution). In solid copper-nickel alloy, it is possible to vary the weight percent copper and still retain the solid phase (one degree of freedom), or to change the temperature and retain the solid phase (the second degree of freedom). Because these two variables can be changed independently while maintaining the same phase, the number of degrees of freedom is two.

Example 5.2A. How many degrees of freedom are there for a liquid copper-nickel alloy in equilibrium with solid copper-nickel? What is (are) the degree(s) of freedom?

Solution: Because there are two components (copper, nickel) and two phases (copper-nickel liquid solution and copper-nickel solid solution), there is only one degree of freedom according to the phase rule: $F = C - P + 1$, or $F = 2 - 2 + 1 = 1$. It is possible to vary the composition (weight percent copper), but once the composition is given, then there is only one temperature for the given composition at which both phases exist: there is only one degree of freedom. On the other hand, it is possible to vary the temperature, but once a temperature is chosen, then there is only one composition* that will still maintain the two phases in equilibrium. It is impossible to change *both* the temperature and composition independently and retain the same two phases. Therefore, only one degree of freedom exists for the given conditions.

Note: A given phase may have a large composition range: e.g., 90Cu-10Ni is the same phase as 10Cu-90Ni at room temperature.

Example 5.2B. How many degrees of freedom are there for a pure metal at its melting point (defined as the temperature at which solid and liquid are in equilibrium)?

Solution: In a pure metal there is one component, and at the melting point two phases are in equilibrium: $F = C - P + 1$, or $F = 1 - 2 + 1 = 0$. Therefore, there are zero degrees of freedom: no variables can be changed without upsetting the equilibrium. Thus, the melting point of a pure metal occurs at one specific temperature. Note that it is impossible to vary the composition of a pure substance (e.g., copper is 100% Cu).

It has been assumed that the systems with which we have been dealing (copper-nickel, silver-copper, etc.) are at equilibrium. The phase rule holds true *only* for systems at equilibrium: it is possible to quench silver-copper alloys fast enough to prevent precipitation of copper. Thus, a one-phase alloy, a supersaturated solution of copper in silver, is obtainable at room temperature, but the phase rule does not apply. If the silver-copper supersaturated solution is annealed at an elevated temperature, copper precipitates and the alloy reaches equilibrium in time—and then the phase rule will apply.

A supersaturated solution is *metastable,* which means it has temporary stability. A block of wood standing on its smaller end is metastable because it has a higher potential energy than when the block is lying down. The latter is the stable state. Metastable states may exist for indefinite periods if they are not disturbed: unless some force is applied to tip the wood over, it will not spontaneously tip over from the metastable to the stable state. Likewise, the silver-copper supersaturated solution (metastable) will remain at room temperature for a very long period. However, when its temperature is elevated several hundred degrees, copper precipitates, and the stable, or equilibrium, state is reached.

---

*One composition for the solid phase, for which there is a corresponding fixed composition of the liquid phase (the two phases have different compositions; see section 5.3).

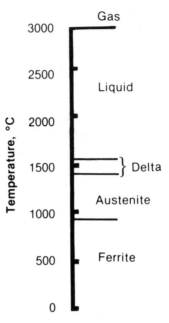

Fig. 5.1   Phase diagram of pure iron

5.3   Phase Diagrams

One type of graph that is most useful to materials scientists is the phase diagram. *Phase diagrams* are usually plots of the stable (equilibrium) phase or phases for various chemical compositions as a function of temperature. A phase diagram for a pure metal, such as iron (see Fig. 5.1), reduces merely to a line because a single pure component cannot vary in concentration. However, if pressure is introduced as an additional variable, the phase diagram of a single component is two dimensional, and the phase rule is $F = C - P + 2$. Variables are temperature and pressure. Most phase diagrams for materials scientists and engineers do not contain pressure as a variable, and hence $F = C - P + 1$. Binary phase diagrams (involving two components) contain a tremendous amount of useful information: melting points, phases present, compositions of phases present, relative amounts of phases present at a given set of conditions, and solubilities. We shall learn how to obtain all of the available information from a given phase diagram.

Copper and nickel are completely soluble in one another in both the solid and liquid state. This is indicated by the phase diagram of copper-nickel, shown in Fig. 5.2. Any point on the phase diagram indicates a unique combination of temperature and composition. The copper-nickel phase diagram shows that there are two phases in equilibrium for all points within the banana-shaped area between the liquidus and solidus curves. The *liquidus* curve is the upper curve, or the curve between the liquid phase and the two-phase region (liquid + solid). The *solidus* curve is the lower curve, or the curve between the solid phase and the two-phase region.

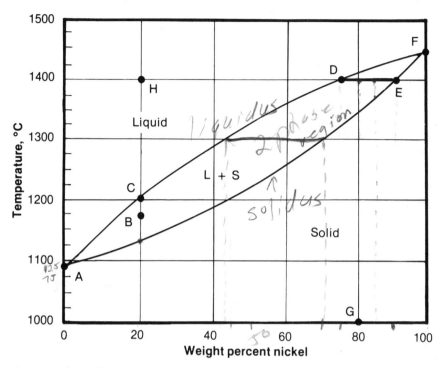

Fig. 5.2   Phase diagram of copper-nickel

Example 5.3A. Locate points on the copper-nickel phase diagram for the following temperatures and compositions: (A) 0% Ni, 1083 °C, (B) 20% Ni, 1170 °C, (C) 80% Cu, 1200 °C, (D) 74% Ni, 1400 °C, (E) 9% Cu, 1400 °C, (F) 100% Ni, 1455 °C, (G) 80% Ni, 1000 °C, and (H) 20% Ni, 1400 °C.

Solution: Composition is given by weight percent nickel or weight percent copper. Note that wt.% nickel = 100 − wt.% copper. On a phase diagram, a given composition is represented by a vertical line, the abscissa (x-axis) value. To locate a point on the phase diagram, first locate the composition line, then travel up this vertical line to the appropriate temperature. The points in Example 5.3A are plotted on the phase diagram in Fig. 5.2.

Example 5.3B. How many degrees of freedom are there at points A, B, G, and H on the copper-nickel phase diagram in Fig. 5.2?

Solution: Note that there are two components, copper and nickel, for all the points except A, where there is only one component. For point A, using the phase rule where F = C − P + 1, F = 1 − 2 + 1 = 0, or there are no degrees of freedom: the melting point of pure copper is fixed at 1083 °C. For point B, F = 2 − 2 + 1 = 1, or one degree of freedom: either temperature or composition may be changed, but not both. If the temperature is changed to another value, then the composition is fixed, or if the composition is changed to a different value, the temperature is determined.

Note: On the solidus curve, there are two phases: solid and an infinitesimal amount of liquid. On the liquidus curve, there is liquid plus an infinitesimal amount of solid. At point G, $F = 2 - 1 + 1 = 2$, or two degrees of freedom: both temperature and composition may be varied while still maintaining only one phase, solid. At point H, $F = 2 - 1 + 1 = 2$, or two degrees of freedom: same as G, only the liquid phase is the stable one.

A *tie-line* is a horizontal line, representing a constant temperature, drawn across a two-phase region until the line intersects its boundaries. In Fig. 5.2, a tie-line has been drawn between points D and E. The line is horizontal because the temperature is 1400 °C all along the tie-line. The intersection of the tie-line (across a two-phase region) with the liquidus curve gives the chemical composition of the liquid in equilibrium with the solid. The composition of the solid is given by the intersection of the tie-line with the solidus curve.

Example 5.3C. For an overall chemical composition of 80% Ni (i.e., 80 g Ni to 20 g Cu) in a copper-nickel alloy, what are the chemical compositions (i.e., the weight percent nickel) of the solid and liquid phases at 1400 °C?

Solution: Location of the point on the phase diagram at 80% Ni and 1400 °C shows that it is within a two-phase region. To determine the chemical compositions of the solid and liquid phases, first draw a tie-line (the line D-E in Fig. 5.2). The chemical composition, or weight percent nickel, for the liquid phase is found by dropping a vertical line from point D to the abscissa. This is found to correspond to 74 wt.% Ni: the liquid phase has a chemical composition of 74Ni-26Cu. To determine the chemical composition of the solid, drop a vertical line from point E. This corresponds to 91% Ni, and this is the composition of the solid phase (91Ni-13Cu). Note that neither the liquid nor the solid phase has a composition of 80% Ni. Also note that *any* overall chemical composition between 74 and 91% Ni at 1400 °C has the *same* tie-line and identical compositions for the two phases: 74% Ni in the liquid phase and 91% Ni in the solid phase.

Example 5.3D. What are the compositions for the solid and liquid phases in the copper-nickel system at 1400 °C when the overall chemical composition is 85% Ni?

Solution: Draw a tie-line at 1400 °C. Because this is line D-E and is identical to the tie-line in Example 5.3C, the compositions of the solid and liquid phases are also the same: 74% Ni in the liquid phase and 91% Ni in the solid phase.

The reader may wonder why an overall alloy content of 80Ni-20Cu can have exactly the same solid and liquid phase compositions at 1400 °C as an alloy of 85Ni-15Cu. The answer lies in a simple fact: the proportions, or ratios, of the two phases are not the same in the above two cases. In the alloy containing 85% Ni, there is much more nickel-rich solid than liquid present. In the 80% Ni alloy, there is slightly more copper-rich liquid than solid.

At this point, the reader must make a clear distinction between two separate concepts. (1) The *chemical composition* of a phase is the weight percent of each element present. The *overall* chemical composition is the average weight percent of each element in the total system. (2) The *relative proportion* of two phases, or the ratio of weights of the two phases, is the weight in

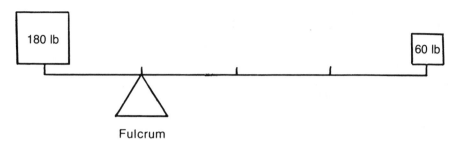

Fig. 5.3 Lever rule
Balance is achieved if the large square on the left represents a weight of 180 lb, and the smaller square on the right weighs 60 lb.

grams of one phase compared with the weight in grams of the second phase. This may be expressed as a ratio or fraction or given as the weight in grams of the two phases. It is frequently convenient to assume a basis of 100 g of alloy and then to calculate the number of grams of phase 1 and the number of grams of phase 2. The reader must understand clearly the difference between chemical compositions and relative amounts of two phases.

The chemical compositions of two phases in equilibrium are read directly from the appropriate tie-line on the phase diagram. The relative amounts of two phases may be calculated by a principle, or equation, called the lever rule (see below).

Point D on the liquidus curve of Fig. 5.2 is 100% liquid and 0% solid, or, based on 100 g of alloy, there would be 100 g of liquid and 0 g of solid. Based on 100 g of alloy, point E in Fig. 5.2 corresponds to 100 g of solid and 0 g of liquid. If we chose a point on the phase diagram very near to E, say 86% Ni and 1400 °C, the two phases would have the chemical compositions given in Example 5.3C, but the relative proportion of the two phases would be comprised of much solid and only a little liquid. Likewise, a point at 75% Ni and 1400 °C would correspond to an alloy comprised of two phases: mainly liquid with a small amount of solid.

If we desire to calculate quantitatively the relative amounts or ratio of the two phases present, it is necessary to know the chemical compositions of three points: (1) the liquid phase, (2) the solid phase in equilibrium with the liquid phase, and (3) the overall composition of the alloy. When one knows these, the relative amount or fraction of liquid phase present is found from the following tie-line; the principle is known as the lever rule (Fig. 5.3):

% Ni in liquid	Avg. % Ni present	% Ni in solid

$$\text{Fraction of liquid phase present} = \frac{\% \text{ Ni in solid} - \text{Avg. } \% \text{ Ni present}}{\% \text{ Ni in solid} - \% \text{ Ni in liquid}}$$

$$\text{Fraction of solid phase present} = \frac{\text{Avg. } \% \text{ Ni present} - \% \text{ Ni in liquid}}{\% \text{ Ni in solid} - \% \text{ Ni in liquid}}$$

The *lever rule* states that the fraction of the phase on the left (the liquid in the above example) is equal to the length of the *opposite* arm of the lever divided by the total length of the lever. The calculation of relative amounts of phases present is analogous to the weights of a father and a small daughter on a seesaw. To balance the seesaw, the father, being heavier, must sit on the short arm of the seesaw, while the daughter, being lighter, must sit on the long arm (Fig. 5.3). The weight of the daughter is proportional to the opposite or short arm of the seesaw, and the weight of the father is proportional to the arm opposite him, namely, the long arm.

Example 5.3E. Calculate the relative amounts or fraction of liquid and solid phases present in Example 5.3C (80Ni-20Cu overall composition, 1400 °C; 74% Ni is the chemical composition of the liquid phase, and 91% Ni is present in the solid).
Solution: Set up the appropriate lever:

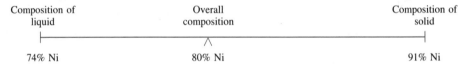

Composition of liquid	Overall composition	Composition of solid
74% Ni	80% Ni	91% Ni

According to the lever rule, fraction of liquid present = (91 − 80)/(91 − 74) = 11/17 = 0.647; fraction of solid present = (80 − 74)/(91 − 74) = 6/17 = 0.353. Note that 0.647 + 0.353 = 1.000, and this must be so: the fraction of liquid phase + fraction of solid phase = 1.000. If there had been 100 g of alloy present and an overall composition of 80% Ni, then the weight of liquid present would equal the fraction of the liquid phase × 100 g = 0.647 × 100 g = 64.7 g liquid. The weight of the solid phase present = the fraction of solid phase × 100 g = 0.353 × 100 g = 35.3 g solid. 64.7 g + 35.3 g = 100 g total.

Example 5.3F. Based on 100 g of copper-nickel alloy at 1400 °C, how many grams of solid phase are present when the overall composition is 86% Ni?
Solution: First set up the appropriate lever:

Composition of liquid		Overall composition	Composition of solid
74% Ni		86% Ni	91% Ni

$$\text{Weight of solid present} = \frac{\text{opposite side of lever}}{\text{total length of lever}} \times 100 \text{ g}$$

$$= \frac{86 - 74}{91 - 74} \times 100 \text{ g} = \frac{12}{17} \times 100 \text{ g}$$

$$= 70.6 \text{ g}$$

Note that the weight of liquid may be calculated either from the lever rule or from the following: weight of liquid phase = 100 g total − 70.6 g solid = 29.4 g.

A *material balance* is defined as the number of grams of each component in each phase. A material balance may be obtained by (1) determining the chemical compositions of the two phases present, (2) determining the relative amounts of the two phases present by the lever rule (usually on the basis of 100 g of alloy), and (3) calculating the weight of *each* component present in *each* phase by combining (1) and (2).

Example 5.3G. Determine the material balance for 100 g of 60Ni-40Cu alloy at 1300 °C.

Solution: First, determine the chemical compositions of the liquid and solid phases by referring to the phase diagram in Fig. 5.2 and by drawing a tie-line at 1300 °C. At the ends of the tie-line, drop vertical lines to the abscissa and read the following compositions from the abscissa: % nickel in liquid = 46.5; % nickel in solid = 63.5. Second, determine the relative amounts of the two phases using the lever rule:

Composition of liquid		Overall composition	Composition of solid
├───────────────────	────────────────	──∧──────────────	──┤
46.5% Ni		60% Ni	63.5% Ni

$$\text{Weight of liquid phase present} = \frac{\text{opposite lever arm}}{\text{total lever length}} \times 100 \text{ g}$$

$$= \frac{63.5 - 60}{63.5 - 46.5} \times 100 \text{ g} = \frac{3.5}{17.0} \times 100 \text{ g}$$

$$= 20.6 \text{ g}$$

$$\text{Weight of solid present} = \frac{60 - 46.5}{63.5 - 46.5} \times 100 \text{ g}$$

$$= \frac{13.5}{17.0} \times 100 \text{ g} = 79.4 \text{ g}$$

Third, calculate the number of grams of nickel and copper in each phase: grams of nickel in liquid = fraction of nickel in liquid × grams of liquid present = 0.465 × 20.6 g = 9.6 g. Grams of copper in liquid = fraction of copper in liquid × grams of liquid = 0.535 × 20.6 = 11.0 g Cu. Note that 9.6 g Ni + 11.0 g Cu = 20.6 g of liquid. Grams of nickel in solid = fraction of nickel in solid × grams of solid present = 0.635 × 79.4 g = 50.4 g. Grams of copper in solid = 0.365 × 79.4 g = 29.0 g. Note that 50.4 g Ni + 29.0 g Cu = 79.4 g solid.

We now have a material balance: we know the compositions of both phases and the number of grams of each component in each phase. The total equals 100 g: 9.6 g Ni in liquid + 50.4

g Ni in solid = 60.0 g Ni present. 11.0 g Cu in liquid + 29.0 g Cu in solid = 40.0 g Cu present. 60.0 g Ni + 40.0 g Cu = 100.0 g of alloy.

5.4   Cooling Curves and the Determination of Phase Diagrams

When phase changes occur, heat is generally evolved or absorbed. For instance, when water freezes, heat is given off (the heat of fusion), and when ice melts, heat is absorbed. Liquid metals have heats of fusion also, and heat is evolved when metals freeze. A pure metal freezes at a specific temperature; for instance, copper freezes (or melts) at 1083 °C, and silver freezes at 960.5 °C. If pure liquid silver is melted in a crucible and allowed to cool slowly (the crucible may be insulated, or it may contain a large amount of silver, such as 1 kg), the cooling rate (temperature plotted as a function of time) will be nearly linear as heat is lost to the surroundings. The temperature may be measured by a thermocouple connected to a recorder.

As long as the silver is completely liquid, the temperature decrease is nearly linear with time; however, at 960.5 °C, silver solid begins to form and heat is given out. As soon as this occurs, the temperature remains at 960.5 °C until all the liquid silver has frozen. This usually requires at least several minutes. Once the silver has completely frozen, then the temperature once again decreases nearly linearly with time until eventually the silver in the crucible attains the same temperature as the surroundings (often room temperature). The time-temperature curve is called a *cooling curve* (see Fig. 5.4). The freezing point of silver is one point on any phase diagram in which silver is one of the components (for example, see Fig. 5.5). In a like manner, the freezing temperature of pure copper may be obtained from a cooling curve.

What would the cooling curve for a composition of 50Ag-50Cu look like? First, the cooling rate for the single liquid phase would be very nearly linear. At 875 °C, a copper-rich phase, called β, begins to solidify. When this happens, heat is given out, and the rate of cooling decreases (i.e., the slope of the cooling curve changes abruptly; see Fig. 5.6). Because we are

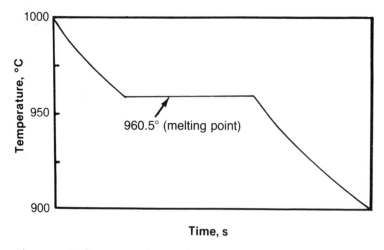

Fig. 5.4   Cooling curve of pure silver
The temperature arrest at 960.5 °C is due to the heat of fusion being given off.

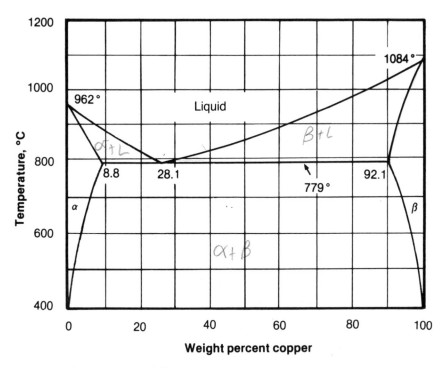

Fig. 5.5 **Phase diagram of silver-copper**

now in a two-phase region, $F = C - P + 1 = 2 - 2 + 1 = 1$ degree of freedom. This means that the temperature can change while the two phases, liquid and solid, coexist. As the temperature decreases, several things happen: (1) the composition of the liquid changes: it takes on a composition higher in silver and lower in copper; (2) the composition of the solid changes: it too becomes slightly enriched in silver with a lower copper concentration; (3) surprisingly, *both* phases shift the *same way* in concentration; this apparent contradiction is resolved when we realize that the ratio of solid to liquid phases also changes with cooling (different tie-lines are involved, and each tie-line at a lower temperature corresponds to a higher fraction of solid and a lower fraction of liquid); (4) the composition of the liquid eventually reaches a minimum of 28.1% Cu.

At 28.1% Cu the composition is called the eutectic composition. In a binary phase diagram, a *eutectic* comprises three phases in equilibrium: one liquid phase, which is the only phase stable above the eutectic temperature, and two solid phases, which are stable at temperatures below the eutectic point. In this system, three phases are in equilibrium at a temperature of 779 °C: liquid, silver-rich α phase (solid), and copper-rich β phase (solid). The liquid of eutectic composition, 71.9Ag-28.1Cu alloy, freezes at a specific temperature, 779 °C, to form a solid, but the solid consists of an intimate mixture of two phases: α and β. Likewise, if a solid 71.9Ag-28.1Cu alloy is heated, it has a sharp melting point at 779 °C (provided the two phases are intimately mixed). At the eutectic point, $F = C - P + 1 = 2 - 3 + 1 = 0$. There are zero degrees of freedom: nature has set the eutectic temperature *and* composition.

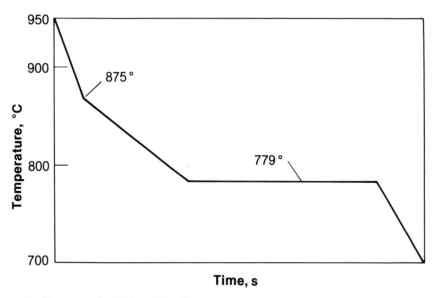

Fig. 5.6   Cooling curve for 50Ag-50Cu alloy
The slope in the cooling curve decreases at 875 °C due to the solidification of copper-rich solid. The temperature arrest at 779 °C is due to the heat of fusion given off by the eutectic.

Once the β phase begins to precipitate, the cooling curve will show an abrupt change of slope, as mentioned above. A plateau in the cooling curve occurs at the eutectic composition because the eutectic liquid phase freezes at a constant temperature, 779 °C (Fig. 5.6). In experimental measurements of cooling curves, the alloy may not always be at equilibrium because the cooling rate may be too rapid for complete equilibration, which requires diffusion over appreciable distances.

When all of the liquid has frozen to α and β solids, the alloy will then cool at a uniform rate; once again the cooling curve is nearly linear, but it may have a slightly different slope from the part of the cooling curve for the liquid only. Note that the important points in the phase diagram are indicated by abrupt changes in slope of the cooling curve: 875 and 779 °C in the case of 50Ag-50Cu alloy. The cooling curve for a liquid having the eutectic composition is similar to the cooling curve for pure silver or pure copper. Cooling curves may be used to obtain points all along the liquidus and solidus curves.

The *solvus curve* is the curve between the single α solid phase and the two-solid-phase region, α and β. Another solvus occurs between the β and α + β phase fields. Because of small heat effects involved in the solid phase transformations, the solvus is difficult or impossible to obtain from cooling curves. However, the solvus curve actually represents the *maximum solubility* of copper in silver-rich α at a given temperature; likewise, the other solvus gives the maximum solid solubility of silver in copper at any temperature up to 779 °C.

The solvus curves may be obtained by metallographic techniques. For example, one may prepare an alloy containing 5 wt.% Cu and anneal it at various temperatures. Samples of the alloy are quenched from various temperatures and investigated metallographically to determine

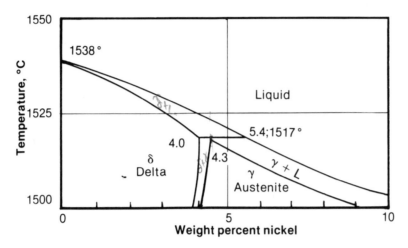

Fig. 5.7   Peritectic in the iron-nickel phase diagram

the temperature above which no β phase appears. This temperature gives a point on the solvus curve. Note that it is not necessary to obtain more than three or four points on each solvus curve to determine the shape of the curve. The shape of the solvus curve is represented by an equation of the form wt.% Cu (dissolved in silver) = $Ae^{-Q/RT}$ (see Chapter 4, section 4.5). Solvus curves may also be determined from electron microprobe analysis of binary alloy diffusion zones (see section 5.7).

5.5   Peritectic Transformation

The eutectic transformation in binary phase diagrams may be summarized as follows:

$$\underset{\text{(higher temperature)}}{\text{liquid phase}} \rightleftarrows \underset{\text{(lower temperature)}}{\text{solid phase 1 + solid phase 2}}$$

At the eutectic temperature and composition, three phases are in equilibrium, and F = 2 − 3 + 1 = 0, or there are no degrees of freedom: both the temperature and composition are fixed (this corresponds to a point on the phase diagram). Because three phases are in equilibrium, this is frequently called a *three-phase reaction*.

Another type of three-phase equilibrium is the *peritectic*, in which an equilibrium occurs between a liquid and a solid phase to form another solid phase:

$$\underset{\text{(higher temperature)}}{\text{liquid phase + solid phase 1}} \rightleftarrows \underset{\text{(lower temperature)}}{\text{solid phase 2}}$$

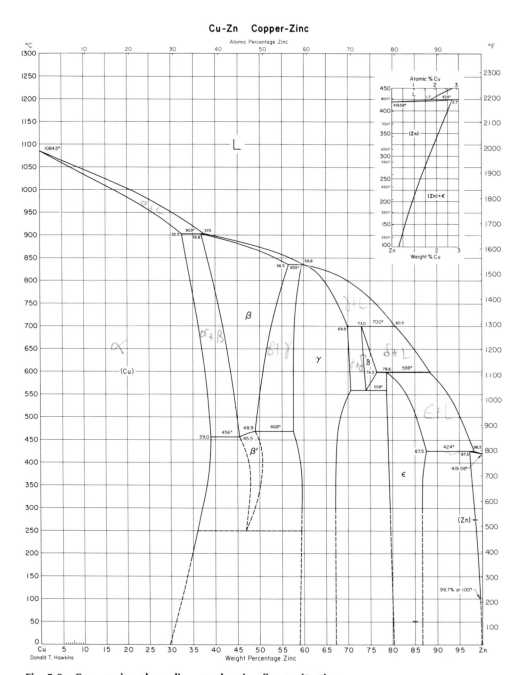

Fig. 5.8   Copper-zinc phase diagram showing five peritectics

An example of a peritectic reaction is shown in Fig. 5.7, which is a portion of the iron-nickel phase diagram. When iron containing 4.3% Ni is cooled slowly from 1600 °C, where the alloy is a single liquid phase, solid δ phase begins to precipitate at 1522 °C. The alloy then comprises two phases: liquid + solid δ. At 1517 °C, the liquid has a composition of 5.4% Ni and the solid contains 4.0% Ni. Upon cooling, these two phases transform to γ solid containing 4.3% Ni.

Example 5.5A. Calculate the relative amounts of liquid and solid δ phases at 1517 °C in the iron-nickel peritectic.

Solution: Set up the appropriate lever:

Composition of solid δ   Composition of solid γ   Composition of liquid

4.0% Ni   4.3% Ni   5.4% Ni

Using the lever rule, the fraction of δ is:

$$\delta = \frac{\text{opposite arm}}{\text{total length of lever}} = \frac{5.4 - 4.3}{5.4 - 4.0} = \frac{1.1}{1.4} = 0.786$$

$$\text{Fraction of liquid phase} = 1.000 - 0.786 = 0.214$$

Based on 100 g of peritectic alloy, 78.6 g δ and 21.4 g liquid transform upon cooling to 100 g of solid γ. Compositions of iron-nickel containing from 4.0 to 4.3% Ni form two solid phases upon cooling: L + δ → δ + γ; on the other hand, compositions containing from 4.3 to 5.4% Ni cool to form L + γ solid: L + δ → L + γ.

Peritectic transformations, or peritectics for short, are common in binary phase diagrams. They occur frequently at the highest temperature of many solid phases. For instance, the copper-zinc phase diagram contains five peritectics (see Fig. 5.8 for this phase diagram).

5.6 Eutectoid and Peritectoid Transformations

Other types of three-phase equilibria are the eutectoid and peritectoid equilibria, in which three solid phases are involved. The *eutectoid* is analogous to the eutectic equilibrium except that all three phases are solid:

$$\begin{array}{ccc} \text{solid phase 1} & & \text{solid phase 2 + solid phase 3} \\ \text{(higher temperature)} & \rightleftarrows & \text{(lower temperature)} \end{array}$$

The most famous eutectoid in metallurgy is in the iron-carbon system, and it has tremendous practical importance (see Chapter 6). Another eutectoid may be found in the copper-zinc phase diagram. The reader should look through the phase diagrams in this chapter and be able to identify all of the three-phase equilibria.

The last three-phase transformation to be described is the *peritectoid* transformation, which involves three solid phases:

$$\text{solid phase 1 + solid phase 2} \underset{}{\overset{}{\rightleftarrows}} \text{solid phase 3}$$
$$\text{(higher temperature)} \qquad \text{(lower temperature)}$$

Peritectoid transformations are comparatively rare.

According to the phase rule, a maximum of three phases may be in equilibrium with each other for a binary alloy unless pressure is involved as an additional variable; then F = C − P + 2. If F = 0, then 0 = 2 − P + 2, or P = 4, or four phases may coexist in equilibrium when both temperature and pressure are variables in binary systems.

5.7   Analysis of Binary Phase Diagrams

The simplest type of binary phase diagram is one in which both *terminal phases,* which are the phases of the two pure components, have the same crystal structure and in which the atoms are about the same size and are chemically similar. Then the solids form a single solution phase, as in the case of copper-nickel. In addition, there is a single liquid phase. Because single phases must be separated from one another by two-phase fields (although they may come together at a point), the liquid phase is separated from the solid phase by a two-phase field: liquid + solid.

The next least complicated phase diagram after one having only a single solid phase, such as copper-nickel, is a diagram exhibited by gold-copper (observe Fig. 5.9 only at the region above 450 °C for the present). In this diagram, there are two two-phase (liquid + solid) lobes, and at the minimum, the liquid has a sharp freezing point: this is called a *congruent point* because the solid ⇄ liquid transformation occurs sharply at one composition. Congruent max-

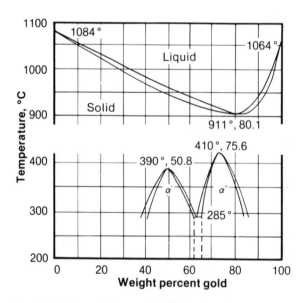

Fig. 5.9   Phase diagram of gold-copper
α′ and α″ are long-range ordered phases.

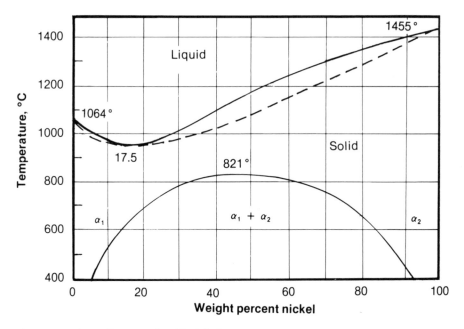

Fig. 5.10 Phase diagram of gold-nickel
The dashed curve is only approximately known.

imum points are common for phase diagrams involving several solid phases. At congruent points, freezing can occur with no change in composition.

Curved boundaries that surround two-phase regions can meet only at congruent points or at compositions of pure components. Some binary phase diagrams contain solid miscibility gaps in which the metals form a single solid solution at higher temperatures but two immiscible (insoluble) solid phases at lower temperatures. An example is the gold-nickel phase diagram. Between 821 and 950 °C, there is only one solid phase, while below 821 °C, there are two solid phases, as indicated on the phase diagram (Fig. 5.10). According to Hume-Rothery, solid phases are usually miscible at room temperature when the atomic diameters of the two atoms differ by less than 15% and when the two metals are similar chemically (electrode potentials are reasonably close) and are near one another in the periodic table (see Chapter 4). Increasing temperature favors greater miscibility, and miscibility is greatly increased in liquid phases. Most liquid metals are miscible, particularly at high temperatures. Miscibility gaps are responsible for eutectics and peritectics.

Terminal phases are defined as single solid phases at the extreme edges of the phase diagram, the phases ascribed to the pure solid components. *Intermediate phases* are solid phases of metallic compounds other than terminal phases. Many intermediate phases exist—look at the copper-zinc phase diagram in Fig. 5.8, which contains four intermediate phases. Intermediate phases may have a broad range of solid solubility, such as the β, γ, or ϵ phases in copper-zinc, or they may have essentially no solid solubility, in which case they are stoichiometric compounds, such as Fe_3C in the iron-carbon diagram (see Chapter 6). In binary metal systems, these

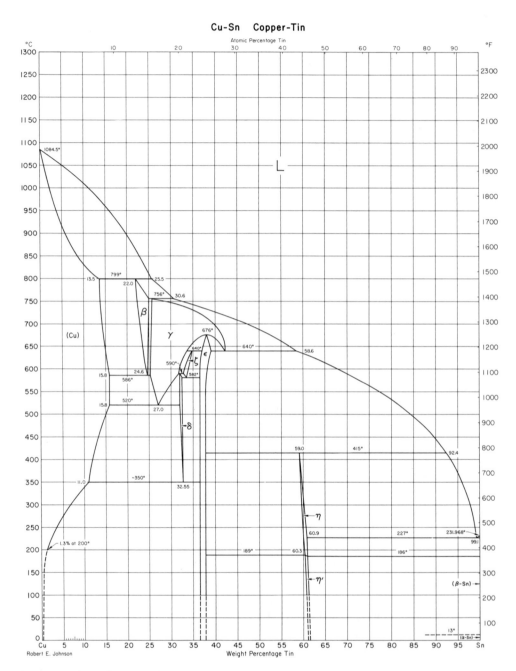

Fig. 5.11   Phase diagram of copper-tin

stoichiometric compounds are often called *intermetallic compounds;* two examples are CaAl$_2$ and NbSi$_2$.

Intermediate phases either have a congruent maximum melting point or a peritectic point at their melting point. A few solid phases disappear upon heating due to a peritectoid point, e.g., the β and δ phases in copper-silicon. Some phases have maximum "congruent" points, but a solid solution rather than a liquid is formed at the higher temperature.

If a horizontal line is drawn completely across any binary phase diagram at any given temperature, and if the line crosses two single and one two-phase fields, the sequence from left to right will be a one-phase field, a two-phase field, and another one-phase field. If intermediate phases are present, the sequence will be 1-2-1-2-1 for one intermediate phase and 1-2-1-2-1-2-1 for two intermediate phases, and so on. For instance, starting from the left of the copper-zinc phase diagram at 650 °C, the sequence of phase fields is α, α + β, β, β + γ, γ, γ + δ, δ, δ + liquid, and liquid, or 1-2-1-2-1-2-1-2-1.

Note that any two-phase region can be identified by the single phases on either side of it. When the components of a phase diagram are pure elements, the terminal phases are always single (unless the terminal point is a melting point or a transformation point, such as 910 °C for iron, at which temperature both BCC and FCC iron are in equilibrium with each other). Phases are usually given names corresponding to letters of the Greek alphabet, α, β, γ, δ, ε, etc., beginning at the left of the phase diagram.

The reader should now be able to interpret completely most binary phase diagrams. Three phase diagrams for reference in conjunction with problems at the end of this chapter are presented in Fig. 5.11 (Cu-Sn), Fig 5.12 (Pb-Sn), and Fig. 5.13 (Al-Nb).

Example 5.7A. List the composition and temperature for each of the following using the copper-zinc phase diagram: (a) five peritectics, (b) two eutectoids, (c) a point in the two-phase field δ + ε, (d) a point in the two-phase field β + L, (e) the maximum solubility of zinc in copper (α), (f) the maximum solubility of copper in zinc, and (g) the maximum concentration of zinc in δ.

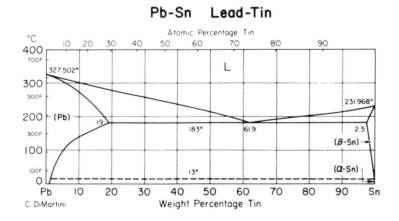

Fig. 5.12   Phase diagram of lead-tin (solder)

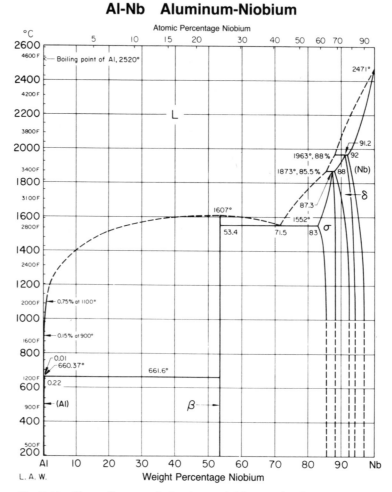

Al-Nb   Aluminum-Niobium

Fig. 5.13   Phase diagram of aluminum-niobium (columbium)

Solution: The following points (temperature, composition) were read from the copper-zinc phase diagram: (a) 903 °C, 36.8% Zn; 835 °C, 59.0% Zn; 700 °C, 73.0% Zn; 598 °C, 78.6% Zn; 424 °C, 97.3% Zn. (b) 250 °C, 47% Zn; 558 °C, 74% Zn. (c) 575 °C, 77% Zn. (d) 850 °C, 55% Zn. (e) 39.0% Zn at 456 °C. (f) 2.7% Cu at 424 °C. (g) 76.5% Zn at 598 °C.

5.8   Problems

1. In the phase diagram of solder (lead-tin), how many phases and how many degrees of freedom are there for the following: (a) 61.9% Sn, 183 °C; (b) 0% Sn, 327.5 °C; (c) 18% Sn, 180 °C; and (d) 50% Sn, 100 °C?

2. In the copper-tin phase diagram (copper-tin alloys are bronze), list the types of three-phase reactions for the upper and lower limits of each of the following intermediate phases: (a) β, (b) δ, and (c) ζ.

3. List the peritectics and peritectoids in the copper-tin phase diagram.

4. At room temperature, 100 g of 50-50 solder (50 wt.% Sn) does not contain exactly 50 g of the lead-rich phase and 50 g of the tin-rich phase. Explain.

5. Draw cooling curves that would be obtained for (a) pure lead, (b) 90Pb-10Sn, (c) Pb-20Sn, and (d) 61.9Sn-38.1Pb.

6. What is the maximum solubility of (a) tin in copper, (b) copper in zinc, (c) zinc in copper, (d) copper in silver, (e) silver in copper, (f) tin in lead, and (g) lead in tin?

7. A bronze alloy (copper-tin) has an overall composition of 20% Sn. At 450 °C what phases are present and what are their compositions? How much of each phase is present based on 100 g of alloy?

8. A bronze alloy has 10% Sn present. How much ε phase would be present in 100 g of alloy at (a) 100 °C, (b) 300 °C, and (c) 400 °C?

9. In the copper-zinc system, what phases are present at 550 °C and at each of the following overall compositions: (a) 40% Zn, (b) 55% Zn, and (c) 90% Zn.

10. What are the chemical compositions of the phases in problem 9?

11. At room temperature, 50 g of γ copper-zinc and 50 g of zinc are mixed and then melted. When the alloy cools, what possible phase or phases could be present? Why?

12. What is the overall composition of an alloy that has 70 g lead-rich phase and 30 g tin-rich phase at 100 °C?

13. Calculate a material balance for a silver-copper alloy having an overall composition of 10% Cu at 700 °C.

14. Compile a list of binary phase diagrams having only one eutectic and no intermediate phases from Volume 8, 8th Edition, of the ASM *Metals Handbook*.

15. Two metals form a eutectic and have no intermediate phases. The eutectic is at 53% metal B and 424 °C. The maximum solubility of B in A is 6.0% and 0.4% of A in B. The melting point of A is 660 °C, and the melting point of B is 938 °C. Construct the phase diagram for these metals. If you want to check your answer, identify the metals from their melting points and find the phase diagram in Volume 8, 8th Edition, of the ASM *Metals Handbook*.

6

The Iron-Carbon System & Heat Treating of Steel

During early civilization, copper was the most important metal of commerce: tools, weapons, armor, mirrors, statues, and many other types of objects were made of copper or copper alloys. Iron and steel emerged only gradually during many hundreds of years beginning about 2000 B.C.; only in the past several hundred years has steel attained the position of foremost metal. Almost literally it is the backbone of civilization: the common availability, excellent strength, and low cost of steel enable it to serve as our primary structural material, especially among metals.

One of the startling aspects of iron is its unique transformation from face-centered cubic (FCC) to body-centered cubic (BCC) as the temperature *decreases*. No other metal undergoes a similar transformation, which is the basis for so many of the useful combinations of properties of commercial steels. If this unique transformation did not exist, many applications would require far more expensive materials.

6.1 Iron-Carbon Phases

Pure iron exists as a gas at temperatures above 2880 °C and as a liquid between 1538 and 2880 °C. In the solid state, pure iron exists as the following phases:

$\alpha-$Fe	BCC	Ferrite	−273 to 912 °C
$\gamma-$Fe	FCC	Austenite	912 to 1394 °C
$\delta-$Fe	BCC		1394 to 1538 °C

Because both α-Fe and δ-Fe are BCC, they are identical in structure but are differentiated by name: α-Fe, or ferrite, is of great practical importance, while δ-Fe, the high-temperature form, is relatively unimportant. Long ago it was believed that another phase, β-Fe, existed, but this was shown to be α-Fe above its Curie temperature, 770 °C (that is, nonmagnetic α-Fe). The *Curie temperature*, which is the temperature above which a metal loses its ferromagnetism, does not represent a phase change because the crystal structure does not change at this temperature. Below 770 °C, FCC iron is nonmagnetic (and unstable), while BCC iron is ferromagnetic. The operation of motors, generators, and electromagnets depends on ferromagnetism;

if iron did not have this property, the only recourse would be to use nickel or possibly cobalt or the magnetic alloys such as Alnico.

Steel is an alloy of iron and carbon. Therefore, the phases of iron and carbon are important in developing an understanding of steel. Carbon is soluble to only a very small extent in α-Fe (up to about 0.022% at 727 °C); because of the larger interstices in the FCC structure, namely the "holes" in the center of each unit cell, FCC iron (γ-Fe) has a much higher solubility for carbon than α-Fe. In fact, the maximum solubility of carbon in γ-Fe is 2.11% C at 1148 °C. Carbon is commonly present in cast iron as graphite, the pure carbon phase in which carbon atoms are bonded covalently in layers by weak Van der Waal's forces between layers. However, graphite does not normally exist in steels: the carbon reacts with iron to form a different phase, Fe_3C (iron carbide), which is frequently called *cementite* or *carbide*. Iron carbide is a metastable phase: iron and graphite are more stable than Fe_3C. However, in normal steels, graphite is formed very slowly from iron carbide (silicon catalyzes the decomposition of Fe_3C into iron and graphite in cast irons). Carbide is stable in steel for a very long time, especially at room temperature. Other iron-carbon phases exist, such as $Fe_{2.2}C$, but these are unimportant in steels because they are uncommon. "Steel" refers to an iron-carbon alloy containing up to about 2% C. Very few commercial steels, however, contain more than 1% C.

The reader should become very familiar with the important phases of iron and carbon (α, γ, and Fe_3C) before proceeding.

When austenite (FCC iron) contains more than 0.022% C, it transforms upon cooling below 727 °C. The transformation products are two phases: ferrite and carbide. The relative amounts of these phases may be calculated from the iron-carbon phase diagram (Fig. 6.1).

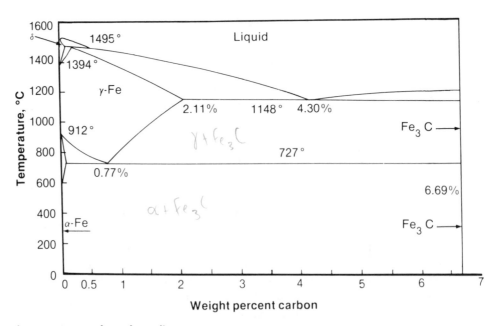

Fig. 6.1   Iron-carbon phase diagram

Below 0.022% C, austenite transforms to ferrite at temperatures ranging from 912 °C for 0% C to 727 °C for 0.022% C.

Example 6.1A. One hundred grams of austenite containing 0.77% C in solution is cooled to below 727 °C. Using the iron-carbon phase diagram, calculate the number of grams of ferrite and the grams of carbide present at 726 °C.

Solution: Use the lever rule, discussed in Chapter 5; first set up the appropriate lever:

The number of grams of α-Fe is proportional to the *opposite* arm of the lever: Grams of α-Fe = (6.69 − 0.77)/(6.69 − 0.022) × 100 g, or (5.92/6.668) × 100 g = 88.8 g. Grams of Fe₃C = (0.77 − 0.022)/(6.69 − 0.022) × 100 g = (0.748/6.668) × 100 g = 11.2 g. Also, because grams of α-Fe + grams of Fe₃C = 100.0 g, then grams of Fe₃C = 100.0 g − 88.8 g = 11.2 g.

Note: The composition of 0.77% C and a temperature of 727 °C is the famous eutectoid of the iron-carbon diagram (*eutectoid*—a solid phase at a higher temperature transforming to two different solid phases at a given temperature: solid phase 1 = solid phase 2 + solid phase 3). The eutectoid transformation in steel at 727 °C is:

$$
\begin{array}{ccc}
 & 727\ °C & \\
\gamma\text{-Fe (0.77\% C)} & \rightleftharpoons & \alpha\text{-Fe (0.022\% C)} + \text{Fe}_3\text{C (6.69\% C)} \\
100.0\ g & & 88.8\ g \qquad\qquad 11.2\ g
\end{array}
$$

When austenite transforms to ferrite and carbide, an intimate mixture of thin platelets of Fe₃C immersed in an α-Fe matrix occurs (see Fig. 6.2); this two-phase mixture is called *pearlite*. Note that pearlite is *not* a phase—it is a two-phase mixture, α-Fe and Fe₃C.

Example 6.1B. One hundred grams of austenite containing 1.00% C in solution is cooled slowly from 1200 °C to room temperature. (a) At what temperature does Fe₃C first form? (b) At what temperature is the formation of Fe₃C completed? (c) What is the chemical composition of the last austenite that transforms? (d) At 726 °C, what is the weight of each phase formed? (e) At room temperature, what are the weights of the two phases formed? (Note that the solubility of carbon in α-Fe at room temperature is essentially zero.) (f) How many grams of pearlite and how many grams of Fe₃C are present at room temperature?

Solution: (a) Using the iron-carbon phase diagram (Fig. 6.1), construct a vertical line at a composition of 1.00%. This line intersects the γ + Fe₃C two-phase region at 820 °C. Therefore, Fe₃C begins to precipitate at 820 °C—this is the temperature at which γ-Fe begins to transform.

Fig. 6.2   Pearlite in 1030 steel
Ferrite is white; pearlite is comprised of platelets of iron carbide in a matrix of ferrite. (Etched, 375×)

(b) The transformation is completed at 727 °C when the last γ-Fe transforms. Note that at 728 °C, most of the steel is γ-Fe with a little bit of proeutectoid Fe₃C present. (*Proeutectoid* carbide refers to Fe_3C formed above 727 °C, the eutectoid temperature.)

(c) The last austenite that transforms has 0.77% C dissolved in it. This is the eutectoid composition. Note that as the alloy is slowly cooled, meaning that the system is in equilibrium and therefore conforms to the phase diagram, the composition of γ-Fe is 1.00% C above 820 °C. Between 820 and 727 °C, the composition of austenite is found by constructing a horizontal tie-line (see Chapter 5) at the given temperature; where the tie-line intersects the curve separating γ-Fe from γ-Fe + Fe₃C, draw a vertical line downward until the line intersects the abscissa (x-axis). The weight percent of carbon in solution is then read from the abscissa. Below 727 °C, the γ-Fe has completely transformed to α and Fe₃C.

(d) Set up the appropriate lever at 726 °C:

α-Fe	1.00% C	Fe₃C
⊢————————————————∧————————————————————————————⊣		
0.022% C	Avg. composition	6.69% C

Using the lever rule, grams of α-Fe = (6.69 − 1.00)/(6.69 − 0.022) × 100 g, or (5.69/6.668) × 100 g = 85.3 g. Grams of Fe₃C = (1.00 − 0.022)/(6.69 − 0.022) × 100 g = (0.978/6.668) × 100 g = 14.7 g. *Note*: 85.3 g of α-Fe + 14.7 g of Fe₃C = 100.0 g total.

(e) At room temperature, the lever is:

```
   α-Fe              1.00% C                                    Fe₃C
   ├─────────────────────────────────────────────────────────────┤
                        ∧
   0.00% C          Avg. composition                          6.69% C
```

Grams of α-Fe = (6.69 − 1.00)/(6.69 − 0.00) × 100 g = (5.69/6.69) × 100 g = 85.1 g. Grams of Fe_3C = 100 g − grams α-Fe = 100 g − 85.1 g = 14.9 g. *Note*: There is only a very small difference between the amounts of the two phases at 726 °C and room temperature.

(f) Because pearlite is formed from eutectoid γ-Fe containing 0.77% C, use this composition in setting up the appropriate lever:

```
   Pearlite           1.00% C                                  Fe₃C
   ├─────────────────────────────────────────────────────────────┤
                        ∧
   0.77% C          Avg. composition                          6.69% C
```

Grams of pearlite = (6.69 − 1.00)/(6.69 − 0.77) × 100 g = (5.69/5.92) × 100 g = 96.1 g. Grams of Fe_3C ("proeutectoid" Fe_3C) = 100 g − 96.1 g = 3.9 g. The proeutectoid carbide is carbide that formed at a temperature higher than 727 °C; proeutectoid carbide is formed *before* pearlite. *Note*: The total number of grams of Fe_3C is equal to the number of grams of proeutectoid Fe_3C plus the number of grams of Fe_3C in the pearlite: grams of Fe_3C = 3.9 + 0.112 × 96.1 = 3.9 + 10.8 = 14.7 g Fe_3C total, which agrees with part (d). The fraction of Fe_3C in pearlite is 0.112, and this is a constant.

It is important to know how to make the above types of calculations because they aid in understanding and interpreting photomicrographs of steels. Pearlite, which is shown in Fig. 6.3, is a distinctive and easily identified constituent. In some cases, particularly at relatively low magnifications such as 200× or lower, the Fe_3C platelets are not resolved, and pearlite simply looks gray. In a given specimen of steel, the approximate weight percent carbon may be calculated from the estimated proportion of pearlite present. If the steel is 100% pearlite, then the composition is 0.77% C. If the steel contains 0% pearlite, the composition is close to 0% C. When steel contains 50% pearlite, the composition is 0.50 × 0.77 = 0.38% C.

Example 6.1C. The microstructure of a given steel reveals 30% pearlite and 70% ferrite as the visible constituents. Calculate the approximate weight percent carbon in the steel.
Solution: Set up the appropriate lever at room temperature:

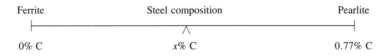

```
   Ferrite          Steel composition                        Pearlite
   ├─────────────────────────────────────────────────────────────┤
                        ∧
   0% C               x% C                                    0.77% C
```

On the basis of 100 g, 30 g pearlite = (x − 0)/(0.77 − 0) × 100 g = 100x/0.77. Therefore, x = (30 g × 0.77)/100 g = 0.23% C.

Note: In general, the approximate weight percent carbon may be calculated by multiplying the fraction of pearlite present times 0.77. The main inaccuracy is the estimation of the percent pearlite present.

Figure 6.3 contains photographs of the microstructures of steels containing various carbon concentrations. Note the correlation of the carbon concentration with the fraction of pearlite present.

A few additional comments need to be made about the iron-carbon phase diagram:

- As mentioned above, graphite actually is the thermodynamically stable phase and Fe_3C is

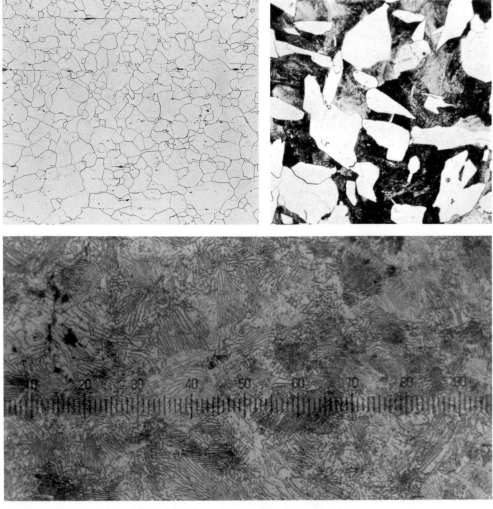

Fig. 6.3   Microstructures of steels with various carbon concentrations
Top left: Steel containing 0.012% C (nital etch, 67×). Top right: Steel containing about 0.35% C; pearlite is dark (nital etch, 250×). Bottom: Eutectoid steel containing 0.8% C (nital etch, 340×)

only metastable. Because Fe_3C decomposes only with great difficulty, it is present in nearly all steels, while graphite is absent (except in cast irons).
- There is a peritectic at 1495 °C and 0.16% C.
- There is a eutectic at 1148 °C and 4.30% C.
- Iron carbide, Fe_3C, is a stoichiometric compound without appreciable carbon solubility; this fact is shown in the phase diagram by a vertical line representing Fe_3C at 6.69% C.
- The part of the iron-carbon phase diagram containing more than 6.69% C is unimportant, practically speaking, and is therefore omitted.

6.2 Transformation of Austenite

According to the iron-carbon phase diagram, austenite containing less than 0.77% C transforms between 912 and 727 °C to form some proeutectoid ferrite. The amount depends on the composition (weight percent carbon) of the austenite.

Example 6.2A. One hundred grams of austenite containing 0.30% C transforms on slow cooling to proeutectoid ferrite and pearlite. How many grams of proeutectoid ferrite are present at 727 °C?

Solution: The proeutectoid ferrite may be determined by using the lever at 727 °C:

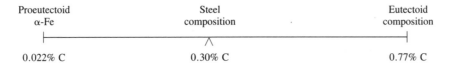

Proeutectoid α-Fe	Steel composition	Eutectoid composition
0.022% C	0.30% C	0.77% C

Grams of proeutectoid ferrite = (0.77 − 0.30)/(0.77 − 0.022) × 100 g = (0.47/0.748) × 100 g = 62.8 g.

Note: When the steel is cooled to room temperature, the proeutectoid ferrite remains essentially the same: about 62.8 g. This proeutectoid ferrite constituent is readily identified from the rest of the ferrite, which is present in the pearlite (α + Fe_3C).

In a similar fashion, austenite containing greater than 0.77% C transforms to proeutectoid carbide at temperatures between 1148 and 727 °C on slow cooling. The proeutectoid carbide forms along the grain boundaries of the original austenite as a thin, white layer; it is easily identified as carbide because pearlite is the only material within the carbide network.

When austenite contains 0.77% C, it transforms completely to pearlite. The overall composition of pearlite is also 0.77% C, because no carbon is lost or gained during transformation. At present, we are discussing the slow cooling of austenite, which assumes that equilibrium is always maintained. When eutectoid austenite transforms to pearlite, about 1 kcal of heat per mole is given off, analogous to the heat of fusion given off when water or other liquids freeze. On slow cooling of the specimen, the heat of transformation is absorbed by the surroundings.

Most phase transformations are time dependent because they involve nucleation and growth of the new phase or phases. In many cases it takes time to remove or add the necessary heat of transformation; usually, diffusion must also occur.

Pearlite is a eutectoid mixture of *two* phases, ferrite and carbide. The carbide is present as fairly parallel platelets in a matrix of ferrite, which is the continuous phase. Pearlite is nucleated in austenite heterogeneously along grain boundaries, and it is nucleated in colonies, or several approximately parallel platelets of carbide very close together.

At a given transformation temperature, such as 727 °C, the spacings between carbide platelets is uniform in pearlite. At lower transformation temperatures (fast cooling to a given temperature and then holding until transformation is complete), the carbide platelets are closer together. *Pearlite nodules,* which are small clumps of pearlite, are not called grains because they are a two-phase mixture; pearlite nodules grow with a constant linear rate.

6.3   Transformation of Austenite at Temperatures Lower Than 727 °C

Because the transformation of austenite to pearlite requires nucleation and growth, time is needed, and it is not surprising to find that the time required for transformation to occur strongly depends on the temperature. Furthermore, the *interlamellar spacing* (which is the perpendicular

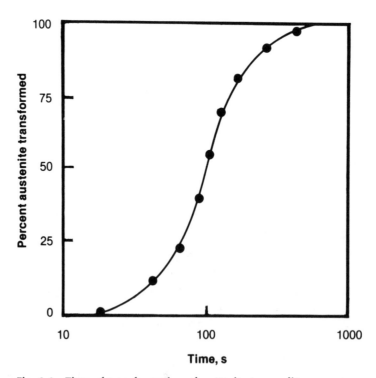

Fig. 6.4   Time of transformation of austenite to pearlite
Isothermal transformation at 680 °C

distance between successive layers of iron carbide platelets in pearlite) depends on the temperature of the transformation. The lower the transformation temperature, the smaller the interlamellar spacing. Small interlamellar spacings cause high hardness and brittleness. Steels cooled moderately fast (not quenched) will transform at moderately low temperatures and hence will be very hard. Slow cooled steel (1) transforms at relatively high temperatures, (2) has relatively large interlamellar spacings, and (3) is relatively soft.

Both nucleation and growth are necessary for transformation to occur. The rate of nucleation, which is the formation of small volumes of pearlite that are capable of growth, is relatively low at temperatures near 727 °C, but the rate of nucleation increases markedly with decreasing transformation temperatures. A nucleus must attain a minimum size before growth can occur. Beneath a given size, the nucleus may either grow larger or smaller. The growth rate of nuclei is rather slow at temperatures just under 727 °C, but growth is faster at lower transformation temperatures. Finally, at about 635 °C, the growth rate remains nearly constant for lower temperatures.

Figure 6.4 shows a typical curve for the time required for austenite to transform to pearlite at a given temperature. Note that the abscissa is a log scale: the time required to start transformation is usually much less than the time necessary to complete transformation.

6.4   Martensite and Tempered Martensite

Martensite is a body-centered tetragonal phase of iron containing carbon. It is metastable, but will remain indefinitely at room temperature. Martensite does not appear in the iron-carbon

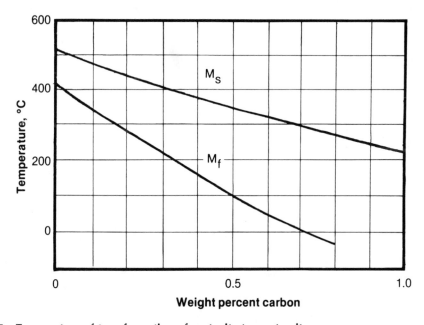

Fig. 6.5   Temperature of transformation of austenite to martensite
M_s represents the temperature at which transformation from austenite to martensite starts; M_f is the temperature at which the transformation finishes.

phase diagram because of its metastability. Martensite is of great practical importance because it is extremely hard and strong, but it is also very brittle. Most important, however, is the ability of martensite to be tempered to a strong and tough material. Martensite is formed by very rapid cooling of austenite, usually by quenching in water, brine (concentrated saltwater), or oil.

The transformation of austenite to martensite does *not* involve diffusion, nucleation, or growth. Rather, the mechanism involves a shear displacement on {111} planes. This occurs very rapidly, and hence the transformation occurs with essentially no elapsed time. However, *the extent of transformation depends only on the temperature*: above a certain temperature, no martensite is formed, and over a rather large temperature range a varying fraction of austenite transforms to martensite. Below a given temperature, the transformation is complete—this is summarized by Fig. 6.5, which also shows that both the start of the transformation to martensite, labeled M_s, and the finishing of the transformation, labeled M_f, depend strongly on the carbon content of the austenite. Any austenite that remains untransformed at room temperature is called *retained austenite*.

Martensite has a body-centered tetragonal structure similar to that of BCC iron; however, the carbon atoms are all placed randomly in $0,0,\frac{1}{2}$ sites, which result in the lengthening of the c-axis (see Fig. 6.6). At the same time, the a-axis is slightly compressed (see Fig. 6.6). There are not enough carbon atoms to be placed in each $0,0,\frac{1}{2}$ site.

The carbon atoms and particularly the lattice distortion undoubtedly impede the motion of dislocations, causing the extreme hardness, high strength, and brittleness characteristic of martensite.

Because martensite is metastable, it only forms when cooled rapidly enough to avoid transformation to pearlite. *Tempering* of martensite is heating martensite to an intermediate temperature of about 350 to 500 °C for about an hour. During tempering, the martensite

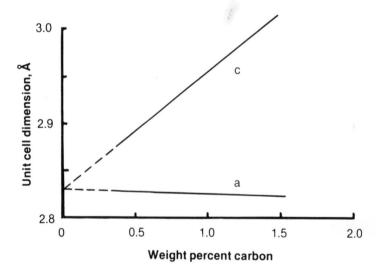

Fig. 6.6   Unit cell dimensions of martensite as a function of carbon content
The unit cell dimension, a, decreases slightly with increasing carbon, while the dimension c increases markedly with carbon.

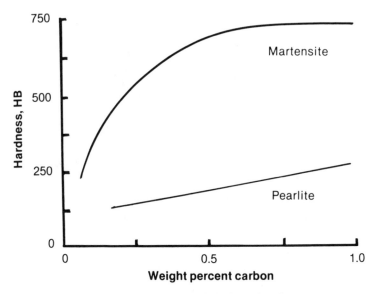

Fig. 6.7   Hardness of martensite as a function of carbon content

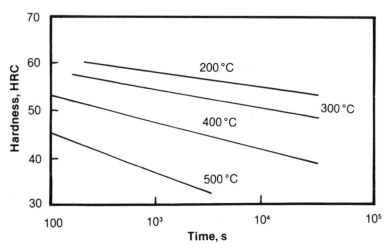

Fig. 6.8   Hardness of tempered martensite as a function of time and temperature of tempering
Note that martensite tempered at 200 or 300 °C is very hard.

decomposes to ferrite and carbide; however, the resulting carbide has a much different shape from carbide in pearlite: *tempered martensite* is a two-phase mixture of ferrite and carbide, with the carbide present as extremely small, chunky particles rather than platelets. Accordingly, dislocations may move through tempered martensite more easily than through martensite or fine pearlite, and the material is stronger and tougher than pearlite or martensite. The finely divided

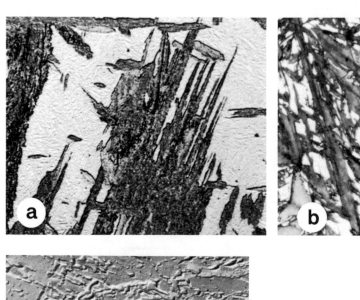

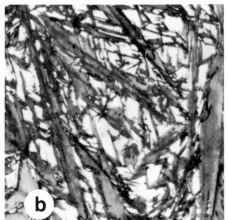

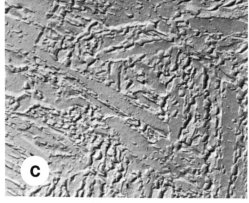

Fig. 6.9 (a) Bainite (750×). (b) Martensite (1000×). (c) Tempered martensite (9130×)

particles of carbide cause tempered martensite to be both strong and moderately ductile, resulting in the ability to absorb a high energy of impact without failure (i.e., tempered martensite is very tough).

The hardness of martensite depends on the weight percent of carbon in the steel and, of course, on other alloying metals (see Fig. 6.7). The hardness of tempered martensite also depends on both the temperature at which tempering occurs and the amount of time the steel is held at this temperature (see Fig. 6.8). The time and temperature of tempering martensite determine the particle size of the resulting carbide, and therefore the strength (increased hardness means increased strength).

6.5 Bainite

If austenite is cooled rapidly to an intermediate temperature to avoid transformation to pearlite and then held at this temperature, it transforms to a different two-phase mixture of ferrite and carbide: *bainite*. Bainite is very similar to tempered martensite in that the particles of carbide are very small and chunky, compared with relatively large platelets of carbide in pearlite. Bainite forms both by diffusion and shear. Thus, the transformation takes time to occur: the steel must be quenched to and held at a temperature in the approximate range of 300 to 500 °C. Although it may sound confusing, bainite, a two-phase mixture of ferrite and carbide, occurs as plates, but the *carbide* within the plates is chunky.

The formation of bainite by quenching to an intermediate temperature and holding is called *austempering*, and it results in a strong and tough steel. Another advantage of austempering is that quench cracking is avoided because the transformation occurs at constant temperature and excessive stresses are avoided. Figure 6.9 compares a photomicrograph of bainite with martensite and tempered martensite.

6.6   Spheroidite

When austenite is cooled to about 700 °C, not far below the transformation temperature, and held for several hours, transformation finally occurs to form *spheroidite,* which is relatively large, chunky (nearly spherical) iron carbide in a matrix of ferrite. Spheroidite has maximum softness for the amount of carbon present in a steel (steels with lower carbon contents are softer than steels with higher carbon contents). Steel that is intended to be quenched to form martensite and then tempered for toughness and strength normally is shipped as spheroidite. Then the manufacturer may machine or shape the steel first before making it hard and tough by quenching and tempering.

6.7   Time-Temperature-Transformation Diagrams

A great deal of useful information may be summarized by constructing a diagram for a given steel in which the time required for isothermal transformation (i.e., transformation at a constant temperature) to begin and to end is plotted versus temperature. This is called a *time-temperature-transformation (TTT) diagram.* TTT diagrams are available for many different steels, and they are a necessity in the work of many engineers.

TTT diagrams are *not* phase diagrams; they do not tell what phases or phase mixtures are stable under given conditions. They are primarily related to kinetics, because they tell how long it takes for a given reaction to occur isothermally (at one constant temperature). Therefore, TTT diagrams must be used with care and interpreted with assistance of the phase diagram.

Figure 6.10 shows the TTT diagram for a eutectoid steel (0.77% C). Above 727 °C, austenite is stable; when it is cooled suddenly to a lower temperature (e.g., 530 °C) by quenching in a salt bath, transformation to pearlite begins in about 0.9 s and ends after holding for 6 s. These data may be determined experimentally by quenching thin steel specimens in a salt bath at 530 °C and holding for various periods before quenching to room temperature. Then the microstructure is investigated to determine the fraction of pearlite present; of course, any austenite still present at 530 °C forms martensite upon quenching. Most data for TTT curves involve times much greater than 1 s and hence may be obtained with greater relative accuracy than that of the above example.

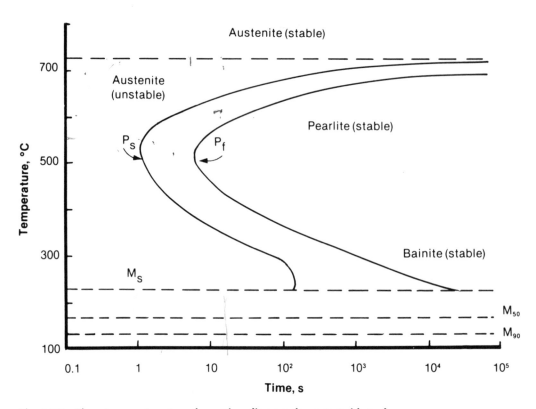

Fig. 6.10   Time-temperature-transformation diagram for eutectoid steel
All transformations are isothermal. P_s represents the start of the transformation to pearlite.

The TTT curves may be used to determine what the microstructure will be following a given heat treatment. Some examples follow.

Example 6.7A. A specimen of eutectoid steel is heated to 800 °C, quenched to 600 °C, held for 20 s, then quenched to room temperature. What phases and constituents are present at the end of this treatment? (A *constituent* is a readily definable region in a microstructure; e.g., pearlite is a constituent.)

Solution: At 800 °C, the specimen is 100% austenite. According to the TTT diagram, after 20 s at 600 °C, the steel would have transformed completely to pearlite (the only constituent present), which is a mixture of α-Fe and Fe_3C (two phases present). Quenching to room temperature does not change pearlite; therefore, the only constituent is pearlite, and the phases present are ferrite and carbide.

Example 6.7B. A specimen of 0.77% C steel is heated to 800 °C, quenched to 550 °C, held for 2.5 s, then quenched to room temperature. What phases and constituents are present?

Solution: After being held 2.5 s at 550 °C, the steel would be about 50% austenite and 50% pearlite (phases: α-Fe and Fe_3C). Upon quenching to room temperature, essentially all of the

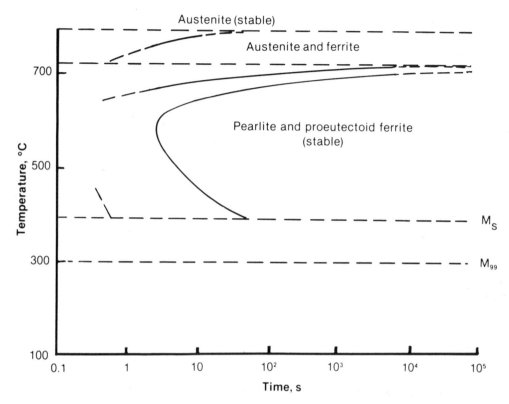

Fig. 6.11   Time-temperature-transformation diagram for 1045 steel
All transformations are isothermal. 1045 steel contains 0.45% C.

austenite would transform to martensite. Three phases are present: α-Fe, Fe_3C, and martensite. Two constituents are observable using a metallograph (metallurgical microscope): pearlite and martensite.

Example 6.7C. A specimen of eutectoid steel is quenched to 600 °C, held for 20 s, quenched to room temperature, and reheated to 700 °C and held for 1 s. What phase or phases and constituents are present at the end of this treatment?

Solution: After holding for 20 s, the steel is completely transformed to pearlite. Neither quenching nor reheating to 700 °C changes the pearlite (unless it is held for several hours at this temperature). Therefore, the phases present are α-Fe and Fe_3C, and the single constituent is pearlite.

Note: The only way to reform γ-Fe once pearlite is formed is to reheat *above* 727 °C. The only way to form martensite is by transformation from austenite: quenching pearlite does *not* form martensite.

Example 6.7D. A specimen of 0.77% steel is quenched to 720 °C and held 10 s. It is then quenched to room temperature. What phase or phases and constituents are present?

Solution: After 10 s at 720 °C, the steel is still essentially 100% austenite: very few nuclei of pearlite have formed. Therefore, when the steel is quenched from 720 °C, it forms martensite completely because pearlite nuclei have had an insufficient time to form and grow. This is an example of how one must be very careful in interpreting TTT diagrams.

Figure 6.11 shows a TTT diagram for a steel containing 0.45% C. There are several differences between this TTT diagram and the TTT diagram for eutectoid steels: (a) the "nose" of the curve occurs at a shorter time than for the eutectoid steel: 0.3 s instead of 0.9 s; (b) both the M_s and M_f temperatures are much higher for the 0.45% C steel; (c) between 727 and 800 °C, the austenite decomposes to form some proeutectoid α-Fe. This is shown on the TTT diagram in the α and γ region at the top. When austenite containing 0.45% C is cooled to form pearlite, some proeutectoid α-Fe always forms. On slow cooling, the constituents present are pearlite and proeutectoid α-Fe.

TTT diagrams are qualitatively understandable in terms of nucleation and growth processes. Just below the equilibrium transformation temperature, both nucleation and growth are slow, and these cause the transformation time to be relatively long. At lower temperatures, both the nucleation and growth rates are fast, and the time to transformation is a minimum at the nose of the curve. At still lower temperatures, the growth rate decreases somewhat, and transformation times are again longer. Additionally, transformation is usually different in that bainite is frequently formed.

6.8   Transformation of Austenite During Continuous Cooling

Up to this point, the discussion of the transformation of austenite to pearlite, martensite, or bainite has dealt only with isothermal transformation: austenitic steel is quenched to a given temperature and held at that temperature until transformation occurs. The assumption has been made that quenching cools the steel instantaneously. In practice, cooling is carried out continuously: the rate of cooling may be several hundred degrees per second, as in water quenching; less in oil quenching; still less, as in air cooling; or as little as 1°/min, as in furnace cooling (the furnace power is shut off). The geometry of the steel object also affects the cooling rate: the interior of a large bar obviously cools much more slowly than the surface of the bar, regardless of the means of cooling. At room temperature, steel objects may vary in their microstructure from one place to another, depending on the cooling rate of each part of the object.

It is useful to know how continuous cooling affects the TTT curves: is the effect of continuous cooling large, small, intermediate, or other? For a eutectoid steel (0.77% C), the nose of the TTT curve occurs at about 1 s at 600 °C. This is the time at 600 °C required for transformation to begin. However, if the steel is continuously cooled to 600 °C during 1 s, transformation will not yet begin because the steel has been at temperatures higher than 600 °C for nearly the whole time, and at higher temperatures it takes longer for transformation to begin. Therefore, in continuous cooling, the nose of the TTT curve will be missed even if the cooling takes slightly longer than 1 s to arrive at 600 °C: martensite will still be obtained.

The net effect of continuous cooling is that the TTT curves are shifted to the right and lowered (see Fig. 6.12). Note that in continuous cooling, martensite may still be obtained if it

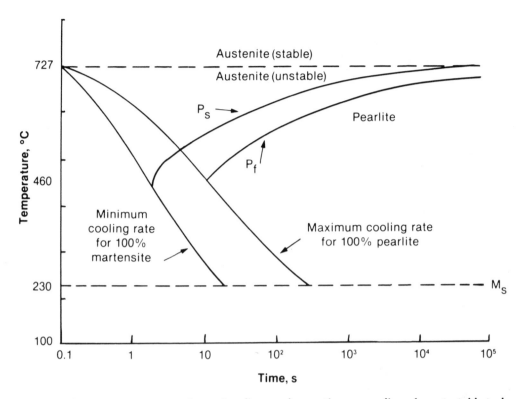

Fig. 6.12   Time-temperature-transformation diagram for continuous cooling of a eutectoid steel

takes 2 s to reach 500 °C, whereas for isothermal transformation, only 1 s at 600 °C results in the beginning of transformation.

Notice that in cooling at various rates, the plot of the temperature versus the logarithm of time gives approximately a straight line. If the nose of the TTT curve is missed during continuous cooling, then only martensite will be formed. If continuous cooling causes the steel to pass through the region of partial transformation, then the steel will comprise a mixture of pearlite and martensite. If the continuous cooling is slow enough for the path to cross into the pearlite region, only pearlite will result. Note that it is impossible to form bainite by continuous cooling. Bainite may only be formed by decreasing the cooling rate once the nose of the TTT curve has been missed.

Two continuous cooling rates are particularly important: (a) the critical cooling rate above which 100% martensite is obtained and (b) the cooling rate below which 100% pearlite is obtained. For a eutectoid steel, these two rates are about 200 °C/s and 50 °C/s through the region 750 to 500 °C.

6.9   Heat Treatment of Steel

Heat treatment of metals often enables one to "have his cake and eat it, too." Metals such as steel may be formed into useful shapes while in a softened condition, and then the final object may be strengthened (hardened) by heat treatment. The net result is a strong object that could not have been easily fabricated in the hardened state.

The first step in nearly all heat treatments of steel is *austenitizing,* which is heating the steel to a temperature sufficient to transform it to γ-Fe (austenite) and to dissolve all carbides present. In industry, the time and temperature for austenitizing are kept to a practical minimum in order to save fuel costs and to minimize oxidation, decarburization, grain growth, and distortion. The temperature for austenitizing is usually about 60 °C above the minimum temperature indicated by the phase diagram for a given composition of steel.

Example 6.9A. What is the lowest temperature at which medium carbon steel containing 0.35% C may be fully austenitized?

Solution: Refer to the iron-carbon phase diagram (Fig. 6.1). Construct a vertical line upward from 0.35%. This line passes into the γ-Fe single-phase region at 790 °C. Therefore, this is the lowest temperature at which the steel could be fully austenitized.

Note: A practical and safe temperature would be 850 °C for this steel. Sometimes hyper-eutectoid steels (>0.77% C) are not heated to a high enough temperature to dissolve all of the excess carbides during austenitizing. If the small amount of carbide dispersed in the austenite does not affect the desired final properties, then austenitizing may occur at temperatures of about 775 °C.

Annealing generally means holding metal at an elevated temperature. However, in the heat treating of steel, ''annealing'' means the furnace cooling of austenitized steel. The furnace containing the steel in coil, bar, sheet, or other form is shut off and allowed to cool over a period of several hours. In this way, transformation of austenite occurs just below 727 °C, and the resulting coarse pearlite is at maximum softness. This is desirable for maximum formability, machinability, and chemical homogenization.

Normalizing refers to the air cooling of austenitized steels. The resulting structures may be coarse to fine pearlite, depending on the geometry and quantity of steel being cooled. Normalizing is used to refine the structure of steel castings and overheated steels.

Spheroidizing is the heating of a pearlitic steel just below 727 °C for several hours. The platelets of iron carbide in the pearlite form small, chunky particles. The approximately spherical shape of the carbide particles gives the spheroidite improved properties over pearlite: the steel has minimum hardness and maximum ductility. Spheroidized steel is very machinable and resists cracking during cold forming. Steel having fine pearlite (relatively small interlamellar spacings) is generally used for spheroidizing. See Fig. 6.13 for a microstructure of spheroidite.

An *interrupted quench,* or *martempering,* is quenching steel below the nose of the TTT curve, holding for a short period, then cooling to obtain martensite. In this process, the surface and interior of the steel transform at about the same time, thus avoiding quench cracking due to differential contractions caused by the transformation.

Ausforming is similar to the interrupted quench, with the steel being plastically deformed at a temperature just above M_s, the start of transformation to martensite. Either martensite or bainite may be subsequently formed. This process is useful for obtaining additional strengthening due to strain hardening.

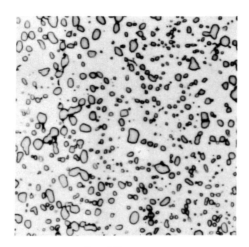

Fig. 6.13   Spheroidite (1000×)

Tempering, discussed previously in section 6.4, is the heating of martensitic steel to a temperature of 200 to 500 °C in order to transform martensite into tempered martensite, which is α-Fe plus chunky iron carbide, Fe_3C. Tempered martensite is tough and strong, with moderate ductility.

Maraging is the formation of relatively soft martensite (usually in stainless steels), machining or cold working the steel, then age hardening by an intermediate heat treatment. This forms very strong steel.

6.10   Hardenability of Steels

Low-carbon steels, containing less than 0.25% C, are useful because of their good formability and weldability. In most applications, such as automobile fenders or doors, good formability is the prime property, and strength is secondary. Hence, low-carbon steels may be heat treated to recrystallize or remove stresses, but they are generally not heat treated with the objective of obtaining high strength and toughness.

Medium-carbon (0.25 to 0.55% C) and *high-carbon steels* (0.55 to about 1% C) are frequently heat treated to maximize their hardness, such as in tool steels, or their strength and toughness, such as in moving, load-bearing machinery parts. In some cases, martensite is the desired end product, but martensite is frequently tempered to obtain maximum strength and toughness. For either type of use, it is necessary to obtain martensite without substantial amounts of pearlite.

To obtain 100% martensite in a given steel object, there is a minimum cooling rate. For unalloyed steels, it is frequently impossible to achieve this high cooling rate, especially at the interiors of rods, bars, or thick sheets. Even if the steel is quenched, the surface may form martensite while the heat transfer from the interior is simply too slow to enable martensite to be formed. Most steels, regardless of their chemical compositions, have about the same heat transfer rate, so little hope lies in achieving martensite throughout thick sections of steel by increasing heat flow. The quenching medium does vary the cooling rate of the surface appre-

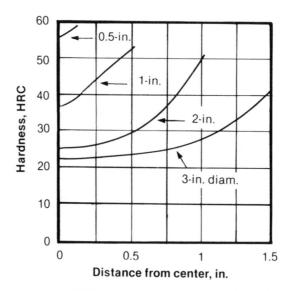

Fig. 6.14   Hardness traverses in 1040 steel specimens (cylindrical specimens of 0.40% C steel)

ciably, but even with instantaneous surface cooling, thick sections of steel transform to pearlite throughout most of their interiors.

If a few steel bars of several diameters but the same composition were quenched in water, each bar would have a different hardness as a function of distance from the surface. Hardness traverses across the steel may be made by cutting the bar in two (and keeping the specimen cool while cutting) and measuring the hardness at several points at various distances from the surface. The hardness traverse may be plotted as in Fig. 6.14. The hardness at the surface is due to the presence of 100% martensite, whereas ½ in. from the surface it decreases to 54 HRC (Rockwell C hardness number). This corresponds to 50% martensite and 50% pearlite. At distances greater than ½ in. from the surface, the hardness is less than 54 HRC due to the presence of more pearlite and the absence or near absence of martensite. The depth at which there is 50% martensite and 50% pearlite may be more accurately determined than most other ratios, so this point is taken as the standard. In Fig. 6.14, all bars with diameters smaller than 1 in. are hardened effectively throughout, whereas larger diameter bars are not.

The *critical diameter* is the maximum diameter at which the center of the bar has 50% martensite and 50% pearlite on quenching in a certain medium. The critical diameter gives a measure of how well a steel responds to hardening by quenching. In Fig. 6.14, the critical diameter, D, is 1.0 in. The *hardenability* of a steel is the *relative* ease with which it may be hardened throughout. (*Beware*: High hardenability has nothing to do with the ability to harden steel—all medium- or high-carbon steels may be hardened to about the same hardness by forming martensite throughout.)

High hardenability means having a relatively large critical diameter. Because the quench medium affects the critical diameter, this variable may be eliminated by using an ideal quench

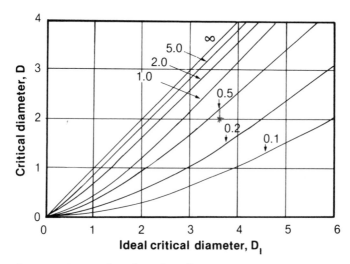

Fig. 6.15   D_I as a function of cooling rate
Each curve represents a specific cooling rate, or H value. See Table 6.1
for meaning of the various H values.

standard. The *ideal quench* is one that would cool the surface of any metal instantaneously to room temperature. The *ideal critical diameter*, D_I, is the critical diameter for an ideal quench. Although an ideal quench medium does not exist, its cooling action may be calculated from heat flow rates and factors derived relative to various quench mediums.

Ideal critical diameters may be obtained from Fig. 6.15 by determining the "H value," listed in Table 6.1 for various quench mediums, and then converting the D value to D_I.

Example 6.10A. What is the D_I of a steel having a critical diameter of 2.0 in. when quenched in well-agitated oil?

Solution: According to Table 6.1, the H value of an oil quench with strong agitation is 0.50. Construct a horizontal line at 2.0 in.; where this line intersects the curve for H = 0.50, draw a vertical line to the abscissa. This line intersects the abscissa at 3.3 in., the D_I value.

Table 6.1   Relative effectiveness (severity of quench) of various quench mediums

Quench medium	Agitation	H value
Oil	None	0.20
	Moderate	0.35
	Strong	0.50
	Very strong	0.70
Water	None	1.00
	Strong	1.50
Brine (H$_2$O + NaCl)	None	2.00
	Very strong	5.00
Ideal		∞

Table 6.2   Nomenclature of steels as related to composition

SAE No.	C	Mn	P max	S max	Si	Ni	Cr	Other
1008	0.10 max	0.40 ± 0.10	0.040	0.050	. . .	. . .	. . .	. . .
1010	0.10 ± 0.02	0.45 ± 0.15	0.040	0.050	. . .	. . .	. . .	. . .
1020	0.20 ± 0.02	0.45 ± 0.15	0.040	0.050	. . .	. . .	. . .	. . .
1025	0.25 ± 0.03	0.45 ± 0.15	0.040	0.050	. . .	. . .	. . .	. . .
1030	0.31 ± 0.03	0.75 ± 0.15	0.040	0.050	. . .	. . .	. . .	. . .
1040	0.40 ± 0.03	0.75 ± 0.15	0.040	0.050	. . .	. . .	. . .	. . .
1050	0.51 ± 0.03	0.75 ± 0.15	0.040	0.050	. . .	. . .	. . .	. . .
1060	0.60 ± 0.05	0.75 ± 0.15	0.040	0.050	. . .	. . .	. . .	. . .
1070	0.70 ± 0.05	0.75 ± 0.15	0.040	0.050	. . .	. . .	. . .	. . .
1080	0.81 ± 0.06	0.75 ± 0.15	0.040	0.050	. . .	. . .	. . .	. . .
1090	0.91 ± 0.06	0.75 ± 0.15	0.040	0.050	. . .	. . .	. . .	. . .
1330	0.30 ± 0.02	1.75 ± 0.15	0.035	0.040	0.27 ± 0.07	. . .	. . .	. . .
1345	0.45 ± 0.02	1.75 ± 0.15	0.035	0.040	0.27 ± 0.07	. . .	. . .	. . .
4012	0.12 ± 0.02	0.87 ± 0.12	0.035	0.040	0.27 ± 0.07	. . .	. . .	0.20 ± 0.05 Mo
4047	0.47 ± 0.02	0.80 ± 0.10	0.035	0.040	0.27 ± 0.07	. . .	. . .	0.25 ± 0.05 Mo
4118	0.20 ± 0.02	0.80 ± 0.10	0.035	0.040	0.27 ± 0.07	. . .	0.50 ± 0.10	0.12 ± 0.04 Mo
4161	0.60 ± 0.05	0.87 ± 0.12	0.035	0.040	0.27 ± 0.07	. . .	0.80 ± 0.10	0.30 ± 0.05 Mo
4320	0.20 ± 0.02	0.55 ± 0.10	0.035	0.040	0.27 ± 0.07	1.82 ± 0.18	0.50 ± 0.10	0.25 ± 0.05 Mo
4340	0.40 ± 0.02	0.70 ± 0.10	0.035	0.040	0.27 ± 0.07	1.82 ± 0.18	0.80 ± 0.10	0.25 ± 0.05 Mo
4419	0.20 ± 0.02	0.55 ± 0.10	0.035	0.040	0.27 ± 0.07	. . .	. . .	0.52 ± 0.08 Mo
4427	0.27 ± 0.02	0.80 ± 0.10	0.035	0.040	0.27 ± 0.07	. . .	. . .	0.40 ± 0.10 Mo
4615	0.15 ± 0.02	0.55 ± 0.10	0.035	0.040	0.27 ± 0.07	1.82 ± 0.18	. . .	0.25 ± 0.05 Mo
4626	0.26 ± 0.02	0.55 ± 0.10	0.035	0.040	0.27 ± 0.07	0.85 ± 0.15	. . .	0.20 ± 0.05 Mo
4720	0.20 ± 0.02	0.60 ± 0.10	0.035	0.040	0.27 ± 0.07	1.05 ± 0.15	0.45 ± 0.10	0.20 ± 0.05 Mo
4820	0.20 ± 0.02	0.60 ± 0.10	0.035	0.040	0.27 ± 0.07	3.50 ± 0.25	. . .	0.25 ± 0.05 Mo
5015	0.15 ± 0.02	0.40 ± 0.10	0.035	0.040	0.27 ± 0.07	. . .	0.40 ± 0.10	. . .
5060	0.60 ± 0.04	0.87 ± 0.12	0.035	0.040	0.27 ± 0.07	. . .	0.50 ± 0.10	. . .
5115	0.15 ± 0.02	0.80 ± 0.10	0.035	0.040	0.27 ± 0.07	. . .	0.80 ± 0.10	. . .
5160	0.60 ± 0.04	0.87 ± 0.12	0.035	0.040	0.27 ± 0.07	. . .	0.80 ± 0.10	. . .
6118	0.18 ± 0.02	0.60 ± 0.10	0.035	0.040	0.27 ± 0.07	. . .	0.60 ± 0.10	0.12 ± 0.02 V
6150	0.50 ± 0.02	0.80 ± 0.10	0.035	0.040	0.27 ± 0.07	. . .	0.95 ± 0.15	0.15 V
8115	0.15 ± 0.02	0.80 ± 0.10	0.035	0.040	0.27 ± 0.07	0.30 ± 0.10	0.40 ± 0.10	0.11 ± 0.04 Mo
8615	0.15 ± 0.02	0.80 ± 0.10	0.035	0.040	0.27 ± 0.07	0.55 ± 0.15	0.50 ± 0.10	0.20 ± 0.05 Mo
8660	0.60 ± 0.04	0.87 ± 0.12	0.035	0.040	0.27 ± 0.07	0.55 ± 0.15	0.50 ± 0.10	0.20 ± 0.05 Mo
8720	0.20 ± 0.02	0.80 ± 0.10	0.035	0.040	0.27 ± 0.07	0.55 ± 0.15	0.50 ± 0.10	0.25 ± 0.05 Mo
8822	0.22 ± 0.02	0.87 ± 0.12	0.035	0.040	0.27 ± 0.07	0.55 ± 0.15	0.50 ± 0.10	0.35 ± 0.05 Mo
9254	0.55 ± 0.04	0.65 ± 0.15	0.035	0.040	1.40 ± 0.20	. . .	0.65 ± 0.15	. . .
9310	0.10 ± 0.02	0.55 ± 0.10	0.025	0.025	0.27 ± 0.07	3.25 ± 0.25	1.20 ± 0.20	0.11 ± 0.04 Mo

Note that agitation always improves the effectiveness of a quench: agitation increases the H value, which is proportional to the cooling rate. Water has a lower viscosity than oil and therefore removes vapor bubbles more rapidly from the surface. On the other hand, brine, a concentrated salt solution, is a very effective quenchant because the salt in the water causes a series of minute explosions to occur at the surface, thereby violently agitating the quench solution.

Factors that affect hardenability are (1) austenitic grain size, (2) carbon content, and (3) alloy content (composition). Hardenability increases with increasing carbon content and with

Table 6.3   Chemical compositions of stainless steels

Stainless steel, AISI type	Typical composition, wt. %
301	17 Cr, 7 Ni, 2 Mn, 1 Si, 0.15 C
304	19 Cr, 9 Ni, 2 Mn, 1 Si, 0.1 C
310	25 Cr, 20 Ni, 2 Mn, 1.5 Si, 0.2 C
316	17 Cr, 12 Ni, 2 Mn, 1 Si, 0.1 C, 2.5 Mo
347	18 Cr, 10.5 Ni, 2 Mn, 1 Si, 0.1 C, 1 Nb
430	17 Cr, 1 Mn, 1 Si, 0.1 C
442	20 Cr, 1 Mn, 1 Si, 0.2 C
410	12 Cr, 1 Si, 0.15 C

increasing grain size (decreasing ASTM grain-size number). The larger the grains, the smaller the grain boundary area for nucleating pearlite, so the greater the hardenability. Every alloying element except cobalt increases the hardenability of steel. Cobalt increases both the rate of nucleation and growth of pearlite. Common alloying elements that increase hardenability include manganese, silicon, nickel, chromium, and molybdenum. These elements are present in relatively low concentrations in high-strength low-alloy (HSLA) steels, which generally have high hardenabilities.

6.11   Nomenclature of Steels

Because carbon is extremely important in steels, usually it is the most important element in describing steels having various properties. Steels are labeled by a four-digit numeral: the SAE (Society of Automotive Engineers) or AISI (American Iron and Steel Institute) number. The last two digits are the number of hundredths of a percent carbon in the steel. Thus, 1010 steel contains 0.10% C and 1080 steel has 0.80% C. The first two digits indicate the presence of elements other than carbon. For instance, steels numbered 10XX indicate *mild carbon* or *plain carbon steels* with nominal concentrations of manganese, sulfur, phosphorus, etc. Other numbers signify the presence of elements as shown in Table 6.2. The nomenclature and compositions for stainless steels are presented in Table 6.3.

6.12   Problems

1. What is the solubility of carbon in (a) δ-Fe at 1500 °C; (b) γ-Fe at 1300, 1100, and 925 °C; and (c) α-Fe at 800 °C?

2. One hundred grams of austenite containing 0.45% C is cooled to 725 °C. Calculate the number of grams of ferrite and grams of carbide present. How many grams of pearlite and how many grams of proeutectoid ferrite are present?

3. One hundred grams of austenite containing 0.60% C is cooled slowly from 1200 °C to room temperature. (a) At what temperature does Fe$_3$C first form? (b) At what temperature is the

formation of Fe_3C completed? (c) What is the chemical composition of the last austenite that transforms? (d) At 725 °C, what is the weight of each phase formed? (e) How many grams of pearlite are present at room temperature?

4. What is the chemical composition of 100 g of steel that contains 5.0 g of proeutectoid carbide?

5. (a) Calculate the weight percent carbon in Fe_3C. (b) What is the mole fraction of carbon in eutectoid steel?

6. The microstructure of a steel indicates 20% pearlite and 80% ferrite as the constituents present. Calculate the approximate weight percent carbon in the steel.

7. What fraction of martensite unit cells contains a carbon atom in martensite formed by quenching 1045 steel?

8. Describe how each of the following may be obtained in a 1040 steel: (a) martensite, (b) tempered martensite, (c) bainite, (d) fine pearlite, (e) coarse pearlite, and (f) spheroidite. Classify each of the above as hard or soft; as tough or brittle. Which of the above have (1) platelets of Fe_3C, (2) chunky Fe_3C, (3) no Fe_3C?

9. What phase or phases are present after each of the following five treatments for eutectoid steel originally at 900 °C: (a) quench to room temperature; (b) quench to 700 °C, hold 1 s, quench to room temperature; (c) quench to 600 °C, hold 10 s, quench to room temperature; (d) quench to 550 °C, hold 100 s, heat to 700 °C and hold 1 s; (e) quench to 600 °C, hold 2 s, quench to room temperature?

10. What constituents are present after each of the five treatments in problem 9?

11. What effect on the position of the "nose" in the TTT curve does each of the following have: (a) increasing chromium content, (b) decreasing carbon content, (c) decreasing cobalt content, and (d) increasing ASTM grain size number?

12. Differentiate among (a) austenitizing, (b) austempering, and (c) ausforming.

13. Give recommended temperatures for austenitizing each of the following: (a) steel containing 0.40% C, (b) 1008 steel, and (c) eutectoid steel.

14. A structural steel part that can withstand a large impact is required. Is it desirable to use a steel of low or high hardenability? Explain.

7

Nonferrous Alloys

Although iron-base alloys are extremely versatile and relatively inexpensive compared with other metals, large quantities of several other metals are used as well. Steel and iron-base alloys are deficient in several properties: corrosion resistance, electrical and heat conductivity, density (too high for some applications), aesthetics, formability, and melting point. Metals of greatest commercial significance are iron, aluminum, and copper; a host of other metals have important applications: zinc, tin, nickel, magnesium, titanium, chromium, silver, gold, and many elements used in relatively small quantities.

7.1   Aluminum and Its Alloys

Toward the end of the 19th century, aluminum was considered something of a miracle metal with a promising future. Indeed, it has lived up to the optimistic predictions. Aluminum combines several highly desirable properties: low density, outstanding corrosion resistance, the ability to be markedly strengthened by alloying and heat treatment, ease of recycling (low melting temperature), and great abundance of ores.

The electron configuration of aluminum is useful in understanding many of its properties: $1s^2 2s^2 2p^6 3s^2 3p^1$ (refer to standard chemistry texts for an explanation of the spectroscopic notation for electrons in atoms). Aluminum has three valence electrons: $3s^2 3p^1$, all in the third shell. In chemical reactions, aluminum always loses these three electrons to form $+3$ ions; in covalent compounds it shares its three valence electrons with other atoms. Although aluminum is more chemically active than iron (for instance, aluminum can reduce iron compounds to the metal), surprisingly aluminum has outstanding corrosion resistance. The reason is that aluminum forms an extremely adherent oxide film whenever exposed to air or water. This oxide film is highly protective and prevents aluminum from corroding. The oxide film reforms quickly and spontaneously when it is broken or removed by scratching or abrading. In fact, it reforms so quickly that one cannot solder aluminum—the oxide film prevents solder from adhering to the aluminum. Because the oxide film is so highly protective, it does not grow very thick. Accordingly, aluminum surfaces always appear like bare metal and usually are very attractive.

The valence electrons, which are in the third shell, are not strongly attracted to the nucleus, similar to Group IA and Group IIA metals. Therefore, the aluminum atom is rather large for an atom with a total of only 13 electrons. Because the atoms are fairly large and the mass is fairly small, the density of aluminum is relatively low: 2.70 g/cm^3 compared with 7.86 g/cm^3 for pure

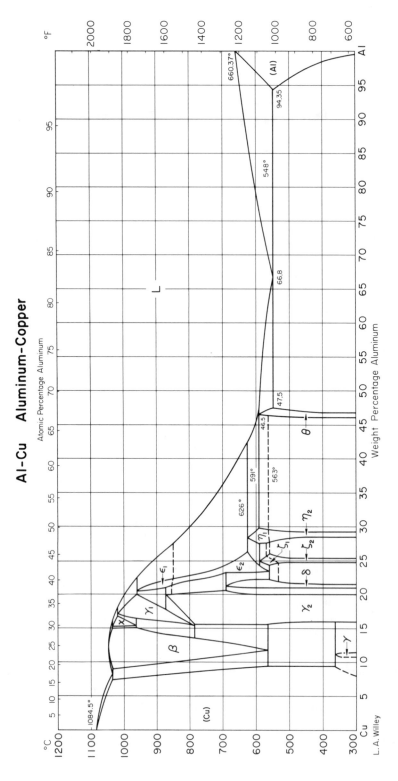

Al-Cu   Aluminum-Copper

Fig. 7.1 Aluminum-copper phase diagram

iron. The melting point of aluminum is 660 °C, which is moderately low compared with iron and other transition metals having bonding involving some d electrons in addition to the usual s and p electrons. The high electrical conductivity of aluminum is due to the fact that it has an odd number of valence electrons and only one p valence electron, leaving many orbitals empty. This means that it is easy to excite valence electrons to slightly higher energy levels, which is necessary for them to become conducting electrons.

Aluminum is face-centered cubic (FCC), which makes it ductile and easy to form. (Recall that FCC crystals have many slip systems usually in favorable orientations for slip to occur.) The coefficient of thermal expansion for aluminum is moderately high because of the large size of the atom for only 13 electrons. Also, the bonding is only moderately strong: although aluminum has no d electrons for strong bonding, similar to many transition metals, it does have three bonding electrons per atom. Accordingly, pure aluminum is moderately strong.

Pure aluminum can be strengthened by work hardening, such as cold rolling, or by solution hardening by alloying with metals such as magnesium or manganese. The great importance of aluminum alloys, however, is due to their ability to be age hardened. Age hardening depends on having a two-phase alloy stable at room temperature with a single solid phase being stable at some elevated temperature. For instance, aluminum containing 4.5% Cu has two phases at room temperature, but only a single phase from about 500 °C to its melting point (see Fig. 7.1 for the Al-Cu phase diagram). When this alloy is heated to above 500 °C to obtain a single phase and then quenched to room temperature, it is still in a single phase—namely, aluminum supersaturated with copper. When this supersaturated Al-Cu alloy is heated to 200 to 300 °C for an hour, some clusters of copper atoms form, and the aluminum is age hardened.

When the resulting material is cooled to room temperature and used at or only slightly above room temperature, it has a strength several times that of pure aluminum and as high as a good steel (see Tables 7.1 and 3.3 for a comparison). The corresponding strength-to-weight ratio is very high, especially when compared with steel. This explains why age-hardened aluminum alloys are used in aircraft.

7.2 Magnesium and Its Alloys

Many of the comments made above for aluminum also apply to magnesium. Magnesium has only two valence electrons, and therefore its bonding is somewhat weaker than in aluminum. Magnesium is a very active metal chemically, but has moderately good corrosion resistance considering its high reactivity. The electron configuration for magnesium is $1s^2 2s^2 2p^6 3s^2$: the last two electrons, $3s^2$, are the valence electrons. The density of magnesium is only 1.74 g/cm^3 compared with 2.70 g/cm^3 for aluminum.

Although pure magnesium metal has few uses except as an incendiary, its alloys are sufficiently corrosion resistant to be useful. Also, the alloys are moderately strong when age hardened with such elements as aluminum or zinc. Magnesium alloys are used for ladders, luggage, lawn mower housings, in aircraft and spacecraft, machinery, and tools. Because magnesium is ''mined'' from seawater, the amount of magnesium available is essentially unlimited.

Table 7.1   Typical properties of aluminum alloys

Alloy No.	Chemical analysis, wt.%(a)	Condition(b)	Tensile strength(c), psi × 10³	Yield strength(c), psi × 10³	% elongation	HB(d)	Applications
Non-heat-treatable wrought alloys							
1100	99.0 min Al	0	13	5	35	23	Sheet-metal
		H18	24	22	5	44	work
3003	1.2 Mn, 0.6	0	16	6	30	28	Aircraft engines
	Si, 0.7 Fe	H18	29	27	4	55	
5052	2.5 Mg, 0.4	0	28	13	25	47	Bus body
	Fe, 0.2 Cr	H38	42	37	7	77	sheets, sheet-metal work
Heat-treatable wrought alloys							
2014	4.5 Cu, 0.8	0	27	14	18	45	Aircraft
	Si, 0.8 Mn, 0.5 Mg, 0.7 Fe	T6	70	60	13	135	structures, truck frames
6061	1 Mg, 0.6 Si,	0	18	8	25	30	Pipelines
	0.2 Cr, 0.3 Cu, 0.7 Fe	T6	45	40	12	95	
7075	7 Zn, 2.5 Mg,	0	33	15	17	60	Aircraft
	1.6 Cu, 0.4 Si, 0.5 Fe, 0.2 Cr	T6	83	73	11	150	structures, cladding
Casting alloys							
296.0	2.5 Si, 1.2	T4	37	19	9	75	Pumps
	Fe, 4.5 Cu	T6	40	26	5	90	
A390.0	17 Si, 0.5 Fe, 4.5 Cu, 0.5 Mg	T6	45	45	<1	145	Die castings
356.0	7 Si, 0.6 Fe, 0.25 Cu, 0.35 Mg, 0.35 Mn, 0.35 Zn	T6	38	27	5	80	Auto transmission casings

(a) Balance aluminum. (b) 0, soft; H18, cold worked; H38, highly cold worked; T4, naturally age hardened; T6, age hardened by heating. (c) Multiply psi by 6.9 × 10⁻³ to obtain MPa. (d) Brinell hardness

When machining magnesium alloys, one must be extremely careful to prevent magnesium turnings from catching fire. Magnesium burns with great intensity, and magnesium fires are very difficult to extinguish.

7.3   Copper and Its Alloys

Copper has been used longer than any other metal. In fact, copper is intimately connected with the evolution of civilization. Undoubtedly, the first use of copper was in the distant past when native copper was first noticed. Indeed, the Bronze Age is named for copper-base alloys. The Iron Age began two or three thousand years *after* the beginning of the Bronze Age. Copper

Table 7.2 Typical properties of copper alloys

Alloy No.	Chemical analysis, wt.%	Condition	Tensile strength(a), psi × 10³	Yield strength(a), psi × 10³	% elonga-tion	Applications
Wrought alloys						
C10100	99.99 Cu	Annealed	32	10	55	Busbars,
		Cold-worked	66	53	4	electronics
C17200	97.5 Cu, 1.9	Annealed	68	25	48	Springs,
	Be, 0.2 Co	Age-hardened	212	195	1	valves
C26800	65 Cu, 35 Zn	Annealed	46	14	65	Radiators,
	(brass)	Cold-worked	128	62	3	lamps,
						hinges
C52400	90 Cu, 10 Sn	Annealed	66	28	70	Wear-resistant
	(bronze)	Cold-worked	147	...	3	bars and
						plates
C61300	92.7 Cu, 0.35	Annealed	70	30	42	Nuts, bolts,
	Sn, 7.0 Al	Cold-worked	85	58	35	tanks,
						tubing
C71700	67.8 Cu, 31.0	Annealed	70	30	40	Seawater
	Ni, 0.7 Fe,	Cold-worked	200	180	4	parts,
	0.5 Be					springs,
	(cupronickel)					screws
Cast alloys						
C80100	99.95 Cu	As cast	25	9	40	Electrical
						conductors
C82500	97 Cu, 2 Be,	As cast	80	45	20	Molds for
	0.5 Co	Age-hardened	160	...	1	plastic parts
C83600	85 Cu, 5 Sn,	As cast	37	17	30	Valves,
	5 Pb, 5 Zn					pumps
C90700	89 Cu, 11 Sn	As cast	44	22	20	Gears,
		Heat-treated	55	30	16	bearings
C92700	88 Cu, 10	As cast	42	21	20	Bearings,
	Sn, 2 Pb					bushings
C96400	69 Cu, 30	As cast	68	37	28	Valves,
	Ni, 1 Fe					pumps for
						seawater

(a) Multiply psi by 6.9 × 10⁻³ to obtain MPa

and its main alloys, brass (Cu-Zn alloys) and bronze (Cu-Sn alloys), are still very important and give every indication of remaining so.

The electron structure of copper is as follows: $1s^2 2s^2 2p^6 3s^2 3p^6 3d^{10} 4s^1$. At first glance, one might expect copper to be a chemically active metal similar to potassium or sodium because it has a single valence electron in its highest shell. Potassium is very active chemically because its $4s^1$ electron is at a much higher energy level than the underlying electrons. Consequently, the $4s^1$ electron of potassium is easily lost, thus accounting for the high chemical activity of the metal. In copper, the $4s^1$ electron has an energy that is only slightly higher than that of the underlying d electrons. The $4s^1$ electron in copper is attracted to the copper nucleus almost as strongly as the d electrons. Therefore, copper, far from being active chemically, is relatively chemically unreactive. The corrosion resistance of the metal and its alloys is outstanding: many copper-based objects from antiquity have survived in fair to good condition (one author of this text has a number of Roman coins in good condition in spite of their being buried in the ground or in river bottoms for almost 2000 years).

High-purity copper has great importance, primarily because of its outstanding electrical conductivity and corrosion resistance. Elements with high electrical conductivities also have high thermal conductivities, because both depend on the easy movement of electrons. The easy movement of electrons in turn depends on the number of energy levels available for conduction electrons, among other things. Elements with an odd number of valence electrons fulfill this requirement best of all. Hence, copper, silver, gold, sodium, and aluminum all have very good electrical conductivities. Sodium is not used for this purpose because it is so easily corroded. Aluminum has been used for electrical wiring, but despite its protective oxide film, aluminum tends to corrode under some conditions, and it is not likely to displace copper in wiring.

Another property of great value is the ductility of copper: it may be drawn down to fine wires. The crystal structure of copper is FCC; FCC crystals have a large number of slip systems (slip planes and directions of slip) that are oriented almost always so that one slip system or another allows slip to occur easily. Copper may easily be purified to high levels by electrolysis. Because copper is relatively inactive, it is relatively easy to reduce copper ores to the metal. Important considerations are the comparative scarcity of copper ores and the relatively low concentration of copper in the ores.

Copper has outstanding corrosion resistance, ductility, and pleasing aesthetics: even when oxidized in air, copper often forms a pleasing blue-green patina; nicely patinated ancient coins often sell for several times the price of a coin without such a coating. Copper is frequently used for architectural applications, such as gutters and downspouts, panels, roofs, and domes.

Brass is an alloy of copper and zinc in which the zinc has dissolved in the copper (see the Cu-Zn phase diagram, Fig. 5.8). In fact, there are many different brasses, ranging from red brass (low-zinc brass) to yellow brasses with high zinc contents. United States pennies used to be made of copper containing 5% Zn. The zinc was used to strengthen the coin, and the coins were cold-worked as well. Brass has been known since antiquity (developed around the first century B.C.), but brass is not nearly as ancient as bronze. The technology required for the formation of brass is sophisticated compared with the production of bronze, which merely requires tin to be added to molten copper. In antiquity, brass could not be made by adding zinc to molten copper, because zinc is so volatile that it could not be made in pure metallic form. Brass used to be made by reducing zinc oxide with charcoal in the presence of copper metal; zinc then diffused into copper, producing brass.

Table 7.2 shows the compositions, properties, and applications of some brass alloys. Brass has a golden color which makes it most attractive. Excellent corrosion resistance and easy formability are other valuable properties. Because at present zinc is appreciably cheaper than copper, brass is frequently less costly than pure copper. Brass alloys with fairly high zinc contents are susceptible to two strange types of corrosion: (1) *dezincification* is the corrosion of brass (both Cu and Zn) with the redeposition of copper because it is less chemically active than the alloy being corroded; thus, the brass corrodes some, and then corrodes more in partially reducing the corrosion product back to copper, which coats the object with a more or less nonadherent layer of copper; (2) *stress-corrosion cracking* is the catastrophic cracking of an alloy that has been stressed below the yield stress while subjected to a specific corrosive environment—ammonia in the case of brass. The mechanism appears to involve the formation of a crack which grows due to corrosion.

Bronze is the general term for alloys of copper that contain tin. Many different bronzes have been developed for a wide variety of applications. Bronze is the first alloy to have been developed, and its superior properties to copper undoubtedly gave a great advantage to its first developers: both tools and weapons were much harder and less ductile, making better swords, knives, etc. Because tin ores, although somewhat scarce, have been readily available in a few areas, and because tin ores are easily reduced, even by charcoal fires, tin has been known in its pure form for thousands of years. Of course, tin ore and copper ores could be reduced separately or together.

Bronze may be hardened to form bearings, gears, or tools. Since ancient China, bronze has been used to cast bells. In fact, the ancient Chinese perfected the composition of bronze that gives the finest of tones. Other applications of bronze include tools, springs, pumps, plaques, etc. The copper-tin phase diagram (see Fig. 5.11) shows that copper may dissolve only 15.8% Sn before a two-phase system occurs.

Nickel has an atomic radius close to that of copper, and nickel is also FCC and similar to copper chemically. It is not surprising, therefore, to discover that copper and nickel dissolve completely in one another. Several Cu-Ni alloys are used commercially; for example, Monel 400 is used to make corrosion-resistant parts (Monel is particularly resistant to corrosion by seawater, and consequently is used for many marine applications). Some Monels contain rather high concentrations of nickel.

7.4   Nickel and Its Alloys

Nickel is an amazing metal with several outstanding properties. Like iron, it is ferromagnetic. Nickel is extremely corrosion resistant: its electron structure is $1s^2 2s^2 2p^6 3s^2 3p^6 3d^8 4s^2$, and the 3d electrons and the 4s electrons have nearly the same energies. Therefore, all 12 of these electrons see a relatively large nuclear charge, and electrons are removed only with difficulty. Several of the d electrons, as well as the two 4s electrons, are bonding electrons, which cause nickel to be strong. Nickel and nickel alloys have excellent resistance to oxidation at elevated temperatures. Surprisingly, nickel alloys also maintain high strength at very low temperatures. In addition, some nickel alloys may be age hardened to produce very strong materials.

Several alloys go by the general name of Inconel, which is an alloy primarily of nickel with chromium and other elements such as iron or niobium. Inconel has outstanding resistance to high-temperature oxidation. Another major application of nickel is in the austenitic stainless steels, but, of course, these are considered iron-base alloys.

7.5   Titanium and Its Alloys

Titanium has the following electron structure: $1s^2 2s^2 2p^6 3s^2 3p^6 3d^2 4s^2$. Titanium has four valence electrons and forms a very stable oxide, TiO_2. Titanium, like aluminum, is active chemically, and it also forms an adherent oxide film that is highly protective. Consequently, titanium is one of the metals with outstanding corrosion resistance and is often used in the chemical industry for that reason. Because titanium has only two d electrons, its density is moderately low: 4.5 g/cm^3 compared with 2.70 g/cm^3 for aluminum and 7.86 g/cm^3 for iron. The four valence electrons experience the attraction of essentially four protons in the nucleus (the remaining 18 protons in the nucleus are countered by the inner core of 18 electrons in titanium).

Because titanium is so chemically active, it is difficult to reduce the metal from its compounds (primarily from TiO_2). Either metallic sodium or magnesium is used commercially to prepare metallic titanium. The high chemical activity of titanium makes it difficult to form or work titanium alloys. Liquid titanium is close to being a "universal solvent": it reacts with nearly everything. Accordingly, it is melted in water-cooled copper crucibles. Of course, a solid titanium shell forms on the copper.

Titanium is hexagonal closest-packed (HCP) at room temperature. This tends to limit its plastic deformation because of the scarcity of good slip planes and slip directions. At 880 °C titanium transforms to BCC. Some elements tend to stabilize the HCP structure (e.g., aluminum), while other elements stabilize the high-temperature structure to lower temperatures (chromium, iron, molybdenum, and vanadium).

Not only are titanium alloys of great use because of their outstanding corrosion resistance, these alloys may be age hardened to form very strong and hard alloys, while maintaining their relatively low density. Hence, their strength to weight ratio is high. Supersonic airplanes generate excessive heat due to friction with air; this heat would cause aluminum alloys to overage (at 200 to 300 °C) and lose their outstanding strength. However, titanium alloys do not overage at these temperatures.

7.6   Lead

Lead is another metal known and used since antiquity. In contrast to several of the above metals, lead is not a transition metal. Accordingly, it does not have bonding due to d electrons, and because of this lead is low melting and not very strong. The atoms are large because there are many electrons to pack around the nucleus, and the valence electrons are held with only moderate strength.

Because lead is not valued for its strength, it does not have many alloys of importance. Lead is soft and malleable, and it is corrosion resistant. In antiquity lead pipes were used, and many fragments still exist, complete with stamps naming the current Roman Emperor. More recently lead has been commonly used in plumbing (which comes from the Latin word for lead, *plumbum*; hence the chemical symbol of Pb for lead).

Because of its relatively low melting point (327 °C), lead is used to make fuses that melt when excessive electric currents pass through them. Solder, the widely used alloy of lead and tin, is important because of its low melting temperature. It may easily be melted and used to adhere certain metals, such as copper and its alloys. Modern pewter is lead-free, although in former times, lead-tin pewter was commonly used for plates, pitchers, etc. Probably many a person suffered unknowingly from lead poisoning due to the widespread usage of lead-base objects. (Wealthy Romans were known to have had a fondness for a syrup made from wine boiled down in lead vessels; it is suspected that some of the madness and infertility problems of the aristocracy may have been due to lead.)

In recent times lead has gained another important use: as a shield from radiation. Lead bricks or lead sheet may be used, depending on the energy of the photons that need to be absorbed. Lead sheet is sufficient to absorb low-energy x-rays, but lead bricks are necessary to absorb high-energy gamma rays (which also are photons).

7.7 Tin

In many respects tin is similar to lead, and this is understandable from the fact that tin is directly above lead in the periodic table. Like lead, tin is soft, low melting, and corrosion resistant. However, there is a major difference: tin is not toxic. This fact is made use of in "tin" cans, which are actually steel cans coated with tin. Tin also is lighter in color than lead, and is thus sometimes used for jewelry and coins (especially fake coins—which is also a use for lead). In fact, tin looks much like silver and is sometimes used as a substitute. Mention has already been made of several alloys of tin: solder, pewter, and bronze.

7.8 Precious Metals

The precious metals have long been valued highly because of their comparative scarcity and especially for their beauty. Gold has an outstanding color and is nearly impossible to oxidize, even at elevated temperatures (references in the Bible to "trial by fire" refer to the testing of gold for its purity by melting it; if impurities are present, they oxidize and form dross that floats on the surface). Gold is FCC and accordingly very ductile. In fact, gold leaf may be formed by beating gold to an extremely thin sheet. Recently gold is valued for its use in the electronics industry because it forms outstanding contacts that do not corrode. Gold fillings and crowns for teeth are important applications.

The greater abundance of silver has made this metal even more popular than gold, although there have been times in history when gold was not much more expensive than silver. The brilliance of silver and its ductility and hardenability when alloyed with copper have added to its usefulness. In the development of coinage, gold was used first (actually a gold-silver alloy), and then silver and finally copper, in the form of pure copper, brass, or bronze coins. Silver and bronze are most frequently used for medals. In recent years the importance of silver compounds in photography has superseded the usage of silver in coins—we literally no longer can afford to make coins from silver because we have more valuable applications for it.

Other precious metals include platinum (also used for jewelry), rhodium, ruthenium, osmium, and iridium. Each of these metals has a few specialized applications.

7.9 Less Common Metals

Many other metals have important, but small, usages. The *refractory metals* are those that are able to serve at high temperatures: molybdenum, niobium, tungsten, tantalum, and rhenium. Notice that all these metals are transition elements with many bonding d electrons. When a transition element contains more than five d electrons, then the electrons begin to pair and consequently reduce the number of unpaired electrons available for bonding.

Most of the refractory metals need to be coated before use at high temperatures. Oxidation-resistant coatings must not only withstand the atmosphere at elevated temperatures, but they must not diffuse too rapidly, or the coating will literally disappear. Also, they should not spall off when the temperature is cycled from room temperature to the elevated temperature and then back again to room temperature. In fact, the coated metal must be able to withstand many heating and cooling cycles.

Several other metals have moderate to small uses, and frequently these elements are called the less common metals; in fact, there is a journal devoted to the less common metals. Group IA metals are examples of these. Sodium metal is manufactured in the millions of pounds annually. However, sodium is made primarily for its use as an active chemical rather than as simply a metal or alloy. Its primary use in the past has been to make tetraethyllead, an antiknock agent for gasoline ("leaded" gasoline).

The properties of sodium may be explained by its electron structure: $1s^22s^22p^63s^1$. The $3s^1$ electron is a "lonesome" electron in that its energy is considerably higher than that of all the other electrons in the atom. Therefore, this electron may easily be lost, which accounts for the high chemical activity of sodium. Accordingly, sodium is an excellent reducing agent and is used to reduce titanium chloride to titanium metal.

Because the $3s^1$ electron in sodium is at a high energy level, it is relatively far away from the sodium nucleus. This causes the metal to have a low density (0.97 g/cm^3, which is less than the density of water!). Also, sodium is weak mechanically: it is possible to bend a bar of sodium about 2 in. in diameter with one's "bare" hands (however, rubber gloves should be worn). Because sodium has just one valence electron, it has excellent electrical and thermal conductivity. In fact, liquid sodium is used in nuclear reactors to conduct heat (sometimes a sodium-potassium alloy is used).

Group IIA metals are used more than Group IA metals. Already we have discussed magnesium and its alloys. Beryllium has limited uses; although it is relatively strong and has a very low density (1.848 g/cm^3), beryllium is rare, expensive, and extremely toxic. Calcium metal has limited uses, primarily as a strong reducing agent: it is one of the best because of the extreme stability of its oxide in particular. One of the authors of this book invented a process for coating carbon steel with a thin diffusion coating of stainless steel by immersing steel in liquid calcium containing chromium and nickel. However, this process is not used commercially.

The lanthanide metals are also active chemically and are used in a few applications. Usually they are used as misch metal, a mixture of lanthanide metals. Because they are very similar chemically, it is difficult to separate the individual metals. In most applications, any lanthanide will suffice, hence the use of misch metal, which is much less costly to produce than pure lanthanide elements.

Mercury is unique among metals in that it is liquid at room temperature. Cesium and gallium melt at temperatures just above room temperature; however, these are expensive metals, and cesium is extremely active chemically. Mercury is corrosion resistant and is commonly used in thermometers and barometers. Although mercury liquid is unreactive (several hundred years ago people drank mercury for the weird feeling it gave in the stomach!), the vapor is extremely toxic, as are the compounds of mercury. Mercury is also valuable to form amalgams with many other metals. Amalgams are safely used as dental fillings. Mercury has been used in the electrolysis of aqueous sodium chloride to prepare sodium hydroxide. It was this process that was responsible for the pollution of a number of lakes with mercury. The low melting point of mercury is due in part to the large size of its atoms and the relatively weak bonding among them. Its 5d shell of electrons is filled (with 10 electrons), and it has two 6s electrons, which have about the same energy as the 5d electrons.

7.10   Summary

Our survey of metals is now complete, and one may marvel at the wide range of properties available in the metallic state. Advances are still being made, such as in glassy metals that have been cooled at exceptionally high rates. The lack of time for crystallization to occur leads to some unusual properties among the metals. Because the structure of metals is relatively simple compared with other materials, much is understood about metals and their properties. However, much remains to be discovered before one can literally calculate properties precisely. Although the field is reasonably mature, future improvements in the understanding of metals are undoubtedly possible.

7.11   Problems

1. Why are aluminum alloys unsuitable for use in supersonic aircraft despite their high strength to weight ratio?

2. Compare the boiling points of sodium, magnesium, and aluminum. Explain the trend that occurs.

3. 2014 aluminum alloy contains 4.5% Cu, 0.8% Si, 0.8% Mn, and 0.5% Mg. Based on 100 atoms of this alloy, how many atoms of each element are present?

4. What are three major ways of hardening aluminum?

5. Is it possible to age harden aluminum by alloying it with a small concentration of magnesium? If you replied in the affirmative, give the specific composition of the alloy and the times at various required temperatures. Is this alloy likely to be important commercially, or does it have a limitation that would prevent its usage as a structural alloy?

6. Referring to the Al-Cu phase diagram (Fig. 7.1), determine the chemical formula for the θ phase. What is the maximum solubility of copper in aluminum? Give the composition, temperature, and phases present in a peritectic; in a eutectoid; in a peritectoid.

7. Is it possible to age harden a bronze alloy containing 10% Sn? If so, is it difficult to overage this alloy?

8. Based on 100 g of alloy, how many grams of ε phase would be present in a 10% Sn bronze at room temperature?

9. What is the maximum solubility of tin in solid copper at any temperature? What is the maximum solubility of copper in solid tin at any temperature?

10. Give the temperature, chemical composition, and phases involved in the following three-phase reactions in the Cu-Sn phase diagram (see Fig. 5.11): (a) eutectoid; (b) peritectic; (c) peritectoid; (d) eutectic (hint: this is difficult to see, but it occurs in the low copper part of the diagram).

11. Identify all the three-phase reactions in the Cu-Zn (brass) phase diagram (see Fig. 5.8).

12. What is the maximum solubility of zinc in copper? Copper in zinc? Why is there such a great difference in these solubilities when the ratio of atom diameters is the same in both cases?

13. Why is copper containing only 2% Be so strong compared with copper containing 2% Zn or 2% Sn? Explain in terms of dislocation immobilization.

14. What unusual property does nickel share with iron and cobalt?

15. Why are copper and nickel completely soluble in each other at room temperature?

16. Why do you suppose that alchemists were convinced that they could chemically transform lead into gold?

17. Why do lead, tin, antimony, and bismuth have relatively low melting points? In general, what is the effect of size of an atom on melting point? What other factor besides size is important?

18. List several properties that work against the use of lead or tin as structural metals.

8

Structure of Ceramic Materials

Ceramic materials can be defined in the broadest sense as substances composed of inorganic, nonmetallic elements or compounds. These compounds usually contain metallic elements combined with strongly nonmetallic elements of the right half of the periodic table, especially oxygen, fluorine, chlorine, and sulfur. Several elements, such as boron, carbon, and sulfur, are also classified as ceramics by materials scientists.

As commonly used, the word "ceramic" brings to mind pottery and china. There is, however, a host of more technical ceramics such as glass, brick, refractories, electronic circuit substrates and insulators, and electronic components themselves. All have related molecular structures and mechanical properties. As with other materials, certain properties of ceramics contribute to their being chosen for specific end-use applications. Typical properties are structural strength and dimensional stability at high temperatures, high dielectric constants, high refractive index, chemical inertness, and hardness. Many of these properties are a result of the type of bonding found in ceramics.

Ceramics, like metals, are usually crystalline. In ceramics, unlike metals, both covalent and ionic bonding of the atoms throughout the crystalline structure contribute greatly to the desirable properties. Because a typical ceramic is composed of several atoms of different size and of differing bonding capability, the crystalline structures of ceramic materials are usually more complex than those of metals. In addition to crystalline ceramics, there is an equally valuable group of ceramic materials which are primarily in the glassy state—e.g., rigid amorphous materials. Ordinary window glass is the most common example of a ceramic glass. The glassy state occurs as a result of bonding between groups of atoms of ceramic materials in the liquid state, which makes rearrangement of the liquid structures into crystals very slow. Ceramic glasses are also closely related in structure to the amorphous polymers discussed in Chapters 10 and 11.

The ceramics industry has evolved from articles fashioned in clay, such as brick, pottery, and decorative objects, into a large array of materials with a wide range of end uses. Many of the current ceramic materials require very high levels of technology to produce and are used for their superior mechanical, thermal, and electrical properties. Future uses range from parts for internal combustion and gas turbine engines to improved prosthetic devices.

Such "high tech" uses are continuing to increase at a rate greater than the economy as a whole.

8.1   Nature of Bonding in Ceramics

Bonding in ceramics is both ionic and covalent. The *ionic bond* occurs when there is *electron transfer* between two atoms resulting in ions of opposite polarity. The *covalent bond* occurs when there is *electron sharing* between two atoms resulting in a fixed orientation between the atoms and a resulting directionality of the bonds. The electrons that are transferred or shared are usually called *valence electrons*. In ionic bonding, the atoms resulting after electron transfer either have an excess of electrons in an orbital about the atom or have a deficiency of electrons and an available, unfilled orbital. Ions with excess electrons are, of course, negatively charged, *negative ions* or *anions*, while those with a deficit of electrons are positively charged, *positive ions* or *cations*.

There is a force of attraction, F_c, set up between oppositely charged ions which follows the well-known coulombic relationship, $F_c = K/a^2$, where a is the separation distance between the two atoms. Note that once a charged ion is formed, it is independent of the other atom which took part in the electron transfer and can associate with or bond with any other ion of opposite charge. The proportionality, K, is a function of the valence, Z, of the charged ion, q, the charge of a single electron, and a constant $k = 9 \times 10^9 \, V \cdot m/C$, which for two ions of valence Z_1 and Z_2 is given by $K = k(Z_1 q)(Z_2 q)$. The attractive force between oppositely charged ions increases exponentially as the separation between them becomes smaller and smaller. As the separation distance becomes very small, however, a repulsive force comes into play which is opposite in direction to the coulombic force. The repulsive force, F_r, occurs from the presence of other electrons in the atomic shells around the atoms which are being forced into closer proximity and from the effect of bringing the two positive nuclei together. The form of the repulsive force is $F_r = Le^{-a/p}$, where L and p are characteristic of a given ion pair, e.g., Na^+ and Cl^-. The net result of these two opposing forces is an equilibrium distance for any two ion pairs, as shown in Fig. 8.1. Note that any movement from the equilibrium position at a_o will result in the expenditure of energy. This energy of separation of the two ions to a very large separation distance (∞) is known as the *bonding energy* and will be different for each different pair of ions. The equilibrium separation distance a_o is called the *bond length*.

Since the bond length is the distance separating two *different* atoms, it is useful to think of it as comprising two different lengths, r_1 and r_2, each of which is the amount of separation attributable to the respective atom of the atom pair. It is as though the atoms were hard spheres with identifiable external surfaces located at the radii r_1 and r_2. Characteristic radii have been determined for most common ions as a result of examining a large number of ionic compounds (see Table 8.1). These radii are of great importance in determining the structures of ceramic crystals. Although the interatomic distances of atoms may be measured with great precision, the atomic radius of a given ion varies somewhat depending on the other ion of the ion pair. Ionic radii are, therefore, less precise than interatomic distances.

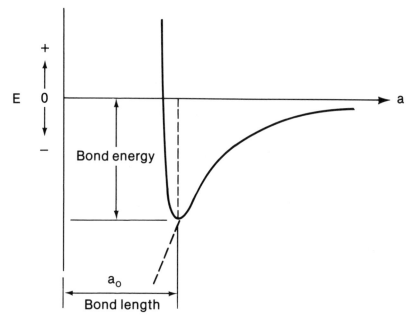

Fig. 8.1 General shape of bond energy curves for covalent, ionic, or metallic bonding
Source: J.F. Shackelford, *Introduction to Materials Science for Engineers*, Macmillan, 1985, p. 43

8.2 Coordination Numbers

Although ionic bonds are not directional in the sense that covalent bonds are considered directional, there are spatial considerations that are important in ionic bonding. Any central charge attracts any other opposite charge. As a result, a central positive ion will be surrounded by a cluster of negative ions. As they draw closer, the closest surrounding negative ions will occupy a definite ordered structure and will vary in number depending on the nature of the central positive ion. Thus a sodium ion, Na^+, will be surrounded by six chloride ions and vice versa. This number of closest neighbors is called the *coordination number*, CN, and is a characteristic number that depends on the relative size of the two ions. The ionic radii have been indicated for the elements in Table 8.1; not only are cations generally smaller in radius than anions of similar mass, but ions that have a higher coordination number are generally larger. The greater number of electrons in the region around an ion with a high CN increases the overall electron repulsion and expands the radius slightly.

The correct coordination number of the ion with the smaller radius of a given ion pair can often be calculated from the relative sizes of the ionic radii of the two ions. This relative size is measured by the *radius ratio*, r/R, where r is the radius of the smaller ion and R is the radius of the larger ion. Figure 8.2 illustrates the minimum radius ratio that can produce three-fold coordination in a planar arrangement. In general, the maximum coordination number for an ion

Table 8.1   Ionic radii of the elements

| | | Bond radii | | | | | | | | | | |
| | | Radius, nm ($\times 10$) | | | | | | | | | | |
Name	Atomic No.	'−4	'−3	'−2	'−1	'+1	'+2	'+3	'+4	'+5	'+6	'+7
H	1				1.54	0.46						
Li	3					0.68						
Be	4					0.44	0.35					
B	5					0.35		0.23				
C	6	2.6							0.16			
N	7					0.25		0.16		0.13		
O	8			1.32	1.76	0.22					0.09	
F	9				1.33							0.08
Ne	10					1.12						
Na	11					0.97						
Mg	12					0.82	0.66					
Al	13							0.51				
Si	14	2.71			3.84	0.65			0.42			
P	15		2.12						0.44	0.35		
S	16			1.84			2.19		0.37		0.3	
Cl	17				1.81				0.34			0.27
Ar	18					1.54						
K	19					1.33						
Ca	20					1.18	0.99					
Sc	21							0.73				
Ti	22					0.96	0.94	0.76	0.68			
V	23						0.88	0.74	0.63	0.59		
Cr	24					0.81	0.89	0.63			0.52	
Mn	25						0.8	0.66	0.6			0.46
Fe	26						0.74	0.64				
Co	27						0.72	0.63				
Ni	28						0.69					
Cu	29					0.96	0.72					
Zn	30					0.88	0.74					
Ga	31					0.81		0.62				
Ge	32	2.72						0.73	0.53			
As	33		2.22					0.58		0.46		
Se	34			1.91	2.32	0.66			0.5		0.42	
Br	35				1.96					0.47		0.39
Rb	37					1.47						
Sr	38						1.12					
Y	39							0.89				
Zr	40					1.09			0.79			
Nb	41					1			0.74	0.63		
Mo	42					0.93			0.7		0.62	
Tc	43											0.98
Ru	44								0.67			
Rh	45							0.68				
Pd	46						0.8		0.65			
Ag	47					1.26						
Cd	48					1.14	0.97					
In	49							0.81				
Sn	50	2.94			3.7		0.93		0.71			
Sb	51		2.45						0.76	0.62		
Te	52			2.11	2.5	0.82			0.7		0.56	
I	53				2.2					0.62		0.50
Cs	55					1.67						
Ba	56					1.53	1.34					

(continued)

Table 8.1   (continued)

		Bond radii										
		Radius, nm (×10)										
Name	Atomic No.	'−4	'−3	'−2	'−1	'+1	'+2	'+3	'+4	'+5	'+6	'+7
La	57					1.39		1.02				
Ce	58					1.27		1.03	0.92			
Pr	59							1.01	0.9			
Nd	60							1				
Pm	61							0.98				
Sm	62							0.96				
Eu	63						1.09	0.95				
Gd	64							0.94				
Tb	65							0.92	0.84			
Dy	66							0.91				
Ho	67							0.89				
Er	68							0.88				
Tm	69							0.87				
Yb	70						0.93	0.86				
Lu	71							0.85				
Hf	72								0.78			
Ta	73									0.68		
W	74								0.7		0.62	
Re	75								0.72			0.56
Os	76								0.88		0.69	
Ir	77								0.68			
Pt	78						0.8		0.65			
Au	79					1.37		0.85				
Hg	80					1.27	1.1					
Tl	81					1.47		0.95				
Pb	82						1.2		0.84			
Bi	83					0.98		0.96	0.74			
Po	84										0.67	
At	85											0.62
Fr	87					1.8						
Ra	88						1.43					
Ac	89							1.18				
Th	90								1.02			
Pa	91							1.13	0.98	0.89		
U	92								0.97		0.8	
Np	93							1.1	0.95			0.71
Pu	94							1.08	0.93			
Am	95							1.07	0.92			
NH$_4$	11−						1.43					

Source: Data from *Handbook of Chemistry & Physics*, R.C. Weast, Ed., 58th Ed., CRC Press, 1977, p. F213–214

pair is determined by a geometric arrangement of spherical ions in which all ions just touch each other. Any attempt to add additional ions around a given center will cause overlap between the electron clouds of similar ions, leading to repulsion (see Fig. 8.3). Usually the radius ratio of the ions falls within specific ranges for each of the maximum coordination numbers, although an ion with a high coordination number will sometimes associate with fewer numbers of ions than the maximum. Table 8.2 gives the radius ratio ranges for coordination numbers between CN = 2 and CN = 12.

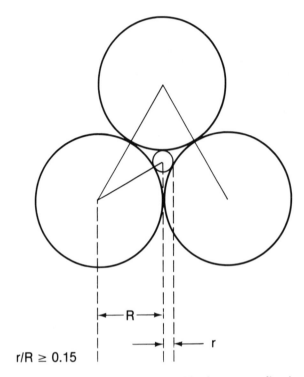

r/R ≥ 0.15

Fig. 8.2   Relative radii for three-fold (planar) coordination

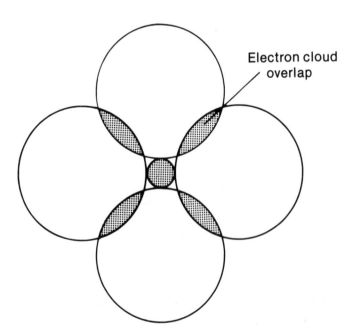

Fig. 8.3   Addition of the fourth ion causes overlap of electron clouds of the surrounding ions—an unstable arrangement

Table 8.2   Coordination numbers for ionic bonding

Coordination No.	Radius ratio, r/R	Coordination geometry
2	$0 < \dfrac{r}{R} < 0.155$	
3	$0.155 \le \dfrac{r}{R} < 0.225$	
4	$0.225 \le \dfrac{r}{R} < 0.414$	
6	$0.414 \le \dfrac{r}{R} < 0.732$	
8	$0.732 \le \dfrac{r}{R} < 1$	
12	1	or(a)

(a) The geometry on the left is for the hexagonal close-packed (HCP) structure and that on the right for the face-centered cubic (FCC) structure.

Source: J.F. Shackelford, *Introduction to Materials Science for Engineers*, Macmillan, 1985, p. 36

8.3   Covalent Bonding

Covalent bonds follow the same type of force/energy relationship as diagrammed in Fig. 8.1 for ionic bonding. In a similar way, a characteristic bond length, a_o, is defined, although the equations relating the attractive and repulsive forces are not the same in covalent bonding as in ionic bonding. The tendency to form ions or to form covalent bonds is a function of the electronegativity of the atoms (see Fig. 8.4 and Chapter 1). Atom pairs having a high degree of difference in electronegativity form ionic bonds, and those of nearly equal electronegativity form covalent bonds. Representative covalent bond lengths and bond energies are given for a number of atom pairs in Table 8.3.

In a covalent bond the electron density about the atoms is not relatively localized about each individual atom as in an ion, but the valence electrons primarily occupy space between the

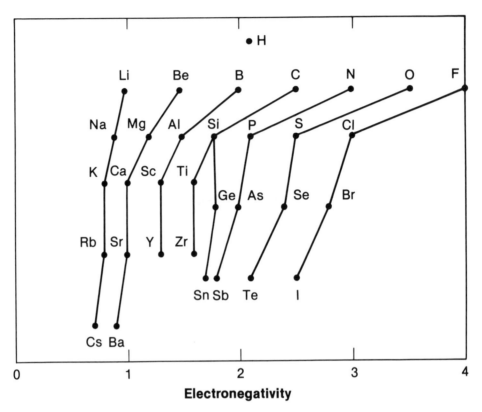

Fig. 8.4   Electronegativity of the elements
Source: Data from W.D. Kingery, *Introduction to Ceramics*, John Wiley & Sons, 1965, p. 109

atoms, forming a directional envelope or localized bond, as illustrated in Fig. 8.5. For two atoms joined by a single covalent bond, the electron density of the bonding or valence electrons is highest along a straight line connecting the two nuclei. The common representation of a covalent bond in terms of electron pairs or lines is also illustrated.

The number of covalent bonds which an atom can form depends on the number of electron pairs that can be accommodated in the outer or valence electron shell. For some atoms, such as carbon and silicon, there is a maximum of eight electrons in the valence shell, which creates the potential for four electron pairs and, therefore, four bonds. Each bond involves the sharing of one electron from each atom to form one electron-pair bond. A maximum of four atoms can be bound to silicon or carbon since they each contribute up to four electrons themselves. When two electrons (an electron pair) form a bond it is a single bond; when four electrons forming two electron pairs bond two atoms together it is a double bond. A triple bond involves three electron pairs, or six electrons. Oxygen atoms often form double bonds (represented by $=$ in Table 8.3), as does carbon, which also can form a triple bond ($\equiv$).

Table 8.3   Bond energies and bond lengths for representative covalent bonds

Bond	Bond energy(a)		Bond length, nm
	kcal/mol	kJ/mol	
C—C	88(b)	370	0.154
C=C	162	680	0.13
C≡C	213	890	0.12
C—H	104	435	0.11
C—N	73	305	0.15
C—O	86	360	0.14
C=O	128	535	0.12
C—F	108	450	0.14
C—Cl	81	340	0.18
O—H	119	500	0.10
O—O	52	220	0.15
O—Si	90	375	0.16
N—H	103	430	0.10
N—O	60	250	0.12
F—F	38	160	0.14
H—H	104	435	0.074

(a) Approximate. The values vary with the type of neighboring bonds. For example, methane (CH_4) has the value shown above for its C—H bond; however, the C—H bond energy is about 5% less in CH_3Cl and 15% less in $CHCl_3$. (b) All values are negative for forming bonds (energy is released) and positive for breaking bonds (energy is required).
Source: L.H. Van Vlack, *Elements of Materials Science and Engineering*, 4th ed., Addison-Wesley, 1980

Electron cloud representation

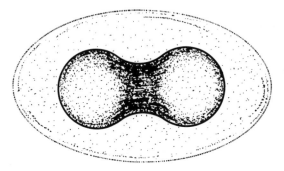

C : C   Electron pair representation

C—C   Single bond representation

Fig. 8.5   Covalent bond
Electrons preferentially occupy space between atoms in covalent bonds.

Covalent bonding is especially important in considering the liquid or glassy state of ceramic materials, although it is also present in crystalline ceramics, as in the case of ZnS, SiC, or diamond (Fig. 8.6).

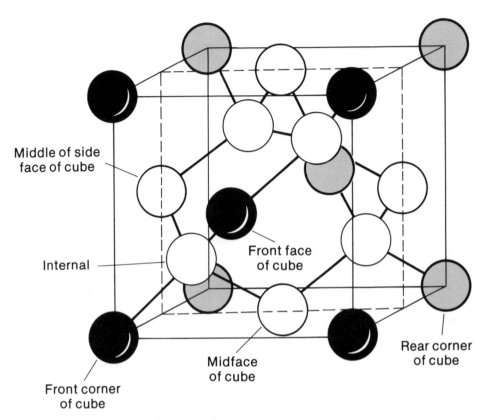

Fig. 8.6   Covalent bonding in diamond crystal structure
Each carbon atom is bonded to four other carbon atoms in the crystal.

8.4   Interstitials

The structure of many ceramic materials, especially oxides which have relatively large oxygen anions, can be understood in terms of an arrangement of close-packed large anions with the cations placed in pockets or holes left in the array of anions. The holes are called *interstices* and the ions in the interstices are called *interstitials* (Fig. 8.7). Each of the following crystal structures can be described in terms of the array of anions and the placement of cations in the various interstitial sites.

For example, the CsCl-type of structure can be seen as a simple cubic array of Cl^- ions with each of the interstices filled with a Cs^+ ion (Fig. 8.8). In the rock salt (NaCl) type structure, the anions are arranged in cubic closest packing, creating octahedral interstitial sites for the cations, consistent with a coordination number of 6 (Fig. 8.9 and 8.7b). (An octahedral site is one that has six atoms arranged in a bipyramid having a common square base.)

The zinc blende (ZnS) structure (Fig. 8.10 and 8.7c) is a cubic closest packing of anions with the four-fold interstitials filled with cations in a coordination number of 4. (A tetrahedral, or four-fold site, has four atoms in a pyramid with all four sides being equilateral triangles.) Other types of nearly close-packed arrays of large anions exist, and the number and types of

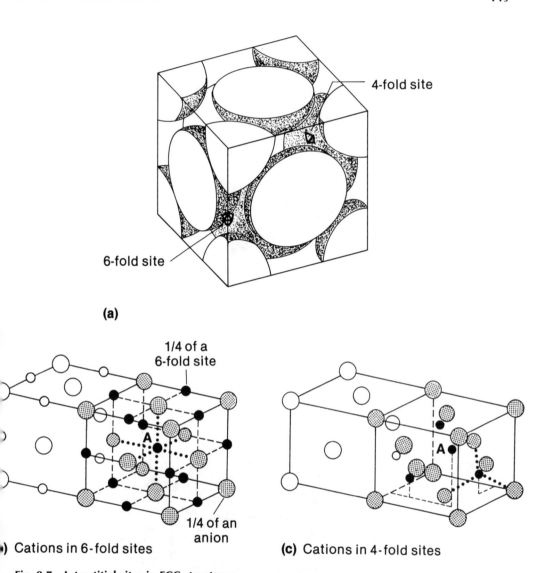

(a)

(b) Cations in 6-fold sites   **(c) Cations in 4-fold sites**

Fig. 8.7   Interstitial sites in FCC structures
(a) There are both four-fold and six-fold sites possible in the FCC structure. (b) The six-fold sites are occupied by Na ions in the NaCl-type structure with CN = 6, while (c) the four-fold sites are occupied by Zn ions in the ZnS (zinc blende) structure. Note that the six-fold sites remain unoccupied in the ZnS structure and can be filled by impurity atoms. Source: Adapted from L.H. Van Vlack, *Materials for Engineering*, Addison-Wesley, 1982, p. 255, 286

interstitial sites vary with each array. Often only a portion of the interstitial sites will be filled, either to achieve a preferred coordination number for the cation or for stoichiometry (valence) considerations. Thus in the face-centered cubic structure of anions, only the six-fold interstitial sites are occupied in the NaCl structure and only the four-fold interstitial sites are occupied in the ZnS structure. There is no reason that the interstitial sites must be filled with identical ions. Substitution of other ions into the interstitial sites of a single type of oxide array can lead to a number of different compounds and even to nonstoichiometric materials.

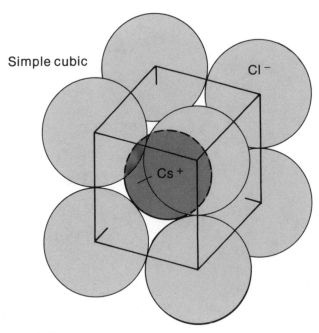

Fig. 8.8   CsCl structure
Source: L.H. Van Vlack, *Materials for Engineering*, Addison-Wesley, 1982, p. 281

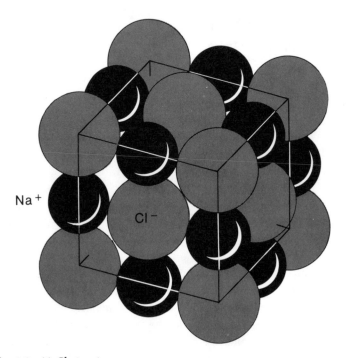

Fig. 8.9   NaCl structure
Source: L.H. Van Vlack, *Materials for Engineering*, Addison-Wesley, 1982, p. 282

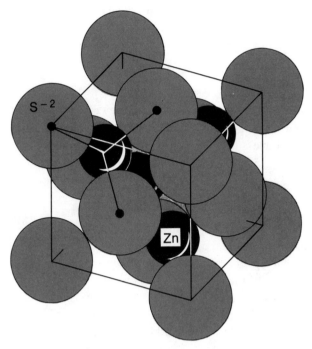

Fig. 8.10 ZnS structure (zinc blende)
Source: L.H. Van Vlack, *Materials for Engineering*, Addison-Wesley, 1982, p. 283

8.5 Crystalline Ceramic Structures

8.5.1 Binary (AB) Compounds

The simplest ceramic compounds are binary, i.e., formed from only two types of atoms although they may have various coordination numbers. It is characteristic of binary compounds that the A ions have only B atoms in their coordination shell, and B atoms are only coordinated with A atoms as nearest neighbors. The resulting crystal structures are highly ordered and usually form cubic crystals. The three principal types of cubic crystals formed are (1) CsCl type, CN = 8; (2) NaCl type, CN = 6; (3) ZnS type, CN = 4. Note that these may be ionic (e.g., MgO) or may have both covalent and ionic character (e.g., ZnS). There are also examples of some AB structures that are not cubic.

CsCl-Type Compounds; CN = 8, r/R ≥ 0.71. The binary compounds that follow the CsCl prototype structure are simple cubic crystals (Fig. 8.8). Although the A ion is located centrally, it is *not* body-centered cubic (BCC), because the atoms at the 0,0,0 and ½,½,½ locations are not identical. Because the A ions are at sites with a CN of 8, they are referred to as being in *eight-fold interstitial sites* or 8-f sites in the cubic pattern.

This type of crystal is relatively uncommon among ceramics because there are few ionic compounds with the relatively high radius ratio of r/R ≥ 0.71 required for a CN = 8 (Table 8.2). In cubic structures, the body diagonal is equal to $\sqrt{3}a$. The length of the body diagonal is fixed by the ionic radii of the A and B ions: diagonal $= 2R_b + 2r_a$. Thus $a = (r_a + R_b)$ (2/$\sqrt{3}$); note that the smaller ionic radius characteristic of the CN $= 8$ must be used. The radius for CN $= 8$ is 0.97 that of CN $= 6$, and the radius for CN $= 4$ is 1.1 that of CN $= 6$ in Table 8.1.

Example 8.5A. Calculate the density of CsCl.

Solution: The density can be calculated from the unit cell volume a^3 and the atomic masses, $\rho = m/a^3$. The atomic mass of Cl is 35.45 and that of Cs is 132.91. (The ionic radius of cesium is 0.97 of the value in Table 8.1 for a CN $= 6$, or $r = 0.162$ nm.) $a = [(0.181 + 0.167)/0.97] \times 2/\sqrt{3} = 0.359 \times 1.1547 = 0.414$ nm. Density $\rho = [(132.91 + 35.45)/6.02 \times 10^{23}]/(0.414 \times 10^{-9})^3 = 3.93 \times 10^6$ g/m^3 or 3.93 g/cm^3. The value determined in the laboratory is 3.99 g/cm^3.

NaCl (Rock Salt) Type Compounds; CN $= 6$, r/R ≥ 0.41. This type of binary compound is most commonly found structure for ceramic materials. It comprises a simple structure in that the two types of atoms alternate along each of the three axes of the cubic cell. Examination of the typical structure (Fig. 8.9) shows that it is face-centered cubic, with each unit cell containing four atoms of each type. Each of the ions is coordinated with six ions of the opposite type. The ratio of the smaller ion to the larger must be greater than 0.41, otherwise the surrounding negative ions would interfere with each other. Even though most ceramics with r/R > 0.71 will form the more stable CsCl-type structure, the NaCl structure is still possible for such combinations. Typical materials with the NaCl structure are MgO, NiO, FeO, CaO, SrO, BaO, CdO, MnO, CoO, CaS, MnS, LiF, TiC, and UC. All the alkali halides (NaCl, SrBr, etc.) except for Cs halides have this structure, as do the alkaline earth sulfides (CaS, BaS, etc.).

ZnS (Zinc Blende) Type Compounds; CN $= 4$, r/R ≥ 0.21. The third example of AB compounds occurs with very small positive ions surrounded by large negative ions: r/R $\geq$ 0.21 (Fig. 8.10). Members of this class comprise several compounds which are important as semiconductors: ZnS, GaAs, InSb, AlP, and CdS. Because the coordination number, CN $= 4$, is consistent with the four-fold covalent bonding preferred by atoms such as carbon, silicon, etc., many of the compounds in this class are covalently bonded. Diamond—crystalline covalently bonded carbon—has this structure. There are, however, only a few ionic compounds with r/R this small, and those compounds with a larger value of r/R will form the NaCl structure (r/R ≥ 0.41), which has a greater coordination number (CN $= 6$) and is energetically more stable. The close-packed arrangement of oxygen atoms in the NaCl and ZnS crystals is similar (see Fig. 8.9 and 8.10). In NaCl, the six-fold sites are occupied by Na$^+$ ions in a CN of 6. In the ZnS structure, none of the six-fold sites and only half of the four-fold sites are occupied by Zn^{+2}. This may be more easily recognized by reference to Fig. 8.7. The unoccupied sites may be filled by impurity atoms.

8.5.2 AB$_2$ Structures

CaF$_2$ Type; CN $= 8$, r/R ≥ 0.73. CaF$_2$ is the model for this type of compound. Other ceramics with this structure include ZrO$_2$ and UO$_2$. In this case, a relatively large positive ion

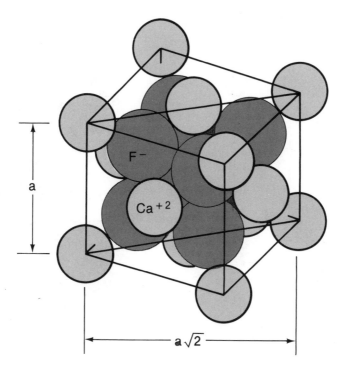

Fig. 8.11 CaF$_2$ structure
Source: L.H. Van Vlack, *Materials for Engineering*, Addison-Wesley, 1982, p. 259

Ca^{+2} (r = 0.099 nm) is paired with one of the smaller negative ions (R = 0.125 nm) to give a ratio r/R of 0.8. Thus each positive ion can be coordinated with eight negative ions. The reverse cannot be true, however, and each negative F$^-$ ion is coordinated with only four positive Ca^{+2} ions, preserving electrical neutrality. The structure is based on a face-centered cubic arrangement of the calcium; the vacancy at the center of the FCC structure, characteristic of all FCC arrangements, is at position "A" in Fig. 8.7b. The F$^-$ ions occupy all of the tetrahedral (four-fold) interstitial sites. Although each calcium ion is situated among eight fluorine ions, half of the eight-fold sites in the cell are vacant. These vacant sites can be seen in Fig. 8.11 to be in the center of the cell and at the center of each edge of the cell.

8.5.3 A$_2$B$_3$ Structures

Cr$_2$O$_3$ Type; CN = 6. The common ceramic alumina, Al$_2$O$_3$, forms a structure of the Cr$_2$O$_3$ type (Fig. 8.12). The unit cell contains the oxygen ions in a hexagonal close-packed structure with the positive ions in each of the six six-fold sites, with CN = 6 relative to the oxygens. Because the stoichiometry requires only two-thirds as many cations as anions, only two-thirds of the six-fold sites are occupied. None of the four-fold sites is occupied.

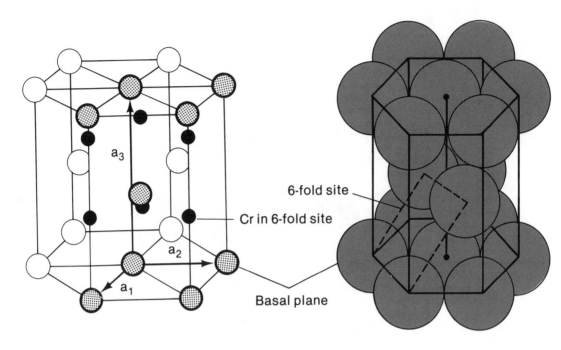

Fig. 8.12   Al$_2$O$_3$ structure
Source: Adapted from A.G. Guy, *Introduction to Materials Science*, McGraw-Hill, 1972, p. 21

8.5.4   AB$_2$C$_4$ Structures

AB$_2$O$_4$ Spinel Types. A typical oxide of this type has a cubic structure in which the oxygen ions are in an FCC close-packing array, with the cations in various arrangements in the interstices. Figure 8.13 illustrates the basic structure of a spinel in which the complete unit cell is constructed from eight more elementary cells. In each elementary cell there are four atoms of oxygen, eight tetrahedral interstices, and four octahedral interstices. Thus in each elementary cell there are a total of 12 spaces to be filled by the two trivalent and one divalent cations. The complete unit cell will contain 32 oxygen ions, 16 octahedral site cations, and 8 tetrahedral site cations, yielding a high degree of complexity.

In the normal spinel structure, the divalent ions are located on tetrahedral sites and the trivalent ions on octahedral sites. This arrangement is typical of the compounds ZnFe$_2$O$_4$, CdFe$_2$O$_4$, FeAl$_2$O$_4$, MgAl$_2$O$_4$, CoAl$_2$O$_4$, NiAl$_2$O$_4$, MnAl$_2$O$_4$, and ZnAl$_2$O$_4$. The more commonly occurring structure is the inverse spinel, which has all of the divalent ions and half of the trivalent ions on octahedral sites and the remaining trivalent ions on tetrahedral sites. This is found for TiFe$_2$O$_4$, MgFe$_2$O$_4$, FeFe$_2$O$_4$, NiFe$_2$O$_4$, SnZn$_2$O$_4$, and other magnetic ferrites.

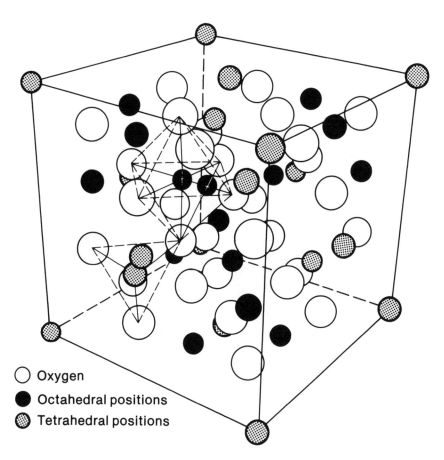

Fig. 8.13   Ion positions in the spinel (MgAl$_2$O$_4$) unit cell
The shaded circles represent Mg^{+2} ions (in tetrahedral or four-coordinated positions), and the black circles represent Al^{+3} ions (in octahedral or six-coordinated positions). Source: F.G. Brockman, *Bull. Am. Ceram. Soc.*, Vol 47, 1967, p. 186

8.5.5   A$_x$B$_y$C$_z$ Structures

BaTiO$_3$ Type; CN = 12. A common ceramic compound that contains more than one type of metal ion is barium titanate, BaTiO$_3$, which is widely used for its electronic properties (see Chapter 14). The structure of the titanate is cubic above 120 °C (Fig. 8.14) and is an example of filling interstitials with more than one cation. Figure 8.14 illustrates a lattice of oxygen anions arranged in a cubic structure in combination with the very large Ba^{+2} ions. The Ti^{+4} ions occupy the octahedral interstices in the center of the FCC cell and are closely associated with the oxygen atoms at the center of each face, with a coordination number of 6. Note that the coordination number for both O^{-2} and for Ba^{+2} is 12, and that the Ba^{+2} ions are less closely associated with the oxygens than is the Ti. This structure is named for the compound CaTiO$_3$ (perovskite) and is also seen in SrTiO$_3$, SrSnO$_3$, CaZrO$_3$, SrZrO$_3$, LaAlO$_3$, YAlO$_3$, and others.

Below 120 °C, a slight rearrangement occurs in the structure and it becomes tetragonal (Fig. 8.15). The Ba^{+2} ions are located as before at the corners of the cell, but the Ti^{+4} has

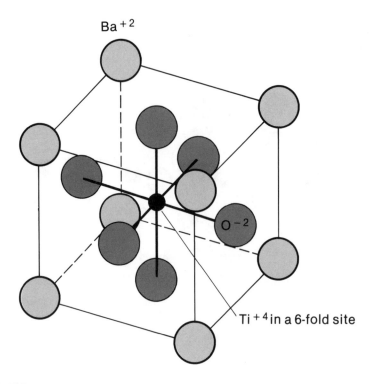

Fig. 8.14   BaTiO₃ structure
FCC arrangement of O^{-2} and Ba^{+2} ions with a Ti^{+4} ion in the center. Source: L.H. Van Vlack, *Materials for Engineering*, Addison-Wesley, 1982, p. 269

shifted upward slightly and the oxygens correspondingly downward so that they are slightly below the center of each face. This umbrellalike structure formed by the Ti^{+4} and the O^{-2} ions has an unbalanced structure so that there is a displacement of the center of the positive charges from the geometric center of the negative charge; i.e., the crystal is polarized. Since the umbrella can be pointed either toward the top or toward the bottom of the cell, a random arrangement among many unit cells leads to little or no macroscopic charge. When placed in a strong field, however, a large number of the cells can be similarly aligned. This results in a strong polarization of the crystal and is the basis for much of its practical applications, such as a transducer material for cartridges for record players and for pressure sensors.

$NiFe_2O_4$ Type; CN = 6. The complex structure of this magnetic oxide ceramic is illustrated in Fig. 8.16. The oxygens are arranged in an FCC structure, which contains two types of lattice sites for cations. There are six-fold octahedral sites, half of which are occupied by the Ni^{+2} ions and half of the Fe^{+3} ions with CN = 6. In addition, there are four-fold sites, one-eighth of which are occupied by the remaining Fe^{+3} ions. The number of six-fold sites is equal to the number of oxygen ions, and the number of four-fold sites is equal to twice the number of oxygen ions.

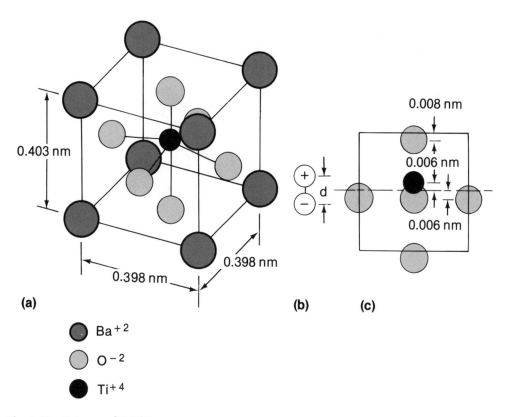

(a)

(b) **(c)**

- Ba^{+2}
- O^{-2}
- Ti^{+4}

Fig. 8.15 Tetragonal BaTiO$_3$
Above 120 °C, BaTiO$_3$ is cubic. Below 120 °C, the ions shift with respect to the corner Ba^{+2} ions. Since the Ti^{+4} and the O^{-2} ions shift in opposite directions, the centers of the positive and negative charges are not identical. The unit cell becomes noncubic. Source: L.H. Van Vlack, *Materials for Engineering*, Addison-Wesley, 1982, p. 532

8.6 Solid Solutions

Solid solutions are formed when the solvent and the solute atoms are of similar size and electron structure. In such cases a solute atom may substitute for one of the atoms in the solvent and produce the solid solution. If such a substitution is random in nature within a crystal structure, the probability of a particular solute atom being in a site is statistically equal to the atomic percent of the solute atoms.

In some types of solid solution, the solute atoms are located in the interstices in the crystal lattice. This is common in ceramic compounds. In these cases the ionic radius of the ion to be inserted must be appropriate for the size of the interstitial site, and the charge on the ion must be appropriate to retain electrical neutrality. This means that either the charge on the new ion must be identical with that replaced or that a balancing charge must be introduced into the structure simultaneously. An example of this type of substitutional solid solution is the substitution of Fe^{+2} for some of the Mg^{+2} in the ceramic MgO. If the Li$^+$ ion is substituted for the Mg^{+2} in MgO and a F$^-$ is substituted for one of the oxygen ions, charge neutrality is maintained and a solid solution of LiF in MgO results.

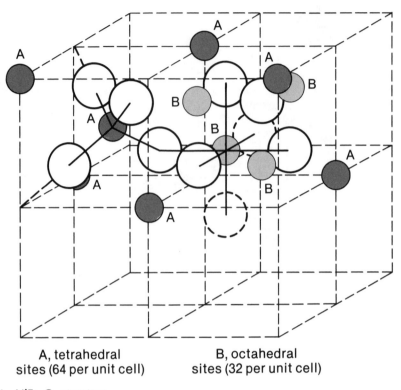

A, tetrahedral   B, octahedral
sites (64 per unit cell)   sites (32 per unit cell)

Fig. 8.16   NiFe$_2$O$_4$ structure
The Ni^{+2} ions occupy the B sites along with one-half of the Fe^{+3} ions. The remaining eight Fe^{+3} ions occupy the A sites; one-half of the A sites remain occupied. See also Fig. 8.13.

The magnetic spinels are often purposely doctored by making a solid solution of certain additives to achieve the best magnetic properties. For example, some of the divalent ferromagnetic Ni^{+2} ions (ionic radius 0.069 nm) are replaced with Zn^{+2} (ionic radius 0.074 nm), in a direct substitution. In some formulations two Ni^{+2} are replaced with an Li^{+}Fe^{+3} pair to form a complex solid solution.

8.7   Defect Structures

Closely related to the phenomena of interstitials and solid solutions are defect structures. Consider a regular lattice, such as the simple array of the FCC NaCl, which normally has a cation in each of the six-fold octahedral sites. The stoichiometry requires an equal number of cations and anions; charge neutrality is preserved if both ions have the same number of charges. If one or more cations is removed from the structure, a *vacancy* is created at one of the interstitial sites; however, the charges are no longer balanced. If the compound were originally NaCl, it would now become Na$_{(1-d)}$Cl, where d is the number of defects per mole of Cl^{-} anions. If we represent the defects or empty sites by the symbol $\widehat{v}$, then the formula for the material containing 10% unoccupied sites (d = 0.1) would be Na$_{0.9}\widehat{v}_{0.1}$Cl.

Many of the structures studied have interstitial sites which are normally unoccupied in the stoichiometric crystal. For example, materials such as GaAs (ZnS structure) have used only the four-fold sites (Fig. 8.7c), leaving the eight unoccupied six-fold sites available (Fig. 8.16). If additional cations are present in these sites, the extra cations are called *interstitials*. These interstitials are defects in the normal structure and can be represented by the symbol ⓘ. Addition of interstitials would be one way of balancing the charges of a defect structure of the type just considered. For example, an equal number of interstitial anions in the tetrahedral sites would restore charge neutrality to the compound $Na_{0.9}ⓘ_{0.1}Cl$. Since an occupied site will have a definite atomic species, the formula will usually indicate that species. For example, if Li^+ ions occupied some of the tetragonal interstitial sites in the NaCl lattice, the structure could be $Na_{0.9}ⓘ_{0.1}Li_{0.1}Cl$. A *defect structure* therefore is any nonstoichiometric compound that contains either vacancies or interstitials in the normal crystalline structure.

An interesting defect structure is produced in the NaCl crystal by the addition of alkali metal. An atom of sodium fills not only one of the Na^+ vacancies but also one of the Cl^- vacancies in the crystal. The metal atom has an extra electron which quickly becomes associated with a Cl^- vacancy, creating the Na^+ ion and a negative entity called an *F center*. Since F centers exhibit an optical absorption which is characteristic of the crystal lattice, they have been used by materials scientists as a method of studying crystal structures.

An example of a compound that frequently has a defect structure is the oxide FeO. The normal compound will have an Fe^{+2} ion at each of the six-fold interstitial sites in the FCC oxygen array (see Fig. 8.8). Usually, however, 10 to 15% of the iron atoms are present as Fe^{+3}. To achieve charge neutrality, two Fe^{+3} ions must replace three Fe^{+2} ions, leaving a vacancy Ⓥ at the site of the missing Fe^{+2} ion. For a 15% replacement of Fe^{+2}, the resulting compound is $Fe_{0.95}O$. Based on 100 of these formula units, or $Fe_{95}O_{100}$, there are 95 Fe ions, 100 oxide ions (O^{-2}), and 5 cation vacancies. From the above, there must be twice as many Fe^{+3} ions as there are vacancies or 10 Fe^{+3} ions, leaving 85 Fe^{+2} ions. The net negative charge is $100 \times (-2) = -200$. The net positive charge is $10 \times (+3) + 85 \times (+2) = 30 + 170 = +200$. Electrical neutrality is maintained in the defect compound.

Example 8.7A. A defect oxide has 11 times as many Fe^{+2} ions as it has Fe^{+3} ions. What percent of the cation sites are vacant and what is the formula of the compound?

Solution: The normal structure will be based on the compound FeO. Again, the number of vacancies is half the number of Fe^{+3} ions. In 100 formula units there will be 100 oxygen, with a net negative charge of $100 \times (-2) = -200$. There will be a net positive charge of $+200$. Let x = the number of Fe^{+3} ions; then the number of Fe^{+2} ions is 11x. Because there are 200 + charges, $x(Fe^{+3}) + 11x(Fe^{+2}) = 200$, or $3x + 22x = 200$, or $x = 200/25 = 8$. The net positive charge is (no. of Fe^{+3}) $\times (+3) + 11 \times$ (no. of Fe^{+3}) $\times (+2) = $ (no. of Fe^{+3}) $\times (22 + 3) = 200$. The number of Fe^{+3} ions is 8, and there are 11x Fe^{+2} ions, or 88 Fe^{+2} ions. The total number of Fe ions is $88 + 8$, or 96, which leaves four vacancies. The formula is written $Fe_{96}O_{100}$ or, more usually, $Fe_{0.96}O$.

A vacancy is an important accelerator of diffusion rates both of ions and of electronic charge. For example, transfer of an electron between adjacent Fe^{+2} and Fe^{+3} is a rapid mode of conduction in a crystal. It is easy to accomplish because it requires no physical movement of ions (the Fe^{+2} becomes an Fe^{+3} and the Fe^{+3} becomes an Fe^{+2}). This is an important mechanism for conduction in special electronic materials and occurs primarily in crystals made from transition elements because they may exist in more than one valence state.

In addition to the example of FeO, similar nonstoichiometric compounds are observed for FeS, CoO, Cu_2O, NiO, and the gamma forms of alumina. Some compounds that have interstitial cations are ZnO, Cr_2O_3, and CdO. Compounds with interstitial anions are more rare; however, UO_2 occurs with interstitial anions.

The defect structures considered above are all point defects, associated with a particular site in the lattice. Ceramic crystals also have line defects or dislocations just as the metals do (see Chapter 4). Dislocations have an important function in the mechanisms for crystal growth. They are also important in plastic deformation of ceramics, as discussed in Chapter 9.

8.8   Silicate Materials

The silicates are the most widely occurring ceramic materials, partly because they are plentiful and inexpensive and partly because of the versatility of their structures, which enables the silicates to have a range of desirable properties. Portland cement is an example of a widely used commercial silicate, and many other construction materials, such as brick and tile, contain silicates. Common sand is SiO_2, or silicon dioxide, and represents the common structural unit of all silicates, the SiO_4 tetrahedron.

Pure SiO_2 consists of oxygen ions in an extensive network of joined tetrahedra in which four oxygens form an interstitial space of CN = 4 for the Si^{+4} ion (see Fig. 8.17). This tetrahedral grouping is the primary structure of the world of silicates. If viewed as a separate entity (Fig. 8.17a), we see that the four oxygens each share one electron of an electron pair with the Si ion and the total charge of the tetrahedron is -4, i.e., SiO_4^{-4}.

The charge on one SiO_4^{-4} tetrahedron must be balanced by a $+4$ charge on one or more cations, e.g., $4\,Na^+$ or $2\,Ca^{+2}$, etc. Any of the oxygen atoms of one tetrahedron, however, may be shared with a second tetrahedron. The shared oxygen (called a *bridging oxygen*) is then part of two tetrahedra, and the resulting formula is $Si_2O_7^{-6}$. By repeated sharing of oxygen atoms, the tetrahedra may be joined into long chains (Fig. 8.17b).

When no metal ions are present, the chaining of silica tetrahedra through all of the oxygens creates a three-dimensional covalent structure. A typical structure, cristobalite (SiO_2), is shown in Fig. 8.18. Note that every oxygen atom joins two tetrahedra, forming a vast network extended through space. The more common form of SiO_2, silica, also is composed of a network of tetrahedral silicon but is of considerably greater complexity.

Fibrous silicate materials, such as asbestos, have chains of strongly bonded tetrahedra separated by cations such as Na^+, Al^{+3}, or Ca^{+2}. Cleavage occurs between the tetrahedral chains, resulting in fibrous materials. Other silicates, such as mica, have planes of SiO_4 tetrahedra separated by the cations. Cleavage occurs between the planes, producing sheetlike materials.

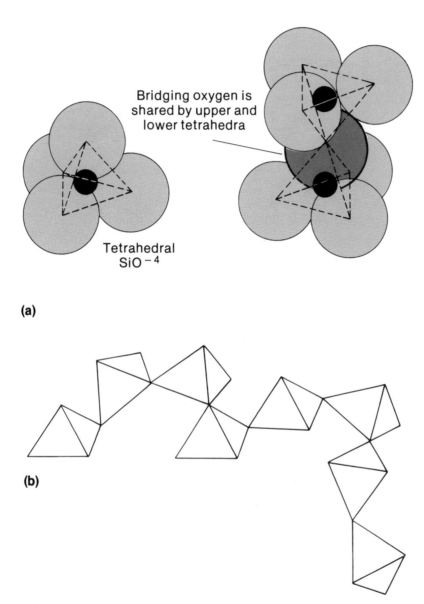

Bridging oxygen is
shared by upper and
lower tetrahedra

Tetrahedral
SiO^{-4}

(a)

(b)

Fig. 8.17   (a) SiO^{-4} tetrahedra. (b) Chains of SiO_4 tetrahedra

8.9   Compounds of Silica With Other Metal Oxides (Silicates)

The silicates may generally be described as compounds in which the silica structure exists in combination with various metal cations and oxygen. Superficially, the silica tetrahedron appears to function stoichiometrically as a large anion relative to the cation, as in the compound sodium silicate (Na_4SiO_4). Usually a complex structure occurs in which two or more tetrahedra are joined by bridging oxygen atoms, as in the $Si_3O_8^{+4}$ anion, which has four bridging oxygen

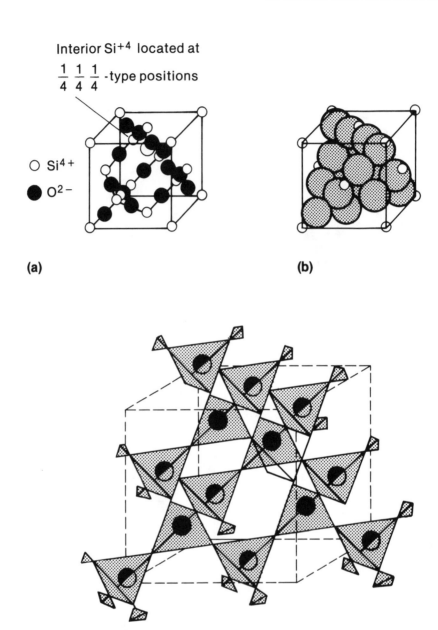

Fig. 8.18   Cristobalite (SiO$_2$) FCC unit cell
(a) Location of Si^{+4} in the unit cell. (b) Ions/unit cell: 8Si^{+4} + 16O^{-2}.
Source: J.F. Shackelford, *Introduction to Materials Science for Engineers*, Macmillan, 1985, p. 79

atoms. *The key factor in all silicate structures is the arrangement of the basic tetrahedral structure.* The tetrahedra can be arranged in a variety of ways by sharing a corner, an edge, or even a common face. The general categories of silicates are (see Fig. 8.19):

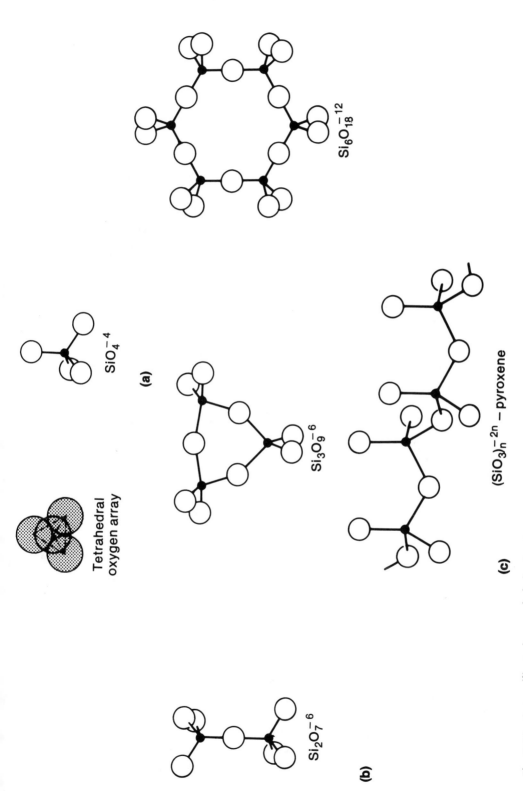

Fig. 8.19 Some silicate ions and chain structures
(a) Orthosilicate. (b) Pyrosilicate. (c) Chain structure

1. Orthosilicates: independent tetrahedra, SiO_4^{-4}
2. Pyrosilicates: two tetrahedra share corners, $Si_2O_7^{-6}$
3. Metasilicates: tetrahedra form chains and rings through shared corners, $(SiO_3)_n^{-2n}$
4. Layered silicates: tetrahedra share three corners, $(Si_2O_5)_n^{-2n}$

In addition, there is a category in which no cations are required:

5. Network silicas: bridging at all four corners, SiO_2

The structure of the naturally occurring mineral feldspar is an example of the modification of the silica structure by introduction of the cations Al^{+3} and K^+. In feldspar every fourth Si is replaced by an Al^{+3}, creating a net deficiency of charge which is remedied by the K^+ ions in interstitial locations. Compare the structure of feldspar in Fig. 8.20 to that of cristobalite in Fig. 8.18. The network structures of the feldspars are similar to the tridymite structure of silica with the alkaline earth or alkali ions in the tetrahedral interstices. Note that smaller positive ions, which are more suitable for octahedral coordination sites, do not form these structures but do form other types of silicate compounds.

The clay minerals are silicate structures in which combinations of silica are made with alumina. The clays are layered structures with alternating silica layers and alumina layers. The basic structure has an Si_2O_5 layer joined at the corners of its tetrahedra alternating with a layer of alumina octahedra (Fig. 8.20). If the oxygen atoms, which are projecting downward from the plane of the silica layer, become part of the alumina plane's octahedra, as in Fig. 8.21, the combined structure is kaolinite, the most common clay mineral. A large variety of clay minerals may be built from different combinations of layers and with various cations. Common substitutions are iron and/or aluminum for silicon in the tetrahedral network, and magnesium and/or iron for aluminum in the octahedral network. Some types of substitutions which lead to a net negative charge (vacancy structures) lead to the absorption of balancing positive ions onto the surface of the clay. These surface ions are easily exchanged, leading to the base ion-exchange character of the mineral clays.

8.10 Phase Diagrams

Phase diagrams illustrate graphically the equilibrium for a given material with respect to a given set of conditions of concentration, temperature, and pressure. In ceramic processing, the time it takes to obtain equilibrium is often very long; consequently, phase diagrams are only part of the information required to determine the structure of a particular ceramic. Nevertheless, as with metals (see Chapter 5), phase diagrams are a useful representation of the equilibrium state of a given ceramic composition.

The equilibrium phase diagram for the SiO_2 system (Fig. 8.22) shows five phases. (The temperatures are for a pressure of 1 atm; they differ only slightly for moderate changes in pressure.) Some of the phase transformations are rapid and reversible (for example, α-quartz to β-quartz), while others are quite sluggish. Several metastable phases regularly occur which are unstable relative to other, lower-energy phases and are only prevented from changing by the slowness of the required transformation. A diagram including such metastable phases is illustrated in Fig. 8.23. Many ceramic materials contain other cations and anions in solid solutions

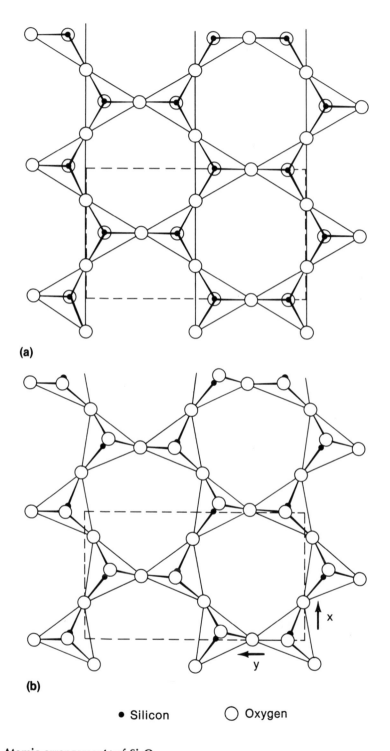

(a)

(b)

● Silicon   ○ Oxygen

Fig. 8.20   Atomic arrangements of Si$_2$O$_5$
(a) Si$_2$O$_5$ layer. (b) Si$_2$O$_5$ layer in kaolinite. Source: G.W. Brindley, in *Ceramic Fabrication Processes*, W.D. Kingery, Ed., John Wiley & Sons, 1958

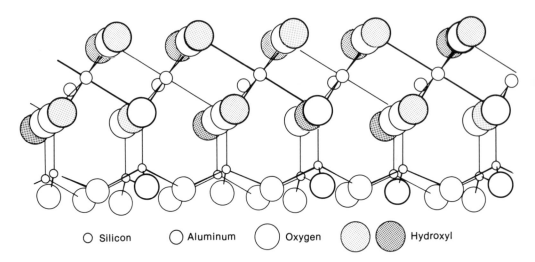

Fig. 8.21 Perspective drawing of kaolinite
Si-O tetrahedra are on the bottom half of the layer and the Al-O, OH octahedra are on the upper half. Source:
G.W. Brindley, in *Ceramic Fabrication Processes*, W.D. Kingery, Ed., John Wiley & Sons, 1958

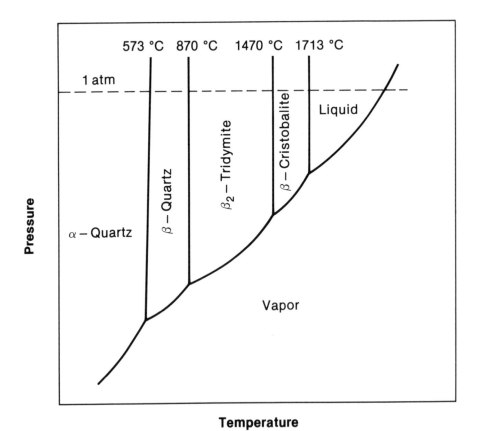

Fig. 8.22   Phase equilibrium diagram for SiO₂
Source: W.D. Kingery, *Introduction to Ceramics*, John Wiley & Sons, 1965, p. 251

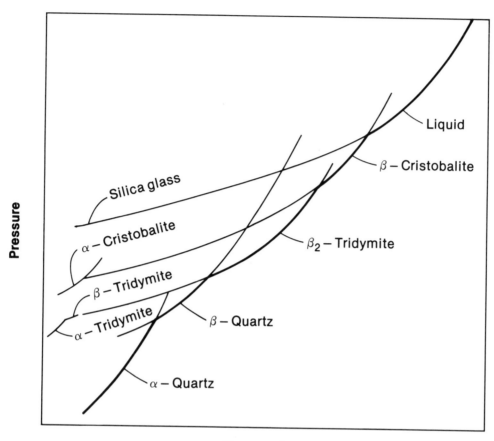

Pressure

Liquid

β – Cristobalite

Silica glass

α – Cristobalite

β_2 – Tridymite

β – Tridymite

α – Tridymite

β – Quartz

α – Quartz

Temperature

Fig. 8.23   Diagram including metastable phases occurring in the system SiO_2
Source: W.D. Kingery, *Introduction to Ceramics*, John Wiley & Sons, 1965, p. 252

or as interstitials, and most ceramic phase diagrams are quite complex, as shown by Fig. 8.24 (Ba_2TiO_4-TiO_2) and Fig. 8.25 (MgO-Al_2O_3).

8.11   Silica Polymorphs

Often a solid element or compound may exist in more than one solid structure—for example, carbon, diamond, and graphite. These structures are called *polymorphs* (see Chapter 2). The FCC cristobalite form of SiO_2 is a polymorph with the common quartz form of silica. Cristobalite is the simplest of the polymorphs; it is cubic, least dense, and stable at the highest temperatures, as shown on its phase diagram (Fig. 8.23). The more common form, quartz, has its tetrahedra joined through their corner atoms into a complex helix structure, resulting in a hexagonal crystal structure.

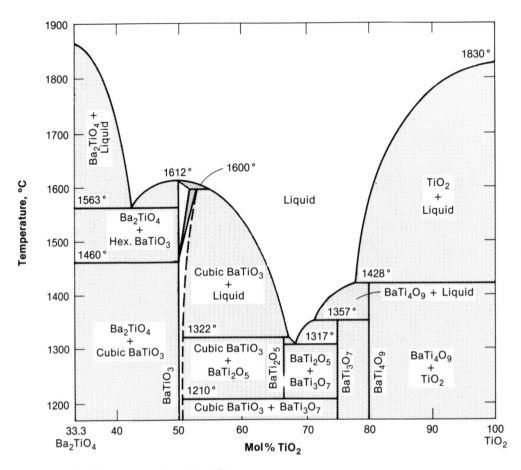

Fig. 8.24   The binary system Ba$_2$TiO$_4$-TiO$_2$
Source: After D.E. Rase and R. Roy, *J. Am. Ceram. Soc.*, Vol. 38, 1955, p. 111

8.12   Silica Glass

If crystalline silica is melted, the crystal structure breaks down only partially. A substantial joining of silica tetrahedra remains in the liquid because of the strong —O—Si—O— bond structure present in the crystal. Although the *long-range*, repeating, ordered structure necessary for crystal structure is no longer present, silica remains as a highly bonded, very large molecular structure in the melt. Consequently, liquid silica is highly viscous and very difficult to process into the desired shapes required for scientific and commercial applications.

When the melt is cooled, it is necessary for a substantial movement and reordering of these large groups of atoms to take place to produce the long- range order required for crystallization. If the melt is held for long periods of time near the melting point, crystallization will occur. Because the melt contains very large molecules and is highly viscous, it is easy to supercool silica to produce an amorphous silica glass, often called *vitreous silica* or *fused silica*. Silica glass has several important properties related to the internal network structure: it accommodates

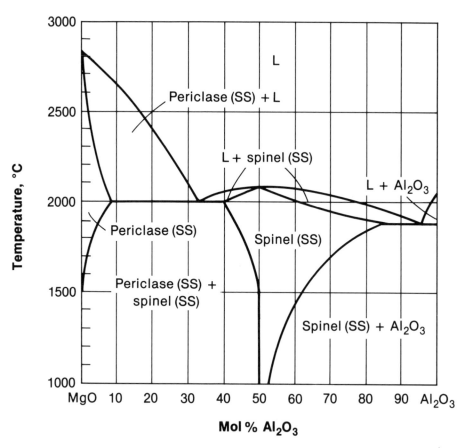

Fig. 8.25 MgO-Al$_2$O$_3$ phase diagram
Spinel is an intermediate compound with ideal stoichiometry MgO · Al$_2$O$_3$. Source: After *Phase Diagrams for Ceramists*, Vol. 1, American Ceramic Society, Columbus, OH, 1964

extra thermal energy through rotation about some of the bonds, and therefore has a low coefficient of thermal expansion, and it is relatively inert to solvents.

8.13 Microstructures in Ceramics

During the processing of ceramics, various methods are used to produce the final ceramic object from its raw materials. The finished ceramic may be crystalline, glassy, or a mixture of both. The crystalline ceramic is usually a polycrystalline material with a grain structure similar to that of metals and alloys described in Chapter 2. The ceramic fabrication process usually comprises heating crystalline and noncrystalline powders at a temperature sufficient to produce the consolidation necessary for a useful object. During the heating a fine-grained, multiphase porous material is initially formed, which subsequently undergoes a decrease in porosity and pore size along with recrystallization and an increase in grain size.

Primary recrystallization refers to the process by which entirely new crystals are formed by nucleation and growth. Grain growth refers to the increase in the average size of existing grains (small homogenous crystalline regions) during heat treatment without a major change in the distribution of grain sizes. A process called secondary recrystallization occurs when a few grains grow at the expense of the smaller or fine-grained structures in a ceramic body. Grain growth and secondary recrystallization are most important in ceramic processing. The properties of ceramics are greatly influenced by the nature and amounts of the phase present in a ceramic body, as discussed in Chapter 9.

8.14   Problems

1. Contrast the general properties of ceramics versus metals.
2. List *all* the types of bonding found in ceramics. List all the types of bonding for each of the following: NaCl, graphite, diamond, limestone ($CaCo_3$), fused silica, and soda-lime glass.
3. What is the coordination number for sodium and chlorine in NaCl, for carbon in graphite, for carbon in diamond, for calcium and fluorine in CaF_2?
4. Calculate the percent of unoccupied space in NaCl and KCl. Are these numbers exactly the same? Why are/aren't they the same?
5. Calculate the density of FeO (FCC structure like NaCl).
6. What is the coordination number of Zr in ZrO_2? What is the coordination number of oxygen in this compound? The structure is similar to CaF_2.
7. Calculate the density of ZrO_2 (CaF_2 structure).
8. What is the ratio of Fe^{+2} ions to Fe^{+3} ions in the compound $Fe_{0.933}O$? What percent of cation sites are vacant? What percent of all sites are vacant?
9. What is the exact formula of ferrous oxide in which there are 8.68 ferrous atoms for each ferric atom? What is the number of vacant cation sites in 100 cation sites?
10. A freshman chemistry lab was directed to prepare a compound by heating copper wire in an excess of sulfur. The excess sulfur burned off, leaving copperII sulfide. The average composition of the compound, however, was $Cu_{1.85}S$, and no students who performed the experiment carefully obtained greater than $Cu_{1.95}S$. Why didn't the careful students obtain Cu_2S?
11. What percent of all oxygens are bridging oxygens in the compound Na_4SiO_4? In $NaSiO_3$? In $NaSi_2O_5$?
12. Clay easily absorbs water molecules in between layers of atoms in its structure. What property or properties does this impart to clay?
13. Speculate on what would happen if a ductile ceramic could be made. Why do ceramics tend to be brittle?

9

Properties of
Ceramic Materials

9.1   Melting Points and Glass Transitions

Thermal motion in ceramics held in the molten state enables the rearrangements necessary for assembly of the constituent species into a more regular order, as in crystallization. We have already seen that some melts contain such large, bulky molecular assemblies that they cannot rearrange rapidly enough to crystallize at normal cooling rates. These materials can be supercooled below their equilibrium melting temperature into a glass. Rearrangement does not stop in these supercooled liquids, and recrystallization continues from the metastable super-cooled liquid to the crystalline solid. On continued cooling of the supercooled liquid phase, there is a point where such rearrangement motion essentially ceases; this point is referred to as the *glass transition temperature*. It is of particular significance in the properties of both ceramic glasses and amorphous polymers. In the case of silicate glasses, however, the glass transition temperature is so high that it is seldom encountered in normal applications.

The phenomenon of the glass transition was first observed in ceramic materials. It is detected by plotting one of several properties of the material as a function of temperature. For example, the volume, or coefficient of expansion, of a liquid plotted against temperature will reveal not only the temperature of phase change to the crystalline state (the *melting point*), but also often shows a deflection point in the metastable supercooled liquid which defines the glass transition temperature (see Fig. 9.1). Note that this transition point is dependent on the cooling rate and does not represent an equilibrium value. The glass transition temperature is discussed more fully in the treatment of polymers in Chapters 10 and 11.

For ceramic materials, several other temperatures near the glass temperature are often specified, due to their significance in processing the material. The *annealing point* is close to the glass transition temperature but is that point at which the material can undergo sufficient molecular rearrangements to reduce stresses within a short time without undergoing significant macroscopic change. Below the *strain point* the glass can be subjected to rapid thermal cycling without inducing any additional internal stresses. In processing a glassy material, it is good practice to anneal the material to remove stress and then to cool the material very slowly from the annealing point to the strain point, at which point the material is sufficiently below the glass transition temperature that rapid cooling can take place. These temperatures are illustrated for a common glass in Fig. 9.2, and are often characterized in terms of the viscosity of the material:

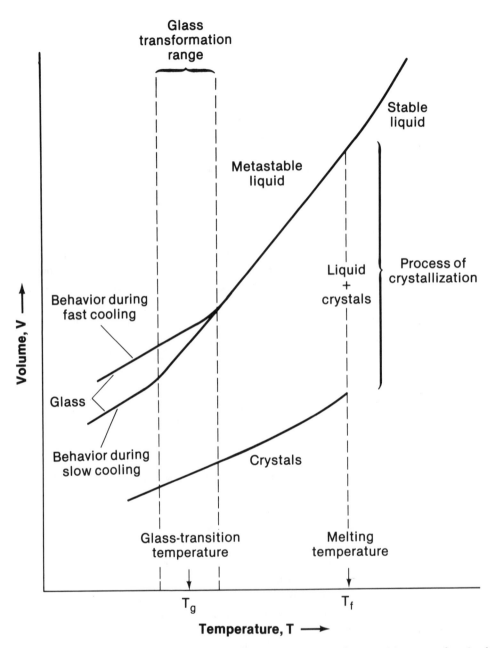

Fig. 9.1   Volume changes in a liquid during crystallization or in the transition to a glass in the range of temperature near T_g
Source: A.G. Guy, *Introduction to Materials Science*, McGraw-Hill, 1972, p. 798

the annealing point occurs when the viscosity is about 10^{12} Pa · s and the strain point corresponds to about 10^{13} to 10^{14} Pa · s. Compare these values to the typical working viscosity at which a glass is usually formed and shaped, about 10^{6} Pa · s.

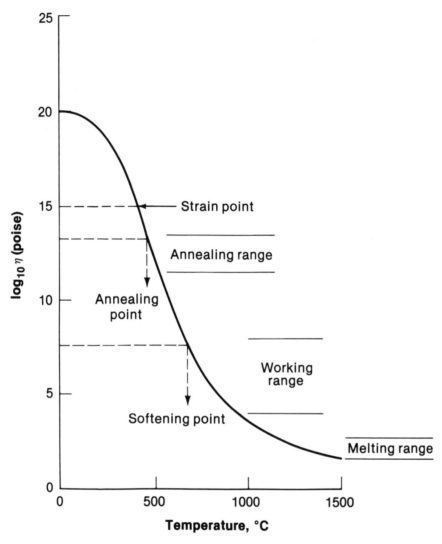

Fig. 9.2 Viscosity of a typical soda-lime-silica glass from room temperature to 1500 °C
Above the glass transition temperature (~450 °C in this case), the viscosity decreases in an Arrhenius fashion.

9.2 Elastic Deformation of Ceramic Materials

Many of the applications for ceramic materials make use of the high degree of resistance to elastic deformation. Both glasses and polycrystalline ceramic materials, but not highly anisotropic single crystals, have primarily an elastic response to stress, which is of the form: $\epsilon = \sigma/E$, where the strain ϵ is proportional to the tensile stress σ and the proportionality constant is called the *Young's modulus*, E. Below the glass transition temperature all glasses exhibit this elastic response.

Shear stress, r, is also directly proportional to the strain, γ: $\Gamma = \gamma/G$. The proportionality constant, G, is called the shear modulus, and the two moduli are related by Poisson's ratio: $\mu = (E/2G) - 1$. For processes where the volume of the material remains fairly constant, such as for creep, plastic flow, and viscous flow, Poisson's ratio (μ) is approximately 0.5. In elastic deformation it is typically between 0.2 and 0.25. Values of Young's modulus for a number of representative ceramic materials are given in Table 9.1. While for some highly crystalline materials E is as high as 42×10^4 MPa (60×10^6 lb/in.^{2}), glasses are in the range of 7 to 10×10^4 MPa (10 to 15×10^6 lb/in.^{2}). Ceramic materials are especially strong in compression or shear; thus the effective strength of glass can be quite high if applications are designed to take advantage of this property.

Example 9.2A. For a borosilicate glass with E = 69×10^3 MPa, the modulus of rupture (strength) is 69 MPa. If Poisson's ratio, μ, for the glass is 0.25, what is the estimated shear modulus, G? The estimated theoretical strength of such glassy materials is between G/4 and G/6. Assuming the value G/5, what is the theoretical strength and what percent of the strength is achieved in practice?

Solution: Applying $\mu = (E/2G) - 1$ for E = 69×10^3 and $\mu = 0.25$, we calculate E/2G = 1.25, or E = 2.5G and G = E/2.5 = 28×10^3 MPa. The approximate theoretical strength is then G/5 = $(28 \times 10^3)/5 = 5.6 \times 10^3$ MPa. The measured rupture strength is 69 MPa. When compared to the theoretical strength, the rupture strength is 69/5600 or 1% of the theoretical strength.

Ceramic crystalline materials have an ordered array of atoms with opposite charge. Plastic deformation occurs in metallic crystals by dislocation movements along slip planes and slip directions (see section 4.9 in Chapter 4). Slip directions are normally the directions with the

Table 9.1 Modulus of elasticity and strength (modulus of rupture) for some ceramics and glasses

Material	E, MPa	MOR, MPa
Mullite (aluminosilicate) porcelain	69×10^3	69
Steatite (magnesia aluminosilicate) porcelain	69×10^3	140
Superduty fireclay (aluminosilicate) brick	97×10^3	5.2
Alumina (Al_2O_3) crystals	380×10^3	340–1000
Sintered alumina (~5% porosity)	370×10^3	210–340
Alumina porcelain (90–95% alumina)	370×10^3	340
Sintered magnesia (~5% porosity)	210×10^3	100
Magnesite (magnesia) brick	170×10^3	28
Sintered spinel (magnesia-aluminate) (~5% porosity)	238×10^3	90
Sintered stabilized zirconia (~5% porosity)	150×10^3	83
Sintered beryllia (~5% porosity)	310×10^3	140–280
Dense silicon carbide (~5% porosity)	470×10^3	170
Bonded silicon carbide (~20% porosity)	340×10^3	14
Hot-pressed boron carbide (~5% porosity)	290×10^3	340
Hot-pressed boron nitride (~5% porosity)	83×10^3	48–100
Silica glass	72.4×10^3	107
Borosilicate glass	69×10^3	69

Source: W.D. Kingery, H.K. Bowen, and D.R. Uhlmann, *Introduction to Ceramics*, 2nd ed., John Wiley & Sons, 1976

highest density of lattice points, or the shortest distance between lattice points, and slip planes are those planes with widest interplanar spacings.

An axial or compressive stress on a crystal is resolved into a shear stress on a slip plane, and movement occurs on the slip plane with the lowest resistance to movement, i.e., the lowest *critical shear stress*. The systems with the lowest critical shear stress are those in the most closely packed directions on the most densely packed planes.

Example 9.2B. In Fig. 9.3 the resolved shear stress, γ, in the slip direction results from the application of a tensile stress, σ, to the axial direction of the crystalline body. The tensile strain is the result of a force F acting on the area A of the material. What is the expression for the resolved shear stress, γ?

Solution: The force F' operating in the slip direction is F' = F cos λ. The effective area for this force is the projection of A onto the slip plane, or A/cos ϕ. The shear stress is the force acting per unit area:

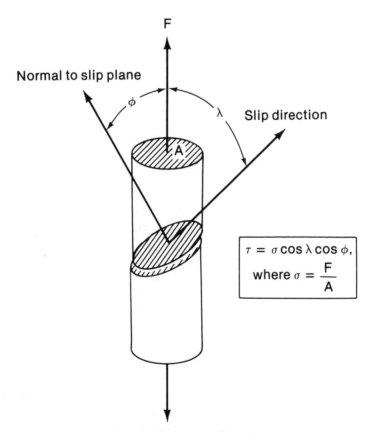

$$\tau = \sigma \cos \lambda \cos \phi,$$

$$\text{where } \sigma = \frac{F}{A}$$

Fig. 9.3   Definition of the resolved shear stress, γ, which directly produces plastic deformation (by a shearing action) as a result of the external application of a simple tensile stress, σ
Source: J.F. Shackelford, *Introduction to Materials Science for Engineers*, Macmillan, 1985, p. 142

$$\gamma = \frac{F \cos\lambda}{A/\cos\phi} = \frac{F}{A} \cos\lambda \, \cos\phi$$

$$\gamma = \sigma \cos\lambda \, \cos\phi$$

The mechanism of slip movement involves the production and movement of a dislocation line. The process requires an energy which is a function of the atomic arrangements. The easiest dislocations to utilize for plastic deformation are those with the shortest unit slip vector. [The displacement of atoms around the dislocation is called its slip, or Burgers vector (Fig. 9.4), and the energy of the dislocation line is proportional to its length, shear modulus of the material, and the square of the slip vector.] The shortest slip vectors are in the direction of greatest linear density of atoms, usually when atoms are touching each other, or $|\mathbf{b}| = 2r$, where $\mathbf{b}$ is the Burgers vector and r is the atomic or ionic radius. The lowest values of the shear modulus, G, are associated with those planes that are the farthest apart and have the greatest density of atoms. In metals this process is greatly aided by the fact that adjacent atoms are generally similar.

In ceramic crystals, plastic deformation is greatly hindered by the dissimilar nature of the atoms in the crystalline array, especially the presence of opposite charges. Figure 9.5 shows a simplified slip process in a typical homogeneous metal and in a ceramic material, a metal oxide. When slip occurs in the metal, the environment of the atoms displaced is identical to that in the original, undeformed crystal. More force is required, however, to displace the ions in a metal oxide than the atoms in a metal. Not only is it necessary to break an ionic bond between the positive and negative ions, but in moving horizontally the strong repulsive force between ions of like charge becomes significant. Both of these add to the required force for shear. In the array illustrated, slip along the horizontal direction is easy for the metal and is highly restricted for the metal oxide. Slip is possible in the oxide along certain other directions (the 45° direction in Fig. 9.5); however, the number of possibilities for slip in the oxide is much more limited than in the metal. In addition, the slip vector length is often greater in the oxide than in the metal; slip in oxides (and ceramics in general) requires movement along greater interatomic distances. These differences are even greater when metals are compared with the more complicated ceramic crystal structures.

Since ceramics cannot easily undergo plastic deformation, they are generally hard and brittle. As a result, many ceramic materials are valuable for use as cutting tools. Those ceramics that are not hard are the fibrous and sheetlike ceramics which are hard within the layers but which have weak interfiber or interlaminar bonding forces.

9.3 Microcracks and Brittle Fracture

All ceramic materials are strong in compression due to the ability to withstand the resolved shear stress. Since the highest shear stress is at 45° to the applied stress for both tension and compression, the tensile strength and compressive strength should be comparable (see Fig. 9.6). In practice, microcracks and other flaws will become the limiting factor in determining the useful strength of ceramic materials. When a tensile stress is imposed on a material with a microcrack (see Fig. 9.7), the stress is more concentrated at the tip of the crack. In a material

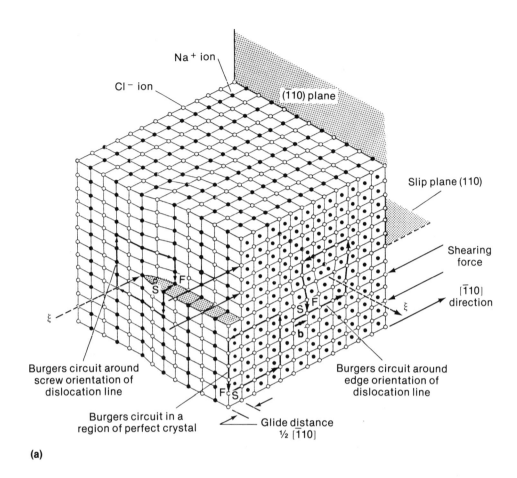

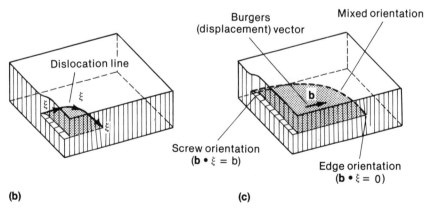

Fig. 9.4   A perfect dislocation forming the boundary of a region of plastic glide in a crystal having the NaCl structure

An m atom by n atom circuit normally forms a closed loop. An n × m circuit around a dislocation is not closed. The vector, **b,** required to close the loop is the Burgers vector. (a) Schematic view of the exterior of a block of crystal lattice containing a curved length of dislocation line. (b) The direction of the dislocation line at any point is described by the unit vector, ξ, tangent to the line. (c) Description of the dislocation line in terms of its orientation relative to the Burgers vector. Source: A.G. Guy, *Introduction to Materials Science,* McGraw-Hill, 1972, p. 171

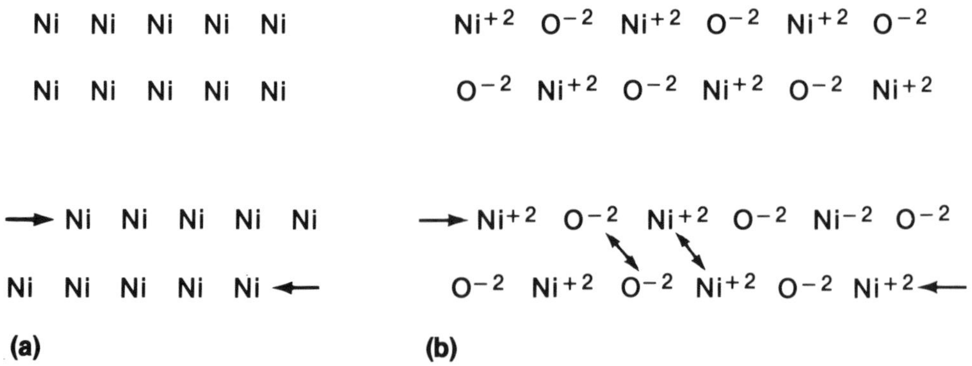

Ni   Ni   Ni   Ni   Ni Ni^{+2} O^{-2} Ni^{+2} O^{-2} Ni^{+2} O^{-2}

Ni   Ni   Ni   Ni   Ni O^{-2} Ni^{+2} O^{-2} Ni^{+2} O^{-2} Ni^{+2}

→ Ni   Ni   Ni   Ni   Ni   → Ni^{+2} O^{-2} Ni^{+2} O^{-2} Ni^{-2} O^{-2}

Ni   Ni   Ni   Ni   Ni ← O^{-2} Ni^{+2} O^{-2} Ni^{+2} O^{-2} Ni^{+2} ←

(a)   **(b)**

Fig. 9.5   Comparison of slip processes (metallic nickel and nickel oxide)
More force is required to displace the ions in NiO than the atoms in nickel. The strong repulsive forces between like ions become significant. Nickel also has more slip systems than does nickel oxide.

of high ductility, the area near the tip will deform slightly, reducing the stress concentration. In glasses and other ceramics which are strong in shear, the crack tip does not deform and the stress is concentrated, causing the crack to lengthen. There is also a high chemical activity in such an area, and as a result, crack propagation is further accelerated by adsorption of water at the active site. Note that under compression the load is transmitted mechanically across the crack so that the effect of the microcrack is reduced or eliminated. The most familiar expression of this ability of surface flaws to reduce the strength of glass is the common technique of scratching a piece of window glass to cause it to break in the exact pattern desired. This type of behavior is characteristic of all brittle materials.

The magnitude of the forces of stress concentration can be estimated by considering the forces surrounding a small elliptical hole (Fig. 9.7). If s is the stress in the bulk of the material at great distance from the hole, then the stress at the small radius end of the hole, s_c, is given by: $s_c = s\,(2\sqrt{c/r})$, where 2c is the length of the hole (c is the depth of an external crack from the surface) and r is the radius of the ellipse at the tip. It is estimated that the tip of a crack in glass has a radius which is of the order of atomic dimensions (0.1 nm).

For a crack to be visible, the length must be about 1000 nm, and the stress concentration increase is $s_c/s = 2\sqrt{(1000/0.1)} =$ a factor of 200. Thus the 70,000 MPa strength of glass cited above must be downgraded to a value less than 350 MPa. Ordinary window glass has a practical strength in tension of about 200 MPa as a result of such flaws introduced in its fabrication.

Example 9.3A. Calculate the radius of curvature of the leading edge of a crack in glass when the stress concentration is 1,500,000 psi, the applied stress is 8000 psi, and the crack depth is 2 μm (0.0002 cm).

Solution: From the equation above, $c/r = s_c^2/4s^2$ and $r = (4s^2 \times c)/s_c^2$. Therefore, $r = (4 \times 8000^2 \times 0.0002)/1,500,000 = 2.28 \times 10^{-8}$ cm, or 0.228 nm.

For the spontaneous growth of a crack in a brittle material, it has been shown that $\sigma = [(2\gamma E)/\pi c]^{1/2}$, where σ is the applied stress necessary for spontaneous crack growth, or

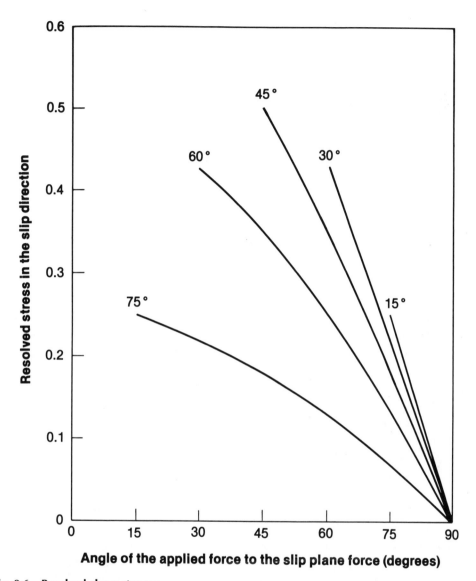

Fig. 9.6 Resolved shear stresses
Resolved stress for unit stress in the axial plane as a function of the angle between the applied force and the resolved force for selected values of the angle between the applied force and the normal to the slip plane

fracture stress; γ is the surface energy of the material in dynes per square centimeter (1 psia = 7×10^4 dyne/cm^2); and c is the crack depth, or radius of an interior crack.

Example 9.3B. Calculate the applied stress needed to propagate a crack in glass having a surface energy of 600 erg/cm^2, a modulus E = 10×10^6 psi, and a surface crack 2 μm deep.
  Solution: $\sigma = [2 \times 600 \times (10 \times 10^6) \times (7 \times 10^4)/\pi \times 0.0002]^{1/2} = [(840 \times 10^{12})/6.28 \times 10^{-4}]^{1/2} = 1.157 \times 10^9$ dyne/cm^2 = 113.7 MPa.

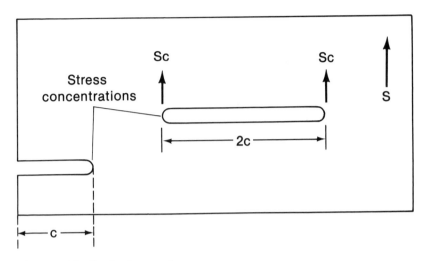

Fig. 9.7   Idealized microcracks

Because many of these flaws occur on the surface of the materials, an improvement in performance can be obtained by treating the surface in such a way that their effect is minimized. Many ceramic materials are strengthened by treatments which subject the surface to compressive forces. This means that the toughened surface area withstands a much greater tensile stress than the normal, untoughened material with stress concentrations in the microcracks. In the case of fine-grained polycrystalline ceramics, the length of the Griffith microcrack is limited to the grain size. These materials are therefore stronger than coarser material of the same composition, and the strength is roughly proportional to the square root of the grain size (diameter of the grain).

Example 9.3C. A ceramic material having an average grain size (diameter) of 50 μm (1 μm = 1000 nm) was found to have an experimental value of average rupture strength of 25 kg/cm^2. If the grain size can be reduced to 10 μm, what is the expected strength in MPa of the modified material?

Solution: Assume the radius of the crack is 0.1 nm as above, and the effective crack length of the original material was 50 μm (the grain size). $s_c = s(2\sqrt{(50,000)/0.1})$ for the original material and $s_c = s'(2\sqrt{(10,000)/0.1})$ for the modified material. Then $s/s' = \sqrt{50,000/\sqrt{10,000}} = \sqrt{5} = 2.236$. The rupture strength of the recrystallized material, s', is multiplied by this ratio and a factor to convert to MPa: $s' = 25$ kg/cm$^2 \times (100$ cm/m$)^2 \times 9.8067$ m/s$^2 \times 1$ MPa/10^6 kg/ms$^2 \times 2.236 = 5.48$ MPa.

Because the observed strength of a material is a function of the presence of flaws of a certain size and shape, each sample tested will display a somewhat different value depending on the distribution of flaws and defects in the material. The value usually quoted for a ceramic material is a *characteristic value* based on the distribution of measured values. The characteristic strengths of some ceramic materials are given in Table 9.1, and examples of the distribution of strength values are shown in Fig. 9.8.

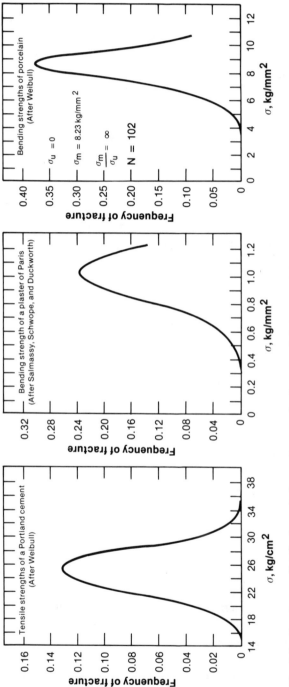

Fig. 9.8 Frequency distribution of observed strengths for some brittle materials
Source: W.D. Kingery, *Introduction to Ceramics*, John Wiley & Sons, 1965, p. 609

Example 9.3D. Assume the strength of the ceramic material was distributed according to a curve whose shape was proportional to that given for Portland cement in Fig. 9.8. What is the value of stress in MPa that you would use to determine a safe application of the material? If you were willing to live with a failure rate of 5%, what cross-section of bar would you need to resist a force of 50 lb?

Solution: The graph indicates that the mean value of strength is 25 kg/cm^2 but that failures occur as low as 14.5 kg/cm^2. (Note: 1 kg/cm^2 = 0.098067 MPa, from the above example.) A value of no more than 14 kg/cm^2 should be used for safety; thus 14 kg/cm^2 × 0.098067 MPa/kg/cm^2 = 1.37 MPa. A value of about 20 kg/cm^2 or 1.96 MPa would be close to the 5% failure rate. To resist 50 lb, it would require a cross-section of: (50 lb/2.2 lb/kg)/20 kg/cm^2 = 22.7/20 = 1.14 cm^2.

Example 9.3E. A ceramic material has a mean value of strength of 1.961 MPa and the distribution of rupture strengths follows a normal distribution with a standard deviation, σ = 0.100 MPa. What is the estimated failure rate at a force of 1.761 MPa?

Solution: The value 1.761 is -2σ values from the mean. The normal distribution gives a probability of occurrence of 2.3% outside the -2σ range; therefore, the estimated failure rate equals 2.3%.

9.4   Fatigue

The measured strength of many ceramics is found to be dependent on: (1) the length of time the material is under stress or (2) the rate of application of the stress. Such materials can resist a very high stress for a short period of time but will fail under low stress that is maintained for long periods. Plastic (polymeric) materials also exhibit this behavior quite markedly. In ceramics, however, this behavior is often related to the presence of atmospheric moisture around the sample. For a soda-lime-silica glass, the time to fracture at constant stress, t, is temperature dependent with an Arrhenius activation energy, E, of about 18 kcal/mol: $t = Ce^{-E/RT}$. This suggests that the mechanism of this failure is chemical in nature and may be from enhanced chemical activity at the apex of the Griffith microcracks.

9.5   Viscous Deformation

Since many ceramic glasses are supercooled liquids, they will exhibit viscous flow or deformation, particularly at temperatures above the glass transition point. The relationship defining this flow is given in terms of the velocity gradient and the force per unit area on a plane parallel to the direction of flow: $dv/dx = F/\eta$. For viscous flow, the *rate of deformation* is directly proportional to the applied stress.

Viscous flow requires a certain minimum energy to overcome the attractive forces between fluid elements and move an individual element to a new position. Because it is a process with an activation energy, viscous flow is temperature dependent. The temperature dependence is given by: $\ln \eta = c + A(1/T)$, where A is the activation energy, E/RT, and c is a constant. Activation energies for some typical fluids are given in Table 9.2. Note that the glasses are

Table 9.2   Activation energies for viscous flow

Type of fluid	Approximate range of values of E (kcal/mol)
Nonpolar liquids, hydrocarbons	½–1
Metals	½–2
Polar liquids	1–3
Hydrogen bonding liquids	2–10
Simple ionic liquids	3–10
Glasses	20–150

Source: W.D. Kingery, *Introduction to Ceramics*, John Wiley & Sons, 1965, p. 572

nearly an order of magnitude higher than other fluids. This large temperature dependence of viscous flow is the basis of glass processing, as described in Chapter 8.

Example 9.5A. A soda-lime glass has a viscosity of 10^6 poise at 1000 °K, 10^9 poise at 833 °K, and 10^{12} poise at 714 °K. At what temperature will it be in the annealing range of $10^{13.4}$ poise? What is the apparent activation energy term A?

Solution: Take the natural logarithms of the viscosities and the reciprocals of the temperatures and plot the values of ln η versus 1/T on graph paper to determine the slope (this may be simplified by using semilogarithmic graph paper). The slope = 34,540. Then, substituting in the equation ln η = 34,540 (1/T) + C for a temperature of 1000 °K gives C = ln η_o = −20.724 so that ln η = 34,540 (1/T) − 20.724. For the annealing range: ln η = ln 2.51 × 10^{13} = 30.85464 = 34,540 (1/T) − 20.724 or (1/T) = (30.854 + 20.724)/34,540 = 0.001493, T = 670 °K. The value 34,540 represents the activation energy term, A. Note that A equals E/RT where R is the molar Boltzmann constant of 1.987 cal/mol · °K.

Example 9.5B. A new glass composition can be worked between η = 10^5 Pa · s and η = 10^2 Pa · s, at 900 and 1400 °C, respectively. Assuming a linear relationship holds between the ln of the viscosity and the reciprocal of the temperature, estimate the form of the relationship and calculate an annealing point for this new composition. What is the corresponding relationship for fluidity, f, of this glass?

$$\ln (10^5) = \frac{A}{900 + 273} + C$$

$$\ln (10^2) = \frac{A}{1400 + 273} + C$$

Eliminating A from these simultaneous equations:

$$(900 + 273) \ln (10^5) = (900 + 273) C + A$$
$$1173 \ln (10^5) = 1173 C + A$$

$$(1400 + 273) \ln (10^2) = (1400 + 273) \; C + A$$
$$1673 \ln (10^2) = 1673 \; C + A$$

$$1173 \ln (10^5) - 1673 \ln (10^2) = -500C$$
$$C = -11.6025, \text{ thus } A = 27,115$$

To achieve an annealing point, we need a viscosity of about $\eta = 10^{12}$ Pa $\cdot$ s; therefore, ln $(10^{12}) = 27,115/T - 11.6025$ and $T = 27,115/39.234 = 691$ °K or 418 °C. Fluidity is the reciprocal of viscosity, so that if:

$$\ln (\eta) = 27,115/T - 11.6025 = \ln (1/f)$$
$$\ln (1) - \ln (f) = 27,115/T - 11.6025$$
$$-\ln (f) = 39.241 - 11.6025 = 27.638$$
$$f = 10^{-12} \text{ Pa}^{-1} \cdot \text{s}^{-1}$$

Example 9.5C. The expression for the dependence of viscosity on temperature is of the classical Arrhenius type, which implies an activation energy for the property being measured. Using the expression for the temperature dependence of viscosity derived in the previous example, calculate the activation energy for viscous flow for this glass.

Solution: Note that $-\ln \eta = \ln (f_o) - E/RT = \ln (f)$, which has the form of $\ln (\eta) = A/T + C$, as in the example above. $-E/RT = A/T$; $E = 27,115 \times 1.987$ cal/mol $= 53,878$ cal/mol.

The viscosity of a ceramic glass is also influenced by its composition. The effectiveness of added ions in changing viscosity is a function of ionic radius and is a function of the relative effectiveness of the ion as a network former/modifier. A table of such additives is given in Table 9.3. The effect of substitution for 8% of the silica in a $74SiO_2$-$10CaO$-$16Na_2O$ glass by several of these modifiers is illustrated in Fig. 9.9. This effect is most pronounced at high temperatures. At lower temperatures the change in viscosity is related to ionic radius within a specific coordination number series; it changes abruptly to a new value as the ions change to a new group. This effect is illustrated in Fig. 9.10 for substitution into the same composition as in the previous figure.

In most ceramics there are one or more crystalline phases, a glass phase, and pores. The glass phase is usually the continuous phase, and its composition and properties primarily determine the overall resistance to deformation. The effect of pores is to decrease the viscosity in direct proportion to the volume fraction of the porosity. Consequently, the high-temperature behavior is closely related to the microstructure and composition. The two extremes of ceramic behavior under stress at high temperatures are: (1) behavior typical of a viscous fluid (glass) with rigid particles dispersed in it, and (2) an aggregate of rigid particles which is bound together by a small amount of liquid (the binding force is due to surface tension). In both cases the creep rate is substantially higher than that which would be observed from a completely crystalline material.

Table 9.3   Network formers/modifiers

Element (M)	Coordination No.	Valence	MO bond strength, kcal/mol	
			Diatomic	MO$_x$(a) (single bond)
Formers				
Boron	3	3	193	119
Silicon	4	4	191	106
Germanium	4	4	158	108
Aluminum	4	3	120	79–101
Boron	4	3	193	89
Phosphorus	4	5	143	88–111
Vanadium	4	5	154	90–112
Arsenic	4	5	115	70–87
Antimony	4	5	89	68–85
Zirconium	6	4	182	81
Modifiers				
Scandium	6	3	161	60
Lanthanum	7	3	191	58
Yttrium	8	3	171	50
Tin	6	4	131	46
Gallium	6	3	68	45
Indium	6	3	86	43
Thorium	12	4	204	43
Lead	6	4	90	39
Magnesium	6	2	91	37
Lithium	4	1	81	36
Lead	4	2	90	36
Zinc	4	2	<66	36
Barium	8	2	135	33
Calcium	8	2	110	32
Strontium	8	2	111	32
Cadmium	4	2	<88	30
Sodium	6	1	61	20
Cadmium	6	2	<88	20
Potassium	6	1	58	13
Rubidium	10	1	61	12
Mercury	6	2		11
Cesium	12	1	72	10

(a) Source: W.D. Kingery, *Introduction to Ceramics*, John Wiley & Sons, 1965, p. 148. Single bond strength = dissociation energy per MO$_x$ divided by the coordination number

Example 9.5D. Assume the materials of Table 9.4 exhibit a temperature dependence of creep that follows an Arrhenius-type behavior expressed by the equation: ln ϵ = A − 74.277 (1000/T). What is the creep for ZrO$_2$ and for Al$_2$O$_3$ at 1800 °C? Note that the data in Table 9.4 are given for a temperature of 1300 °C.

Solution: For the two temperatures, calculate 1000/T. 1000/T = 0.4824 at 1800 °C and 1000/T = 0.6357 at 1300 °C. For ZrO$_2$: ϵ from Table 9.4 = 30 × 10^{-6} mm/(mm · h) at 1300 °C.

$$\ln \epsilon = -10.414 = A - 74.277 \, (0.6357)$$
$$A = 47.218 - 10.414 = 36.804$$

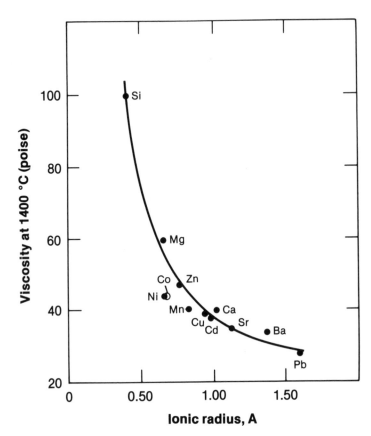

Fig. 9.9   Effect on viscosity of substitution of 8% of silica in a 74SiO₂-10CaO-16Na₂O glass by other divalent oxides on a cation-for-cation basis
Source: After A.F.G. Dingwall and H. Moore, *J. Soc. Glass Technol.*, Vol 37, 1953, p. 337

At 1800 °C:

$$\ln \epsilon = 36.804 - 74.277 \, (0.4824) = 0.99408$$
$$\epsilon = 2.7 \text{ mm/(mm} \cdot \text{h)}$$

For Al_2O_3: from Table 9.4, $\epsilon = 1.3 \times 10^{-6}$ mm/(mm · h)

$$\ln \epsilon = -13.553 = A - 74.277 \, (0.6357)$$
$$A = 33.665$$

At 1800 °C:

$$\ln \epsilon = 33.665 - 74.277 \, (0.4824) = -2.166$$
$$\epsilon = 0.11 \text{ mm/(mm} \cdot \text{h)}$$

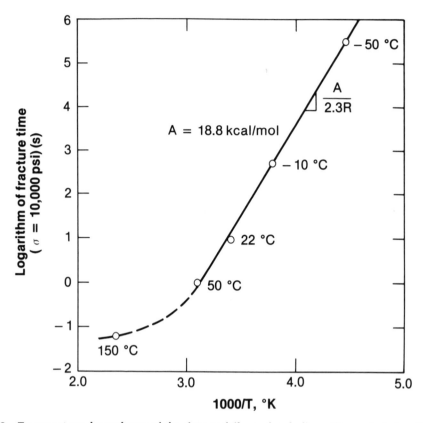

Fig. 9.10 Temperature dependence of the time to failure of soda-lime-silica rods in bending test

Table 9.4 Creep rates for various ceramic materials at 1300 °C

Material(a)	Rate, $\epsilon \times 10^6$ mm/(mm · h)
Al_2O_3	1.3
BeO	300
MgO (slip cast)	330
MgO (hydrostatic pressed)	33
$MgAl_2O_4$ (2–5 µm)	263
$MgAl_2O_4$ (1–3 mm)	1
ThO_2	1000
ZrO_2 (stabilized)	30

(a) Polycrystalline

Source: W.D. Kingery, H.K. Bowen, and D.R. Uhlmann, *Introduction to Ceramics*, 2nd ed., John Wiley & Sons, 1976

Example 9.5E. An Al_2O_3 bar used at 1300 °C and 12.4 MPa stress has a total strain of 1% during its service life of 320 days. Using the data of the preceding example, calculate the maximum operating temperature if ZrO_2 is substituted for the Al_2O_3 if it must not exceed 1% strain during 320 days of service.

Solution: A 320-day service life is consistent with the value of $\epsilon = 1.3 \times 10^{-6}$ mm/(mm · h) given in Table 9.4. Using the data of the preceding example, for ZrO_2:

$$\ln (1.3 \times 10^{-6}) = -13.553 = 36.804 - 74.277 \ (1000/T)$$
$$1000/T = 50.357/74.277$$
$$T = 1000/0.678 = 1475 \ °K \ or \ 1202 \ °C$$

9.6   Optical Properties

Until recently ceramics, and especially glasses, were the primary source of materials for optical uses such as windows, lenses, filters, optical fibers, etc. Today some of these applications are served by specialty polymers, although ceramics retain the bulk of the applications. In addition, the value of a ceramic material in many other applications is directly related to an optical property of the material such as color, transparency, and gloss. This is true for artware, tableware, porcelain enamels, and tiles. The primary properties that determine these many complex applications are the refractive index and the dispersion of the ceramic material.

Light travels through a particular medium at a velocity that is characteristic of the medium. The ratio of the velocity of light in vacuum to that in any other medium is the *refractive index* of that medium. The refractive index of a material is also a function of the wavelength, usually decreasing as the wavelength increases. The rate of change of refractive index with wavelength at any designated wavelength or frequency is called *dispersion*. Dispersion is usually expressed in terms of the reciprocal relative dispersion, v, a finite fraction evaluated at several common single wavelengths:

$$v = \frac{n_D - 1}{n_F - n_C}$$

The subscripted indices of refraction refer to measurements at specific frequencies: the sodium D line, the hydrogen F line, and the hydrogen C line (58,930, 48,610, and 65,630 nm, respectively). Typical values of refractive index for common materials are given in Table 9.5.

9.6.1   Reflection and Refraction

The passage of light from one medium to another is accompanied by a change in its velocity. The change in velocity when light passes from one medium to another is accompanied by two other phenomena: a change in its direction in the transmitted direction and a reflected wave originating at the interface between the media. The change of direction is related to the

Table 9.5   Refractive index of common substances

Material	Index of refraction (20 °C), 589.3 nm
Air	1.0003
Water	1.333
Optical glass (flint)	1.650
Optical glass (crown)	1.517
Fused silica	1.458
Silica, quartz	1.553
Silica, cristobalite	1.48
TiO_2, anatase	2.55
TiO_2, rutile	2.76
ZnS	2.37
ZnO	2.02
Lithopone	1.84–2.08
Clay	1.56
$Al(OH)_3$	1.58
Polycarbonate	1.59–1.60
Polystyrene	1.59–1.60
Polyvinylchloride	1.50–1.55

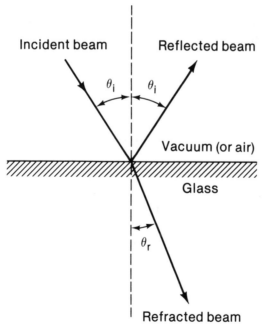

Fig. 9.11   Reflection of light at the surface of a transparent material occurs along with refraction

angle which the entering light makes to the plane bounding the media. Thus in Fig. 9.11 if the angle made to the boundary surface is i and the new angle the wave makes in the transmitted direction after refraction is r, then they are related by the refractive index:

$$n = \frac{\sin i}{\sin r}$$

This assumes that one of the media is either air or vacuum. A correction factor would be needed for other combinations.

The reflected light wave makes an angle of i with the normal to the boundary surface, exactly equal to the incident angle. The amount of light reflected increases with the value of the refractive index and is given by Fresnel's formula:

$$R = \left(\frac{n - 1}{n + 1}\right)^2$$

Although we like the sparkle caused by this reflection in decorative ceramics and glass, it is undesirable in many other ceramic objects—for example, lenses—and a number of techniques have been sought to minimize reflections.

9.6.2 Absorption

The second important optical property is absorption. As light passes through a medium, a certain fraction is absorbed in each layer of the medium through which it passes and converted into heat. This change in transmitted intensity due to absorption follows an exponential law:

$$T = \frac{I}{I_o} = \exp(-\alpha x)$$

or

$$\ln\left(\frac{I}{I_o}\right) = -\alpha x$$

where I_o is the initial intensity and I is the intensity after passing through the distance x. The absorption coefficient α is usually expressed in terms of reciprocal centimeters.

For the case illustrated in Fig. 9.11, the overall transmission loss is the combination of reflection and absorption losses. For normal incidence this becomes:

$$T' = \frac{I}{I_o} = (1 - R^2)\exp(-\alpha x)$$

The absorption coefficient is a strong function of the wavelength. Selective absorption (Fig. 9.12), which removes certain bands of frequencies from white light, gives the color which we perceive in most objects.

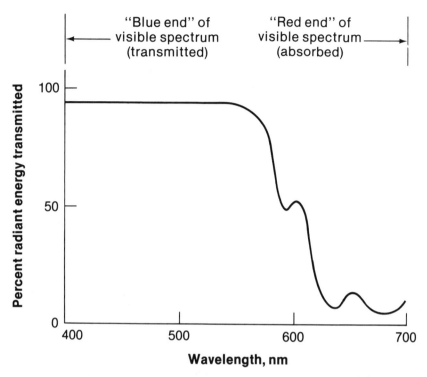

Fig. 9.12 Typical "absorption curve" for a silicate glass containing about 1% cobalt oxide
The characteristic blue color of this material is due to the absorption of much of the red end of the visible light spectrum. Source: J.F. Shackelford, *Introduction to Materials Science for Engineers*, Macmillan, 1985, p. 341

9.7 Optical Scattering

Very small particles exhibit a behavior similar to the phenomena of refraction and reflection; a portion of the incident beam is changed in direction, or *scattered*, some of the beam is absorbed, and the intensity of the transmitted beam is reduced by a combination of the two. In this case, however, the change in direction for the beam is no longer uniform and does not obey the simple expression derived above. It is, in fact, a complex function of the refractive index, absorption coefficient, and the size and shape of the particle.

The most familiar example of scattering is the reflection of light by the pigments used in paints. The fact that a white paint on a wall helps to illuminate the room with a pleasant diffuse light is due to the scattering of the incident light backwards into the room over a multitude of angles. Careful control of the particle size is needed to optimize this scattering. Because the relative index of refraction is directly related to the scattering coefficient, the particle size of pigment needed for "flat" paints, where the pigment is primarily interfacing with air, is different from that needed in "gloss" paint, where the pigment is completely embedded into a polymer matrix.

For thick pigment coatings, the reflectance is determined almost equally by the refractive index and the absorption coefficient. In porcelain enamels this opacifying scattering is often achieved by adding insoluble particles of zirconia or tin oxide to the glaze. In most house paints, the scattering pigment is titanium dioxide. These pigments provide the bulk of the scattering intensity and, in the case of house paints, reflect (scatter) back nearly all of the incident light. Addition of colored pigments provides the selective absorption necessary to achieve the desired colors, but at some sacrifice of intensity.

9.8 Problems

1. What are the differences between the melting point of a ceramic and a glass transition temperature? May a given ceramic material exhibit both a glass transition temperature and a melting point? Either? Neither? Explain.

2. Is the yield stress an important mechanical property for ceramic materials? Explain.

3. Why do ceramic materials have a much greater problem undergoing plastic deformation compared with most metals?

4. In a specimen of soda-lime glass, the stress at the tip of a 1.6 μm microcrack at the surface is 680 times as much as the nominal stress. Calculate the radius of curvature of the tip of the microcrack in nanometers.

5. A stress of 98.4 MPa is required to propagate a crack that is 3.4 μm long; the modulus of elasticity of the glass is 1.0×10^7 psi. Calculate the surface energy of the glass.

6. Why do measured tensile strengths of similar glass specimens vary rather widely?

7. Explain the toughening of glass in terms of the Griffith crack theory.

8. A specimen of glass has a viscosity of a million poise at 800 °C, a billion poise at 630 °C, and a trillion poise at 520 °C. At what temperature will the glass be in the annealing range, i.e., having a viscosity of 2.5×10^{13} poise?

9. Using the Arrhenius method of plotting viscous flow (the reciprocal of viscosity) versus 1/°K, calculate the activation energy for viscous flow from the slope of the curve in Fig. 9.2. Note that activation energies are always positive.

10. What is the difference between reflected and refracted light?

10

Structure of
Polymeric Materials

10.1   Introduction to Giant Molecules; Nature of Polymers

The technology of metals and of ceramic materials has been heavily dependent on the study and understanding of the arrangement of their atoms in crystalline structures and the imperfections of these structures. The class of materials called polymers is composed of giant or ''macro'' molecules, and their properties are heavily dependent on the configuration or structure of the individual molecules themselves, as well as on the ordering of the molecules into regular structures such as crystallites.

Polymers, less accurately known as plastics, are among the most commonly occurring materials in nature. For the past hundred years they have been synthesized by technologists from basic chemical raw materials. As the useful properties of these materials became widely known, synthetic polymers such as nylon, teflon, polyethylene, and vinyl (polyvinyl chloride, or PVC) have become familiar names. They are, in fact, used more extensively than metals in many common household objects.

The science of these materials, whether natural materials or the latest of the synthetic plastics, is concerned with the characteristics and behavior of large molecules. These large molecules are, in general, composed of long chains formed by joining many small molecules together. These giant molecules are composed of many (*poly*) repeated units (*mers*), i.e., *polymers*. The common name for the single (*mono*) molecule of these repeated units is *monomer*. In the case of some monomers, the molecule is repeated almost exactly in the polymer, with only slight changes due to the rearrangement of bonding as the units are joined end to end with each other. For example, the ethylene molecule

$$
\begin{array}{ccc}
\text{H} & & \text{H} \\
| & & | \\
\text{C} & = & \text{C} \\
| & & | \\
\text{H} & & \text{H}
\end{array}
$$

is the monomer from which the vinyl unit

$$
\{-\overset{\displaystyle \underset{|}{\overset{|}{H}}}{C} - \overset{\displaystyle \underset{|}{\overset{|}{H}}}{C} -\}
$$

in polyethylene is derived. The number of such repeating units, or mers, in a macromolecule or polymer chain is known as the *degree of polymerization*. A typical polymer will have hundreds or thousands of such units.

Relatively few elements are commonly found in polymer molecules (see Table 10.1). These comprise atoms of elements such as carbon, nitrogen, silicon, sulfur, and oxygen, which are capable of forming multifunctional covalent bonds (more than one bond per atom). Hydrogen, chlorine, fluorine, and similar atoms have only one covalent bond; these atoms bond to the above chain-forming atoms. The large number of atoms in a single polymer molecule provides opportunity for an exceptionally large variety of chemically different monomer and polymer molecules. Table 10.2 gives an example of some common monomers that are related to the simple ethylene (CH_2=CH_2) monomer by substituting one of the hydrogens of ethylene with another atom or group of atoms. Each of these derived monomers is capable of being joined (polymerized) into a long chain polymer. Thus the common plastic, polystyrene, is formed from the styrene monomer, polypropylene from the propylene monomer, polyvinyl chloride from the vinyl chloride monomer, and so on.

The nature of the polymer material can be intuitively related to the behavior of other chemical molecules by taking into account the effect of the very large molecular weight. For example, small organic molecules such as methane, ethane, and propane are normally gases,

Table 10.1   Elements commonly found in polymers

Element	Atomic weight	Typical bonds
Hydrogen	1.0079	—H
Carbon	12.001	=C⟨ or ⟩C⟨
Nitrogen	14.0067	=N⟨ or —N⟨
Oxygen	15.9994	=O or —O—
Fluorine	18.9984	—F
Silicon	28.0855	=Si⟨ or ⟩Si⟨
Sulfur	32.06	=S or —S—
Chlorine	35.453	—Cl

Table 10.2 Structures of selected vinyl polymers

$$\begin{bmatrix} 1 & 3 \\ C-C \\ 2 & 4 \end{bmatrix}_n$$

Monomer	1	2	3	4(a)	International abbreviation
Polyvinyls (general)	H	H	H	—R	
Ethylene	H	H	H	—H	PE
Vinyl chloride	H	H	H	—Cl	PVC
Styrene	H	H	H	—⬡	PS
Propylene	H	H	H	—CH_3	PP
Vinyl acetate	H	H	H	—$OCCH_3$ (O‖)	PVAc
Acrylonitrile	H	H	H	—CN	PAN
Vinylidene chloride	H	H	—Cl	—Cl	PVDC
Methyl methacrylate	H	H	—CH_3	—$COCH_3$ (O‖)	PMMA
Isobuytlene	H	H	—CH_3	—CH_3	PIB
Tetrafluoroethylene	F	F	F	F	PTFE

(a) R represents an aliphatic chain. ⬡ represents an aromatic ring derived from benzene.
Source: L.H. Van Vlack, *Materials for Engineering*, Addison-Wesley, 1982, p. 201

whereas larger hydrocarbon molecules, like those produced by petroleum refineries for gasoline and heating oil, are typically liquids. When the chains become longer still, the materials begin to take on the properties of solids. They may crystallize and become a material with a true melting point, as is the case with some very highly purified paraffin waxes. In many cases, however, the process of arranging into a highly regular crystalline structure is impeded by the need to move these extremely long molecular chains around in space to form the structures. This process may be visualized by comparing the ease with which a collection of marbles can be jostled into a well-ordered structure with a minimum number of voids and defects to the difficulty of making an orderly arrangement of the strands in a bowl of spaghetti. This is a fundamental difference in the behavior of polymers compared with other materials.

10.2 Natural Polymers

Many of the materials occurring in nature are polymeric—for example, proteins and starches, and the cellulose fiber of cotton. Because they are derived from relatively complex monomers, often by complicated polymerization reactions, their polymeric nature is not always recognized. Often they do not fit the image which is ordinarily held of polymers—namely, they do not appear to be plastics.

Typical of such natural polymers is the starch molecule. This is a long-chain polysaccharide; it is composed of a combination of many molecules of a simple sugar monomer. The materials which constitute our muscles are polymers formed from simple amino acid monomers.

The fibrous nature of the muscle tissue is itself aided by the length of the slender protein molecules.

10.3   Man-made Polymers

The relatively abundant raw materials occurring in petroleum can be transformed into a variety of monomers which are subsequently polymerized. Because of the control over the nature of the monomer and of the polymerization process, many different polymers have been produced, each with its own individual properties. This has led to the production of plastics and other polymers which are custom designed to fit specific end needs.

Probably the most commonly known man-made polymer is nylon. This polymer was synthesized in the early 1940s by the E.I. du Pont Company and found wide use during World War II as a substitute for silk, which was no longer available from the Orient. The development of nylon illustrates a common occurrence: the replacement of a naturally occurring polymer by a man-made polymer.

An even earlier man-made polymer used in fiber for clothing was the chemical modification of the naturally occurring polymer in cotton. By a chemical process the polymer was put into solution, modified, and regenerated as the material known as rayon. This process was known as the viscose process, hence the name viscose rayon.

At the present time there are over 10 billion pounds of man-made fibers produced in the United States, and nearly 30 billion worldwide. A listing of some representative materials is given in Table 10.3.

Similarly, the production of synthetic polymers for structural and mechanical applications has led to a proliferation of materials. Sometimes the same polymer types used in fibers for textiles have found widespread applications elsewhere. In other cases, unique properties of the polymer have been responsible for its widespread use. For example, the polycarbonates have been widely used as a tough, unbreakable, optically clear material for window and other glazing applications. A number of common synthetic engineering polymers are listed in Table 10.4, along with representative application areas.

Table 10.3   World noncellulosic fiber capacity

Fiber type	Production capacity (1988–89), 1000 tons
Acrylic-modacrylic	3,099
Nylon-aramid	4,746
Polyester	9,941
Olefin (estimated)	2,192
Other fibers	265
Total fibers	
Capacity	20,243
Estimated production	17,206

Source: *Textile Organon*, June 1988

Table 10.4   Selected engineering polymers

Polymer type	Estimated U.S. sales (1989), million lb
Nylon	600
Polycarbonate	650
Polyacetal	150
Epoxy	500
Urethane (excludes flexible and insulating foam)	650
Acrylics	750
Polyester (unsaturated)	1350
Polyester (thermoplastic)	2100
Polymer/alloys	350
Others	1050
Total (excludes glass content of reinforced materials)	**8150**

10.4   Importance of Polymers to Modern Society

The reader no doubt already has an appreciation of the impact of synthetic polymers on everyday life. These materials also are extremely important in improving productivity and raising the standard of living.

Polymeric materials are used in many commercial products for two very important reasons: (1) they allow special designs to be fabricated, and (2) they can be processed with great ease and speed to highly decorative finished products that need few extra fabrication steps. Many polymers become fluid when heated. This allows the material to fill a mold of very complex shape; upon cooling, the polymer becomes a structurally strong object that reproduces the shape of the mold to close dimensional tolerances. When complex shapes are constructed of metal, they are usually made of several subassemblies that must be fastened together. The molding of plastics has allowed designers great freedom to create new designs which can be produced economically. For example, the ability to design plastic moldings of great complexity has allowed automotive manufacturers to reduce the number of assembly steps in the modern automobile, creating a great cost saving. In addition, the reduction of weight from substitution of plastic for metal has been a major contributor to the fuel economy of current automobiles.

These advantages have become such a strong driving force behind the introduction of polymers in engineering applications that polymers have almost completely replaced metal in household appliances, office equipment, and many other manufactured goods. The total use of polymers in the United States has grown immensely and totals over 58 billion pounds, distributed over 24 major types of materials, as shown in Table 10.5. Because of the large difference in the average densities of polymers and metals, the total volume of polymers in cubic feet probably exceeds the volume of all metals used each year.

10.5   Structure of Polymer Chains

Polymers can be formed by linking together either synthetic or natural monomers into macromolecules. When only a single identical monomer is used, the repeating units (mers) form a *homopolymer*. If more than one kind of monomer is combined into each chain of the polymer,

Table 10.5   U.S. polymer sales (1988–89)

Material	Million lb	
	1988	1989
ABS	1,279	1,243
Acrylic	697	739
Alkyd	320	325
Cellulosics	90	91
Epoxy	468	492
Nylon	574	595
Phenolic	3,053	3,162
Polyacetal	138	141
Polycarbonate	589	622
Polyester, thermoplastic (PBT, PCT, PET)	2,043	2,101
Polyester, unsaturated	1,373	1,325
Polyethylene, high density	8,084	8,115
Polyethylene, low density(a)	10,947	10,636
Polyphenylene-based alloys(b)	172	176
Polypropylene and copolymers	7,088	7,246
Polystyrene	5,027	5,184
Other styrenics(c)	1,100	1,170
Polyurethane	3,234	3,245
Polyvinyl chloride and copolymers	8,287	8,307
Other vinyls	950	960
Styrene acrylonitrile (SAN)	145	137
Thermoplastic elastomers	491	542
Urea and melamine	1,515	1,367
Others	288	307
Total	**57,952**	**58,228**

(a) Includes LLDPE. (b) Includes modified phenylene oxide and modified phenylene
ether. (c) Excludes ABS and SAN
Source: *Modern Plastics*, Jan. 1990

then the material is referred to as a *copolymer*. Copolymers can have their monomers arranged in a regular alternating structure, arranged randomly, or arranged in a block structure. A *block copolymer* will have its monomers arranged as though short segments of homopolymer were attached together.

For a given set of monomers—monomer A, monomer B, and monomer C—we can visualize these polymer types as follows:

- *Homopolymers*
  AAAAAAAAAA or BBBBBBBBBB or CCCCCCCCCC
- *Alternating copolymers*
  ABABABABABAB or ACACACACACAC or BCBCBCBCBCBC
  ABCABCABCABC or ACBACBACBACB
- *Random copolymers*
  AAABAAAAABAAAA or AAAAAACAAACACAA or AAACAAABABACCAAAA
- *Block copolymers*
  AAAAAAAAAABBBBBAAAAAAAAAABBBBBAAAAAAAAAABBBBB
  AAAAAAAAAACCCCCCCCCBBBBBBBBBBAAAAAAAAAAACCCCCCCC

These structures have been illustrated in terms of simple linear molecules only. It is easily possible to have intricate treelike structures formed in each of these categories. All of the above molecules with the individual mers arranged in a single sequence (as though they could be stretched in a single line) are called *linear* (e.g., linear homopolymer or linear copolymer).

* *Branched copolymer*

```
                          BBBBAAAAAABBBBB
                          B
      AAAAAABBBBBAAAAAABBBBBAAAAAA
                          B
                          B
                          BBBAAAAAABBBBBAAAAAA
                                  A
                                  AAAAABBBBB
```

If the complexity of the arrangement of monomers in a copolymer is considered in light of the number of chemically different monomers available, it is obvious that many thousands of chemically different macromolecules can be formed. Each of these combinations can be slightly different in its properties because the chemical structure is slightly different. This great versatility of polymer science is the basis for the customization of materials for specific end-use properties.

For a given pair of monomers combined in an alternating linear molecule, there are only three possibilities, depending on which groups are on the end of the chain:

$$A \, . \, . \, B \, . \, . \, . \, A \, . \, . \, B \, . \, . \, . \, A$$
$$B \, . \, . \, . \, A \, . \, . \, B \, . \, . \, . \, A \, . \, . \, B$$
$$B \, . \, . \, . \, A \, . \, . \, B \, . \, . \, . \, A \, . \, . \, B \, . \, . \, . \, A$$

Since the chain itself comprises hundreds or thousands of repeating units, these differences are unlikely to be reflected in the material properties. On the other hand, if we have ten varieties of A monomer to choose from and ten varieties of B monomer to choose from, we can synthesize over 100 chemically different polymer molecules, each of which will have different material properties. For random copolymer structures, many thousands of chemically different polymer molecules are possible.

Further diversity is possible in branched polymers due to the arrangement of the branching along the primary polymer chain. If the branches are randomly positioned along the chain, the polymer is *atactic* (Fig. 10.1a); if they are regularly positioned in the same plane along one side of the primary chain they are *isotactic* (Fig. 10.1b); while if they alternate from one side of the primary chain to the other they are called *syndiotactic* (Fig. 10.1c).

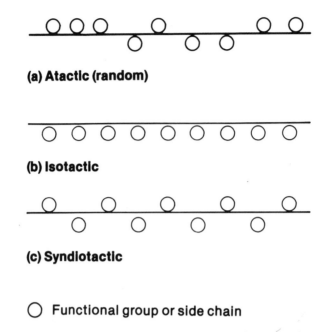

(a) Atactic (random)

(b) Isotactic

(c) Syndiotactic

○ Functional group or side chain

Fig. 10.1   Atactic, isotactic, and syndiotactic polymer configurations

10.6   Synthesis of Polymer Chains

The field of polymer synthesis is concerned with the chemical reaction of monomers in such a way that these various repeat structures are formed in a controlled and known manner. Normally chemical reactions are a statistical process (that is, dependent on the statistics of two reacting components coming together). The task, however, of causing a controlled, orderly combination, such as illustrated by a regular copolymer, requires special conditions.

The synthesis of polymers, i.e., the controlled reaction of monomers to form polymer molecules, is called *polymerization*. There are two basic types of reaction used to form polymers: *addition polymerization* and *condensation polymerization*. Other commonly used synthesis conditions are usually a combination or modification of these two basic types.

10.6.1   Addition Reactions

Addition reactions are those in which the monomers under suitable conditions react by adding one new unit on the end of the growing polymer molecule until reaction ceases. The process can be visualized as a moving group of skaters joined hand in hand in a line; the last person in the line reaches out his or her free hand to any other free skater. Little by little, all of the skaters on the pond are joined into one big linear chain. The process ceases when there are no more free skaters (monomers) to pick up. It effectively stops when the time it takes the long chain to contact another free skater (monomer) becomes very long.

In actual polymerization, the polymer chain maintains a highly reactive, unsatisfied bond on the end of the chain, similar to the free hand of the skater. It is this reactive site, called a *free radical*, which keeps the growth of the chain going. As each monomer is added, the new unit becomes the one with the free radical (unsatisfied bond) and growth continues. Occasionally other impurity molecules in the reactor can meet the free radical and react with it completely without themselves regenerating an active site. This is analogous to someone handing an object to the skater on the end of the chain. His or her free hand is now occupied and the chain no longer grows.

The probability of continued chain growth, compared to the probability of stopping the chain by termination with impurities, is simply related to the relative concentration of impurity molecules compared to monomer molecules. If there is one active growing chain in a group of 1000 monomer molecules, the chain will grow to become 1000 units long (degree of polymerization of 1000). If there is one impurity molecule per every 100 monomer molecules, the chains will *on the average* become only 100 units long (degree of polymerization of 100). It is important that ultrapure monomers are used in polymerization reactions in order to achieve the desired molecular weight. It is also true that careful metering of just the right concentration of a known impurity into an ultrapure monomer can be used to predetermine the degree of polymerization.

10.6.2 Condensation Reactions

Condensation reactions generally refer to those types of polymer formation when two, usually different, molecules react together to form a bond plus a third small molecule. When organic acids react with organic alcohols to form an ester, the formation of the ester is accompanied by elimination of a molecule of water.

$$
\begin{array}{ccc}
& O & O \\
& \parallel & \parallel \\
R\ldots C-OH + R \ldots C-OH \rightarrow R \ldots C-O-C \ldots R + HOH
\end{array}
$$

Such a reaction can only lead to a polymer if there is a reactive site on both ends of the monomer. Thus many of the monomers used in condensation polymerization are *bifunctional*.

There are several choices for the bifunctional mers used in the formation of linear condensation polymers. For example, the reaction may involve (1) a molecule having an acid radical on each end with a molecule having an alcohol or amine group on each end, or (2) it could be with a monomer having an acid group on one end and an alcohol or amine group on the other end of the same molecule.

$$
\begin{array}{l}
HOOC \ldots R1 \ldots COOH + HO \ldots R2 \ldots OH \rightarrow \\
\quad HOOC \ldots R1 \ldots COO \ldots R2 \ldots COO \ldots R1 \ldots COO \ldots R2 \ldots OH + \text{water}
\end{array}
$$

or

$$
\begin{array}{l}
HOOC \ldots R \ldots OH \rightarrow \\
\quad HOOC \ldots R \ldots COO \ldots R \ldots COO \ldots R \ldots COO \ldots R \ldots OH + \text{water}
\end{array}
$$

Some of the monomers used in polycondensation reactions and their typical polymers are given in Table 10.6.

Table 10.6   Condensation polymers

Type	Link unit	Applications
Polyacetal	$\begin{array}{c} \text{H} \\ | \\ -\text{O}-\text{C}-\text{O}- \\ | \\ \text{R} \end{array}$	Molded parts for industrial equipment, business machines, and appliances
Polyamide	$\begin{array}{c} \text{O} \\ \| \\ -\text{C}-\text{NH}- \end{array}$	Nylon, silk for fabrics; molded parts for appliances, business machines, and industrial equipment
Polyester	$\begin{array}{c} \text{O} \\ \| \\ -\text{C}-\text{O}- \end{array}$	Molded parts for high-temperature electrical applications; automotive parts for under-hood applications
Polyurethane	$\begin{array}{c} \text{O} \\ \| \\ -\text{O}-\text{C}-\text{NH}- \end{array}$	Elastomeric tubing and foam, fibers
Cellulose	—O—	Cotton for fabrics
Phenol-formaldehyde	—CH— and Ar—CH—O—CH—Ar cross-links (Ar = o or p substituted aromatic ring)	Electrical equipment
Polysiloxane	$\begin{array}{ccc} \text{R} & & \text{R} \\ | & & | \\ -\text{Si}-&\text{O}-&\text{Si}- \\ | & & | \\ \text{R} & & \text{R} \end{array}$ and $\begin{array}{ccc} \text{R} & & \text{R} \\ | & & | \\ -\text{O}-\text{Si}-&\text{O}-&\text{Si}- \\ | & & | \\ \text{O} & & \text{R} \\ | \\ \text{R}-\text{Si}-\text{R} & \text{cross-links} \\ | \\ \text{O} \end{array}$	Silicone rubbers for gaskets, adhesives, where resistance to severe chemical and temperature environments is needed

Both homopolymers and copolymers can be formed by condensation reaction. However, for practical reasons the production of condensation polymers usually forms copolymers, as illustrated in the first reaction sequence.

10.7   Effect of Polymer Molecule Size on Physical Properties

The properties of polymeric materials are highly dependent on the length of the molecule. For example, simple hydrocarbon molecules such as propane and butane are volatile liquids, whereas the longer molecules present in paraffin result in low volatility and higher melting points for these hydrocarbons. Similarly, the properties of polymeric materials that have short chain lengths vary in melting points, mechanical properties, and even stability from those with very long chains. For relatively small molecules, many physical properties are related to molecular weight. For polymeric materials, however, other measures of polymer chain lengths are useful besides molecular weight.

10.8   Chain Lengths, Molecular Weight, and Degree of Polymerization

Because polymeric materials are composed of a series of small molecules attached together to form a long chain, an important measure of the length of the chain is the number of units that have been joined together. It is impractical or impossible literally to count the number of repeating units or mers in an individual polymer molecule, which is called the *degree of polymerization* (n). The degree of polymerization is obtained, however, from two quantities that can be easily measured—the molecular weight of the polymer and the molecular weight of the mer (sometimes called mer weight):

$$n = \frac{M}{m} = \frac{\text{Molecular weight (polymer)}}{\text{Molecular weight (mer)}}$$

If we know that a polyethylene molecule has a molecular weight of 42,000 g/mol and that the mer unit (CH_2CH_2) has a molecular weight of 28, then the degree of polymerization, n, of the polyethylene is 42,000/28 or 1500 units (mers) per molecule.

Note that in this case the molecular weight of the mer unit is the same as the molecular weight of the monomer molecule. This is not always the case. For example, condensation polymers have a different chemical formula in the repeating units in the chain from the formula of the separate monomers used to form the chain: $NH_3(CH_2)_6COOH$ is the monomer for a certain polyamide (variety of nylon polymer); $-NH_2(CH_2)_6CO-NH_2-(CH_2)_6CO-$ is the repeat structure for two units in the condensation polymer. Note that the molecular weight of the monomer is 208 g/mol, while the weight of a single mer unit is 190 g/mol due to the loss of a molecule of water during the formation of the polymer.

10.9   Number Average and Weight Average Molecular Weights

The use of a variety of types of mers and the random or statistical nature of chain growth in polymerization processes produces many different polymer chains in a material. Differences always include the differences in the size of the molecules, but the structure of the individual molecules may vary as well. Even in the simplest linear polymer, the polymer material consists of a variety of chain lengths. Each chain is, of course, a separate molecule with its own unique molecular weight. But for the material as a whole, there is no one unique molecular weight which describes its molecules. It is necessary somehow to count the molecules of differing length and determine an average molecular weight to characterize a polymer.

If a given polymer consists of 1500 units of ethylene mers, CH_2CH_2, each of which has a unit weight of 28 ($2 \times 12 + 4 \times 1$), then the molecular weight of the polymer chain is 42,000 g/mol. Now suppose the material was actually composed of 150 molecules with degree of polymerization of 1000, 700 molecules with degree of polymerization of 1500, and 150 molecules with degree of polymerization of 2000. The molecular weights of the three components are 28,000, 42,000, and 56,000, respectively. An average of these molecular weights is 42,000; however, it is very time consuming to determine the average molecular weight by summing the weight of each single molecule and dividing by 1000. A shortcut may be made by taking the numerical fraction of each class times its molecular weight to determine a *number average molecular weight*. In this example, there are a total of 1000 molecules, and the number fraction of molecular weight 28,000 is 0.15, that for molecular weight 42,000 is 0.7, and for molecular weight 56,000 is 0.15. The number average molecular weight is then ($0.15 \times 28,000$) + ($0.7 \times 42,000$) + ($0.15 \times 56,000$) = 42,000. Note that this last calculation was made on the basis of exactly 1 mole of polymer, whose weight is 42,000 g.

An alternative method of calculation takes into account the *weight* of the various fractions, rather than the *number* of molecules in each fraction. The calculations for *weight average molecular weights* are given by the generalized formula:

$$M_w = \sum W_i M_i$$

where W_i is the weight fraction and M_i is the molecular weight of each component i. In the example above, on the basis of a total of 1 mole, the weight of the first fraction is 0.150 mole $\times$ 28,000 g/mol = 4200 g. The weights of the other fractions are 0.700 mole $\times$ 42,000 g/mol = 29,400 g and 0.150 mole $\times$ 56,000 g/mol = 8400 g, respectively. The weight of the sample of 1 mole of polymer is 42,000 g. Dividing the weight of each group by the total weight of the sample gives the *weight fraction*, W_i, for each group. We can now average them by taking the weight fraction of each group times its molecular weight to achieve a weight average molecular weight. In the example, the weight fractions are 0.100, 0.700, and 0.200, respectively. The weight average molecular weight of the sample is then ($0.100 \times 28,000$) + ($0.700 \times 42,000$) + ($0.200 \times 56,000$) = 43,400. Notice that the weight average molecular weight gives *greater* emphasis to the *heavier* molecules and produces a *higher* average molecular weight than does the number average. Note also that $\Sigma W_i = 1.000$.

Figure 10.2 illustrates this difference graphically where there is a greater range of sizes. If a similar calculation is carried out, the number average molecular weight is found to be less than the weight average molecular weight. *This is always true*, except for the ideal case where all molecules are the same size, in which case the two are equal. One can learn something about the range or dispersion of molecular sizes by comparing these two numbers. The greater the difference between them, the more disperse are the molecular sizes. An index of this difference, which is calculated as the ratio of the weight average to the number average molecular weight is called the *polydispersivity index* (PDI):

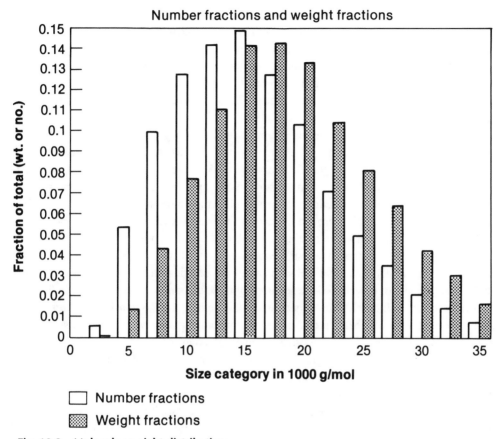

Fig. 10.2   Molecular weight distributions

$$PDI = \frac{M_w}{M_n}$$

The PDI typically ranges from near 1 for highly monodisperse polymers to 5 or greater for typical commercial polymers. A large value of PDI represents a larger quantity of relatively small molecules which may adversely affect properties such as tensile strength or heat distortion.

10.10   Spatial (Shape) Considerations

The C—C bond is the basic structure of most polymers. The nature of the C—C bond determines much of the way a polymer behaves spatially. An understanding of the shape of carbon-based polymer chains helps in understanding how differences in the basic bond in other polymers—e.g., Si—Si bonds—influences the behavior of those polymers.

A carbon atom has a maximum of four pairs of shared electrons, hence four bonds. Carbon atoms are capable of sharing electrons with other carbon atoms in C—C bonds, which enables the growth of extended chains of C—C bonds in polymers. Normally the other bonds about the carbon atom are formed with nonmetallic elements, such as hydrogen, chlorine, and fluorine, which form only a single bond with the carbon atom. Other elements that are frequently found in polymers are oxygen, sulfur, and nitrogen. These elements can form multiple bonds with carbon and other atoms.

The bonds of the carbon atom are arranged such that they point toward the corners of a tetrahedron with the carbon atom in the center (see Fig. 10.3). As a result, the angle between any two bonds of a given carbon atom is normally 109.5°, a fact which influences the molecular geometry of polymer molecules. Furthermore, if we look at a pair of carbon atoms joined together, we find that the tetrahedron of the second carbon atom is not fixed in space relative to the first but can rotate freely around the axis of the bond between them, as illustrated in Fig. 10.4. Thus the positions in space of a large number of carbon atoms bound to each other in a chain are not rigidly predetermined, but are randomly placed, within the constraints of remaining attached. The effect of thermal agitation causes a continual redistribution of these positions. The greatest amount of ordering of the positions of these carbon atoms is found when the chains are arranged (usually linearly) into crystallites, but even here there is considerable randomness. This somewhat statistical behavior of polymeric materials uniquely influences many of their properties and sets them aside from other, simpler materials.

To determine the length of a large polyethylene or other vinyl polymer chain of n units we could use the length of the individual C—C bonds in the individual unit and multiply by the number of units, n. The length of a polyethylene chain of 1500 units would be calculated to be 462 nm since each of the 3000 C—C bonds is 0.154 nm long (see Table 10.7). This calculation, however, is based on the false assumption that each of the mers is attached in a simple straight line, much as we have written them on paper. In the real world, the C—C bonds at a single atom are at an angle of 109.5° (Fig. 10.4). This results in a zigzag orientation for the atoms in the chain, and the resulting length of the molecule is shorter than 462 nm.

$$C — C — C — C — C — C — C \text{ (incorrect assumption)}$$
$$1$$

becomes

a   b   c 2d

The zigzag configuration is taken into account by a simple trigonometric correction whereby the distance, 2d, from a to c is $2 \times \sin(109.5/2°) \times 0.154$ nm $= 0.252$ nm. Thus a single C—C length is 0.126 nm and the length of the 3000 bonds in the polyethylene chain is now calculated to be 377.0 nm—nearly 20% shorter than for the naive in-line assumption.

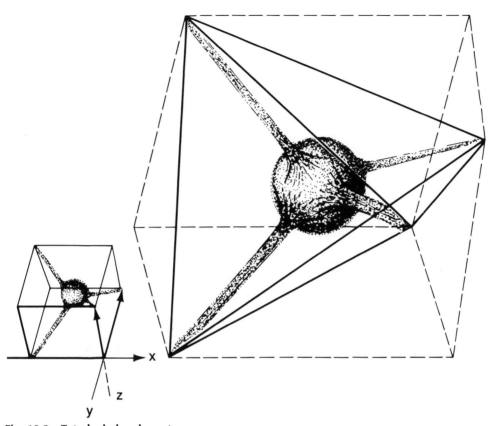

Fig. 10.3 Tetrahedral carbon atom
The four bonds of the carbon atom define a regular tetrahedron. The four corners of this tetrahedron occupy one-half of the positions of a rectangular lattice, as shown by the dashed lines.

Note that even this length is a highly idealized one. Because the single bonds in most molecules are free to rotate, the chain only occupies this length in space when completely stretched out. Normally the freedom of rotation will allow each bond to assume a different rotational angle, and the polymer chain will actually be highly kinked. This results in a more tangled shape as with a ball of yarn and a much shorter end-to-end distance. These differences are significant in determining the mechanical properties of various polymers. For example, the capability of rubber to undergo a high degree of elastic stretching is the result of uncoiling these kinked and coiled molecules under stress. When the stress is removed, the original shape of the material is regained due to the recoiling of the chains. Examples of this random orientation of carbon atoms in a polymer chain are given in Fig. 10.5.

The length of the freely rotating polymer chain can be calculated using statistical methods to take into account the random distribution of bond rotation. The maximum distance between any three carbon atoms in the chain is the 0.252 nm shown above. Statistics show that on the average a shorter length, defined as the *root-mean-square length* (L), represents the distance

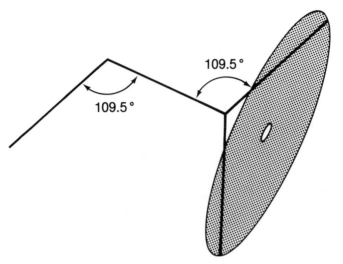

Fig. 10.4   Rotating carbon-carbon bond

Table 10.7   Lengths of commonly occurring bonds

Atom pair	Length, nm	Atom pair	Length, nm
C—C	0.154	C═C	0.13
C—H	0.13	C≡C	0.12
C—N	0.15	C═O	0.12
C—F	0.14		
C—Cl	0.18	O—H	0.10
		O—O	0.15
H—H	0.074	O—Si	0.18
N—H	0.10		
N—O	0.12		

between the ends of the chain. L is calculated from the number of units, n, and the length of a single unit, l, as:

$$L = l \times \sqrt{2n}$$

We use 2n because two bonds are associated with each mer unit (—C—C).

In the example above, the average (root mean square) end-to-end distance of the polyethylene molecule with a bond length of 0.154 nm and 1500 units is calculated to be L = 8.435 nm. Compare this with the 462 nm for a completely linear molecule in the above example or 377 nm for the stretched-out kinked molecule!

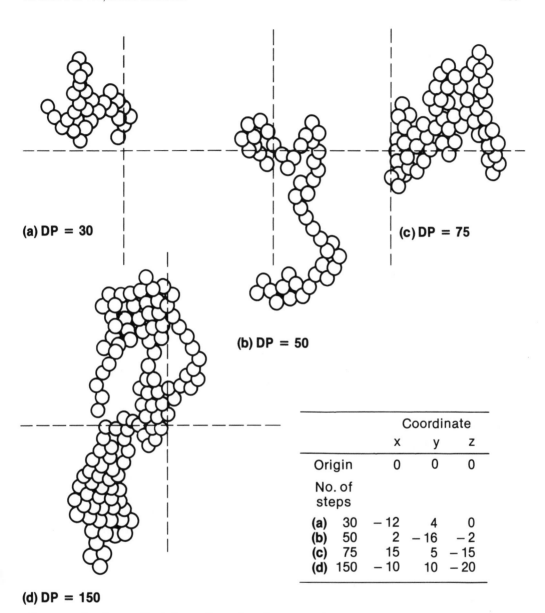

(a) DP = 30

(b) DP = 50

(c) DP = 75

(d) DP = 150

| | | Coordinate | |
	x	y	z
Origin	0	0	0
No. of steps			
(a) 30	− 12	4	0
(b) 50	2	− 16	− 2
(c) 75	15	5	− 15
(d) 150	− 10	10	− 20

Fig. 10.5 Random walk chain configurations for 30 to 150 steps

10.11 Branching

The calculations just described apply to a polymer chain that is relatively simple in configuration—a linear arrangement of mers in a giant molecule, with freely rotating bonds. However, real polymers usually have branched chains. Obviously the calculation of end-to-end distance as described above will no longer apply since there are more than two ends in a branched polymer. Branching is significant in determining the level of order (e.g., crystalline order) which the molecules can assume, and therefore branching influences the mobility, crystallinity, and other physical properties of polymers. The ultimate branched polymer is one in which each branch is terminated by being attached to another portion of the chain, a process called *cross-linking*. When there are a sufficient number of cross-links, a giant three-dimensional network is established. In these highly cross-linked or *network* polymers, the end-to-end distance has no simple meaning since the whole material is one molecule.

The concepts of end-to-end distance and average number of units between branch points are useful even in considering cross-linked and network polymers. Sometimes the material can be considered as though it were a collection of long linear molecules with a given number of attachments or cross-links per chain. Cross-linking modifies the physical properties which would normally be expected for a polymer of a given type. Often the degree of cross-linking may be correlated with physical property changes. Cross-linking is a valuable way of modifying polymeric materials to alter their properties in a controlled manner. Frequently the degree of cross-linking or degree of branching can be estimated from the chemistry of the original synthesis mixture. For example, specific chemicals can be added to a polymerization mixture to cause cross-linking, and the degree of cross-linking is proportional to the mole fraction of the cross-linking agent.

10.12   Microstructures Formed From a Combination of Many Polymer Molecules

Polymeric materials can simultaneously exhibit both the properties of a conventional ''solid,'' such as a metal below its melting point (for example, elasticity), and those of a liquid, such as viscous flow. This results from the various types of microstructures that are developed in polymeric solids as well as the nature of the chains themselves. This property of exhibiting both elasticity and viscous flow in response to stress at a given temperature is called *viscoelasticity*.

As shown in Fig. 10.5, a single chain of a linear thermoplastic polymer may assume many configurations in space, ranging from the completely extended chain to a tighter configuration resembling a ball of yarn. In a dilute solution of a polymer in a suitable solvent, the individual polymer chains can be widely separated. Under such conditions the individual chains do tend to take on the configuration illustrated. The idea of the freely rotating bond led to the calculation of the average end-to-end distance and the generally spherical shape with a characteristic radius. It was assumed that there was enough kinetic energy from surrounding molecules of gas or liquid to enable the chain to reach this equilibrium shape. In a solution there is not just one single shape for a polymer with a corresponding end-to-end distance as calculated above, but a distribution of shapes depending on the nature of the solvent in which the polymer is dissolved, the nature of the polymer chain itself, and the temperature of the system.

As in all substances, polymer chains tend to assume a configuration of minimum energy. When dissolved in a solvent with little interaction between the solvent and the polymer molecule, the polymer chain tends to form itself into the idealized shape illustrated in Fig. 10.6. If, for example, there is a great difference between the chemical nature of the polymer and the solvent, energy is minimized when the materials are separated into two phases. In such case, the polymer does not dissolve at all. When the difference is not so great, only very dilute solutions

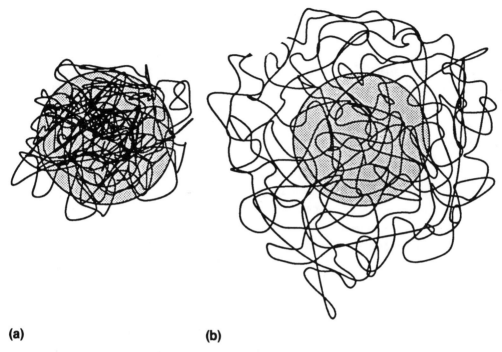

(a)   **(b)**

Fig. 10.6   Polymer chains in solution
(a) In dilute solutions, in poor solvents, or in solvent mixtures near the φ temperature, the chain occupies an "idealized" configuration, as though it were not in solution. The end-to-end distance is that which would be calculated for a molecule in free space. (b) In a good solvent the interaction with the solvent may lead to an extended configuration.

can be formed; this is called a "poor solvent." In poor solvents the individual polymer molecules behave as though they were free in space, and the overall dimensions of the chains are determined by bond angles and bond lengths. This occurs at a fixed temperature, called the φ temperature, because the solvent-polymer interaction energies are highly temperature dependent. Measurements of properties made in dilute solutions at the φ temperature will be those of the normal undisturbed molecule as if it were a statistical entity free in space. At this temperature, interactions between solvent and chain segments are near zero.

It requires a much greater energy of polymer-solvent interaction for the chains to achieve a more extended configuration. The configuration of the polymer is, therefore, a function of the temperature. When a solution is heated, the higher thermal energy allows the chains to uncoil a bit, increasing the apparent diameter of the coiled shape. One may therefore expect a dramatic change in the properties of a polymer solution on heating. Solutions of polymers, for example, have a much lower viscosity when hot. This is useful in processing—for example, in the casting of plastic films from solution, where the low viscosity characteristic of a hot solution of the polymer allows it to flow easily through the casting equipment and form a very thin layer.

Obviously various solvents will interact differently with a given polymer. There is little energy difference between the tightly coiled state and the more loosely coiled structure for a polymer in a solvent that resembles the polymer chain itself. For example, polyethylene feels right at home in a hydrocarbon solvent, such as hexane, and the polymer chains extend even beyond the idealized statistical size. In the same way, a polymer chain with a more polar chemical structure, such as polyvinyl alcohol, will feel more at home in a solvent such as methanol and will become more tightly coiled as increasing amounts of nonpolar solvent are added to the solution. Eventually, at a certain critical point the polyvinyl alcohol polymer will form a second macrophase and precipitate out of solution.

If a given polymer were comprised of chains of only a single length, then the behavior of the polymer in solution could be treated in exactly the same way as has been established for binary solutions of small molecules. A temperature-composition phase diagram could be drawn, and most compositions would lie in a two-phase region having compositions indicated by the tie line. The maximum in this curve defines a critical point for the system, and at temperatures above the critical temperature the system is homogeneous regardless of composition (Fig. 10.7).

For solutions of polymers that have a distribution of molecular weights, a similar phase diagram may be constructed. Because the system has more than one component, the maximum temperature for the coexistence of two phases is not a true critical point since it will vary with the composition of the molecular sizes in the polymer. Whenever precipitation of polymer is caused by addition of a second solvent, at least three components are present in the system; it must then be represented by a conventional type of ternary diagram, as shown in Fig. 10.8.

A linear polymer in solution may be thought of as existing as a dispersion of very tiny balls. For a given polymer, the size of the individual balls is influenced by the temperature and the

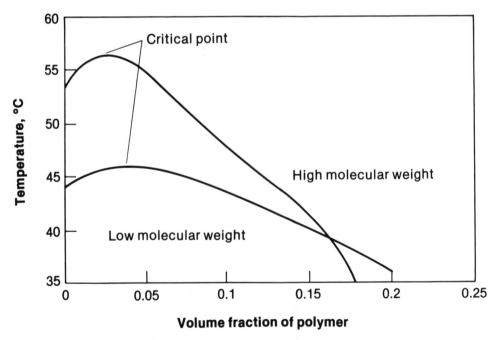

Fig. 10.7 Variation of critical point with molecular size
High and low molecular weight fractions in the same solvent

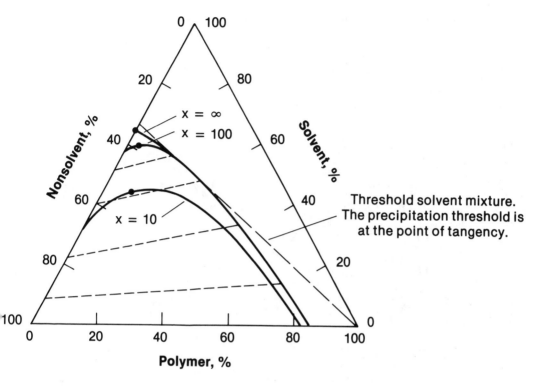

x = Number of segments per molecule

• = Critical points

Fig. 10.8   Ternary phase diagram for polymer in a solvent/nonsolvent mixture
Source: Adapted from P.J. Florey, *Principles of Polymer Chemistry*, Cornell University Press, 1953, p. 552

nature of the solvent in which the polymer is suspended. A similar behavior occurs when the polymer is branched.

Usually the configuration of a copolymer in solution is quite complex. If the copolymer comprises a long chain segment of nonpolar ethylene and a long chain segment of a polar polymer group such as polyvinyl alcohol, then in a hydrocarbon solvent the polyethylene segments would have a relatively open structure, while the polyvinyl alcohol segment would be more tightly coiled. The molecules of this copolymer would resemble a string of pearls when in solution in a hydrocarbon.

10.13   Orientation in Solution

If a solution of a polymer in a "good" solvent is subjected to mechanical action, such as flow through a pipe, a preferential orientation will occur. The polymer molecules that are somewhat extended in the solvent will experience an orienting force due to the net forces of solvent flow in a specific direction. This will cause the individual chains to line up somewhat in the stream direction. This effect is small for polymer solutions in which the chains remain tightly coiled, but is large for those solvent-polymer combinations where the chains are highly extended.

Because of their unique chain structure, some polymers can form weak bonds between chain segments. If these weak bond sites are arranged regularly along the chain of a linear polymer, it can cause an orientation of a substantial number of chains into a regular, often three-dimensional structure, a so-called liquid crystal (see Fig. 10.9). The "ordering" of

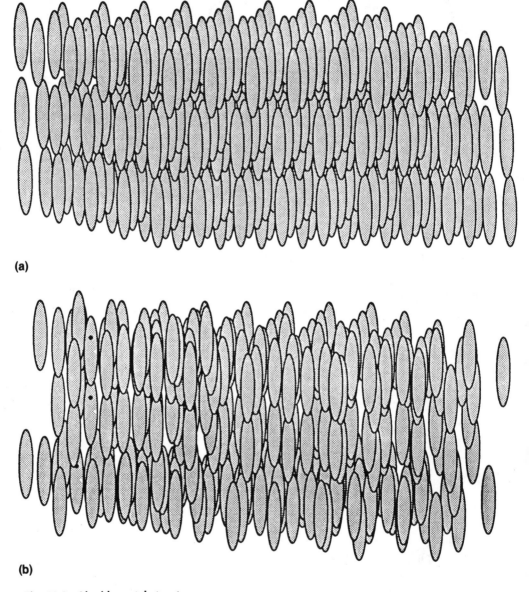

(a)

(b)

Fig. 10.9   Liquid crystal structures
(a) Smectic. (b) Nematic

structures will of course increase as the temperature of the solution decreases; if the temperature is increased, the "order" will decrease until it disappears entirely at some higher temperature.

10.14   Structures in the Solid

A polymer molecule may have a number of configurations in the so-called solid phase. Because of its large molecular size, it does not lend itself easily to the formation of crystals; nevertheless, polymers do form crystals under the right conditions. When a thermoplastic polymer is held at a high enough temperature, it assumes a somewhat mobile liquid state. If it is kept liquid long enough, the various individual molecules will disentangle and will resemble the state of the molecules in solution. As the temperature decreases, the resulting material may become a supercooled liquid. If there is sufficient time and the chain structure is favorable, some of the molecules may become ordered into a crystalline structure. Many polymers comprise a mixture of microcrystals and supercooled solution.

Whenever partial crystal formation occurs, the crystallite may incorporate only a portion of the chain of a given molecule. This is especially likely if the chain is a block copolymer (see Fig. 10.10). The presence of these crystallites in an otherwise amorphous or glassy matrix influences the properties of this type of material. In thermoplastic elastomers, for example, the crystalline regions can function as cross-links to build a type of network polymer that is cross-linked at lower temperatures and is not cross-linked at temperatures above the melting point of the crystallites. The tendency of highly branched structures to crystallize or assume ordered regions is diminished due to the interference of the branching. Similarly, a highly cross-linked polymer that forms a network may have some of the characteristics of uncross-linked materials if the distance between cross-links is sufficiently large. The loss of mobility due to the network precludes large molecular movements or rearrangements; thus the most commonly known network polymers, the thermoset polymers, are generally rigid amorphous materials.

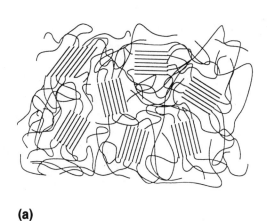

(a)

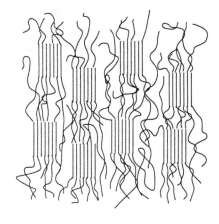

(b)

Fig. 10.10   Crystallinity in polymers
(a) Partially crystalline, randomly arranged. (b) Oriented

Because polymers have so many segments to arrange in order, and because they are spatially quite extended, on cooling from the melt, polymer molecules tend to coil, entangle, and only partially crystallize. Nevertheless, at any given temperature there is some mobility among chain segments and therefore among chains themselves. Over a span of time, chains may entangle or disentangle or even rearrange into crystalline regions. *Time* is of very special importance in describing the state of a polymeric solid, and many of the properties to be discussed in the next chapter are time dependent.

10.15   Crystallinity in Polymers

Although metals usually crystallize when they solidify, most polymers do not crystallize under normal conditions of cooling. There are some polymers that do tend to crystallize, the most common being linear polyethylene.

Usually the atoms that are arranged into unit cells are considered to be approximately spherical in shape. Although these atoms often differ in "diameter" and sometimes have a preference for their nearest neighbor due to charge differences, the treatment of crystallinity has generally been how one can arrange a collection of "balls" into a repeating, long-range pattern. Polymer molecules do not fit this spherical model in the least; in fact, they are almost always extended in space along one axis.

In molecular crystals, however, there are relatively strong bonds between the various atoms comprising a given molecule and weaker bonds between molecules. This weaker inter-molecular bonding may be enhanced by various interactions between polar groups, the most notable being hydrogen bonding where a single hydrogen atom is "shared" between oxygen or nitrogen atoms on two separate molecules.

An example of a simple molecular crystal is that of iodine, which crystallizes in an orthorhombic unit cell. The two atoms of iodine that comprise this molecule have a strong covalent bond between each other but a much weaker bond between two molecules. Thus the individual units that are ordered in the lattice are not spherical, but are dumbbell shaped. It is not possible to assemble dumbbells into an efficient cubic lattice. However, distortion of the lattice sufficiently to accommodate the molecular structure of I_2 forms an orthorhombic unit cell, as shown in Fig. 10.11. Note that the three corner angles are all 90°, but the dimensions are all unequal (a = 0.727 nm, b = 0.979 nm, c = 0.479 nm).

Polyethylene is a long, linear molecule with strong covalent bonding along the chain and weaker interchain bonds (*Van der Waal's dispersion forces*). These interchain bonds are of sufficient strength, however, to cause a regular ordering into a crystalline form, as shown in Fig. 10.12. Note that a single long molecule simultaneously occupies many unit cells. When only a portion of the atoms in one molecule are in unit cells, and the remaining molecules are not in any unit cell, there is a high degree of imperfect crystallization. Because the chains are of unequal length, they must form an imperfect structure. The tendency for chains to coil and kink causes some segments of one polymer molecule to be partially in one or more unit cells and other segments of the same molecule to be in amorphous regions. Even in the most regular of polyethylene crystals, a folding of the chain apparently occurs in which the chains line up in the crystalline lattice, but with folds at the edge of the crystal, much the way a fire hose is kept in

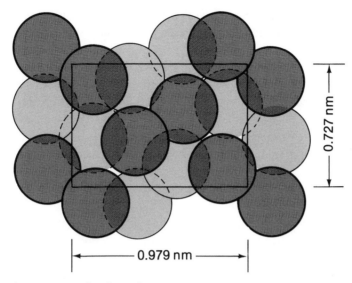

Fig. 10.11   Molecular iodine crystal

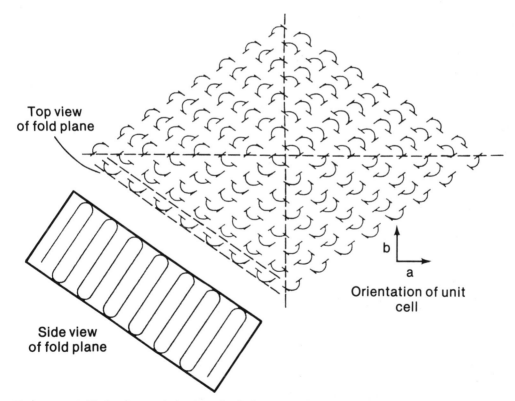

Fig. 10.12   Folded polymer chains in polyethylene crystals
Source: D.J. Williams, *Polymer Science and Engineering*, Prentice-Hall, 1971

its case (see Fig. 10.12). The more regular the chain structure of a polymer, the more easily it crystallizes, and the more disorder in chains, such as branching, the less easily a given polymer crystallizes.

It is possible to have short-range ordered portions of a polymeric solid while there is little or no long-range order. For example, as liquid polyethylene is cooled it is relatively easy to form a small number of reasonably good crystals. However, the long polymer chains extend not only into amorphous regions but even into other crystalline portions. As the crystallites grow, it soon becomes difficult to relocate segments of chains into crystals due to constraints of the long chains. Eventually, a certain fraction of the polymer remains amorphous.

10.16   Networks

When polymer chains are joined by cross-links in a material, the initial effect is to form new chains that are in effect branched (see Fig. 10.13). A common way of producing such cross-linked chains is to add some trifunctional mer into a condensation polymerization between two types of bifunctional mer. The reaction forms a highly branched structure, such as that shown in Fig. 10.14.

As the number of branches increases, the concept of a polymer "chain" vanishes. For the purpose of discussing network polymers, we will refer to a chain as the portion of a molecule between two branch units, or between a branch unit and a terminal end group. The lengths of

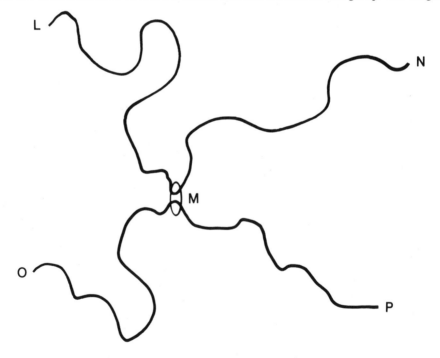

Fig. 10.13   Formation of branched polymers through cross-linking
Two linear polymer chains, LN and OP, become a single chain LMP with two branches, MN and MO, after formation of the cross-link at M.

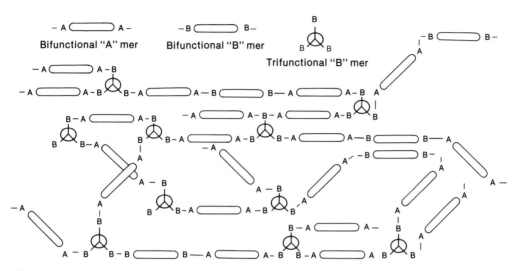

Fig. 10.14   Condensation polymerization with cross-linking through trifunctional mers

these chains varies, of course, depending on the statistics of the polymerization. An important parameter of network polymerization is α, the *branching coefficient*, which is the probability that any particular branch unit (the trifunctional mer BBB) is connected by a polymer chain to a second branch unit. The value of α is related to the number of functional groups in the polymer, and α can be determined for any polyfunctional system from suitable experimental measurements. For trifunctional cross-linking mers (B mers) in a polymerization of bifunctional A mers with bifunctional B mers:

$$\alpha = \frac{rP_b^2\, p}{1 - rP_b^2(1 - p)} = \frac{P_a^2\, p}{r - P_a^2(1 - p)}$$

where r is the ratio of B groups to A groups; p is the ratio of all B functional groups on trifunctional mers to the total number of B groups on all mers, both bifunctional and trifunctional; P_a is the fraction of all A groups which have reacted; and P_b is the fraction of all B groups which have reacted. Note that the probability that a given A group has reacted with a branch is $P_a p$ and the probability that it is connected to a bifunctional B—B group is $P_a(1 - p)$.

When the number of branch groups are so connected that an infinite network is formed, the polymer mass is in effect one giant molecule. The value at which this takes place for the trifunctional example above is $\alpha > \frac{1}{2}$ (see Example 10.16A). This type of polymerization will lead to an unlimited branched structure completely connected into what is called an *infinite network*. Thus, $\alpha = \frac{1}{2}$ is the critical value for formation of an infinite network in the trifunctional system of the example.

Example 10.16A. A polymerization is carried out between an equal number of moles of a bifunctional mer A—A and a trifunctional mer B—(B)—B. What is the branching coefficient when the reaction of A mer is carried to 90% completion?

Solution: In this special case where there are no bifunctional B—B mers, the ratio of trifunctional B groups to all B groups (bi- and trifunctional) is 1, since there are only trifunctional groups present. $p = 1$; $\alpha = rP_b^2 = P_a^2/r$; $r = 3A/2B = 3A/2A = 1.5$; $P_a^2 = 0.90^2 = 0.81$; and $\alpha = P_a^2/r = 0.81/1.5 = 0.54$. This composition has just exceeded the point at which an infinite network is formed.

Example 10.16B. To the reaction of Example 10.16A is added an amount of bifunctional B mers equal to 10% of the trifunctional B mers. What is the branching coefficient when the reaction of A mers is 90% complete?

Solution: The relative number of A groups is 2A, the relative number of B groups on trifunctional mers is 3A, and the relative number of B groups on bifunctional B mers is $2 \times 0.01 \times A$. Therefore, $r = (3A + 0.2A)/2A = 1.60$. $(3B = 3A$ since there are equal numbers of moles of each type of mer.) $p = 3/(3 + 0.2) = 0.9375$; $P_a = 0.90$; $P_b = 0.90/r = 0.34$; $\alpha = P_a^2 p/[r - P_a^2(1 - p)] = (0.81 \times 0.9375)/[1.60 - (0.81 \times 0.0625)] = 0.759/1.55 = 0.49$. This composition, with only 10% of bifunctional mer, has not quite reached the point at which an infinite network exists.

If the branching units are not trifunctional, but tetrafunctional, for example, a similar treatment can be employed. In a system consisting of bifunctional B—B units and f-functional B units $X—(B)_f$ where the B groups can react with other B groups:

$$\alpha = \frac{Pp}{1 - P(1 - p)}$$

A general condition for the formation of critical networks can be stated to include any degree of multifunctionality f as follows: When the final number of chains resulting from the branching of some of the n chains in a polymeric material is greater than the original number of chains, n, an infinite network will be formed. (Note: We refer to a chain as the portion of a molecule between two branch units, or between a branch unit and a terminal end group.) The critical value of α for this condition is:

$$\alpha = \frac{1}{f - 1}$$

The nature of a polymer that is an infinite network is that it is no longer soluble; i.e., the molecules are no longer separable. Solvent can, of course, penetrate the net and cause some swelling. This type of structure is commonly called a *gel*, and the point in a condensation reaction when the infinite network first forms is called a *gel point*.

When a reaction proceeds beyond the gel point, not all of the mers will have reacted into "infinite" molecules. Because the reaction is of a statistical nature, there will be many molecules that are not part of the network as long as the value of α is between ½ and 1. The material that is not part of the infinite network is called the *sol*.

Polymers that are formed into insoluble "infinite" networks by condensation reactions are called *thermoset polymers*. They represent an important class of commercial materials because of their excellent resistance to heat and solvents. Some common thermoset polymers are listed in Table 10.8.

Table 10.8 Some common thermosetting polymers

Name	Monomers	Typical applications
Phenolics (example: phenol formaldehyde)		Electrical equipment, kitchen utensils, laminates
Amino resins (example: urea formaldehyde)		Dishes, decorative laminates
Polyesters		Composites, coatings, automotive parts
Epoxies	(R is an aromatic polyfunctional molecule; R' is a polyfunctional polyamide)	Adhesives, composites, coatings, aircraft parts
Polyurethanes	OCN—R—NCO + HO—R'—OH (diisocyanate) (R is a complex polyisocyanate; R' is a polyol)	Sheet, tubing, foam, elastomers, fibers, coatings
Silicones	Trichlorosilane Trihydroxysilane	Gaskets, adhesives, sealants

In the case of cross-linked structures, we are not dealing with a specific multifunctional mer but the formation of relatively short cross-links in otherwise linear polymers. These are usually presumed to occur at random along the primary chain. If we regard the uncross-linked polymer material to consist of a number, n, of primary molecules (number of molecules before any cross-linking), only one cross-link for two primary molecules (or n/2 total cross-links) is needed to form an infinite network, e.g., for incipient gel formation. When comparing polymers of the same number average molecular weight—i.e., the same number of primary molecules— the broader the molecular weight distribution, the fewer the number of cross-links needed to produce gelation.

10.17 Problems

1. Using values for the electronegativities of various elements and the elements commonly found in polymers (see Table 10.1), determine the largest electronegativity difference found in common polymers. Is this bond still mainly covalent? Determine the smallest electronegativity difference for two different elements and give the elements involved.

2. List all the types of bonding in each of the following: polyethylene, polyvinylchloride, and Lucite (polymethylmethacrylate). Why does none of these contain hydrogen bonding? Nylon contains N—H bonds as well as C=O bonds (see section 10.8); does nylon have hydrogen bonding? How does hydrogen bonding affect the properties of polymers?

3. A specimen of polyethylene contains 0.1 mole having a degree of polymerization, n, equal to 1000; 0.15 mole with n = 1500; 0.4 mole with n = 2000; 0.2 mole with n = 2500; 0.1 mole with n = 3000; and 0.05 mole with n = 3500. Calculate the number average molecular weight.

4. Calculate the weight average molecular weight for the specimen in problem 3. Calculate the polydispersivity index for the answers to problems 3 and 4.

5. Calculate the average distance between ends of polyethylene molecules having a molecular weight of 52,000 g/mol.

6. Describe viscoelasticity in your own words.

7. Describe the effect of a solvent on the coiling of a polymer when the solvent becomes less and less like the polymer due to the addition of a second material to the solvent.

8. Describe the effect of increasing temperature on the degree of coiling of a polymer that is only a little like the given solvent.

9. Discuss the relative ease (or difficulty) of the formation of a crystalline material upon slow cooling of liquid iron, soda-lime glass at a very high temperature, and liquid polyethylene.

10. Why are crystalline polymers stronger than the same polymers (same composition, same molecular weight) that are amorphous or largely amorphous?

11. Discuss the effects of solvents on infinite network polymers. What is a common name for many infinite network polymers?

11

Properties of Polymers

11.1 Basic Mechanical Properties of Polymeric Materials

A description of the mechanical properties of polymers uses much of the same terminology developed for the characterization of metals. Tensile strength and modulus of elasticity are important figures of merit for polymers, just as they are for metals. Because polymers are often supercooled liquids or glasses at the temperatures at which they are used, other properties must also be considered. Most thermoplastic materials exhibit viscous flow as well as an elastic response to stress. Because of its complex structure, a polymeric solid may exhibit both an elastic response and a viscous response to stress at any given temperature. This combined elastic and viscous response of a polymer is referred to as *viscoelasticity*. Amorphous polymers are related to other glasses in having viscoelastic behavior. The measurement of viscoelasticity in various ways is more important in the characterization of polymers than in metals.

For every amorphous polymer there exists a temperature band, called the *glass transition temperature*, in which it changes its response to mechanical stress from that typical of a viscous or rubbery material to that of a glass. A glassy material is an amorphous solid and is defined as a rigid supercooled liquid. This change is not a phase transition or melting point, but is the temperature region in which the polymer is transformed from a liquid to a glass. Not only do brittleness and hardness properties change, but also other properties such as thermal expansion, dielectric constant, heat capacity, and apparent viscosity. This temperature is an important characteristic of thermoplastic materials.

The time duration of the mechanical stress is very important in determining the type of response exhibited by a polymeric material. For thermoplastics, a short-time constant stress will elicit the maximum of elastic response, while a long-time constant stress will exhibit more viscous response. The effects of time can also be duplicated by increasing the temperature: a greater amount of viscous flow is observed at higher temperatures. These effects are often very important in the applications for polymeric materials. Therefore, a great emphasis is placed on mechanical properties relating to impact strength and heat distortion. A typical set of properties of interest to the mechanical designer is given in Table 11.1 for several common polymers.

11.2 Nature of Viscosity in Polymer Liquids and Solutions

Long-chain molecules present in polymers create a resistance to flow in the liquid state. Unexpectedly, they also greatly increase the resistance to flow of solutions even though they are

Table 11.1　Mechanical properties data for selected polymers

Polymer	Young's modulus (E)(a), MPa	Tensile strength, MPa	% elongation at failure	Rockwell hardness, R scale(b)	Impact energy(c), ft · lb	Poisson's ratio, μ
Thermoplastics						
Polyethylene						
High-density	1,100	28	10–1,000	40	0.4–4	0.46
Low-density	170–280	8–30	100–900	10	>>22	0.46
Polyvinyl chloride	2,800	36–90	2–30	110	1.4	0.42
Polystyrene	3,300	36–90	1–2	75	0.4	0.51
Polypropylene(d)	1,100–10,000	24–104	10–700	90	1.4–15	0.54
ABS	2,400	18–55	20–80	95	1.4–14	0.50
Polymethylmethacrylate	3,100	70	3–5	130	0.7	0.50
Cellulosics(d)	3,400–28,000	14–55	5–40	30–115	1–11	0.65
Acetals (homopolymer)	3,600	67	50	120	3	0.39
Polyamides 66(d)	2,900	62–260	60	118	1.4	0.41–0.47
Polycarbonates	2,400	110	110	118	19	0.49
Polytetrafluoroethylene (Teflon)	410	17	100–350	70	5	0.48
Polyesters	3,000	57	50–300	117	1.4	0.46
Thermosets						
Phenolformaldehyde (glass filled)	>48,000	45–55	0.2	125	0.4	0.50
Urea-melamine (filled)	10,000	48	1	115	0.4	0.50
Polyesters (unfilled)	4,400	4–90	1–6	100	1–7	···
Epoxies (unfilled)	2,500	27–90	3–6	90	0.2–1.1	0.50
Epoxies (filled)	20,000	34–38	4	···	···	···
Silicones	···	2.4–27	100 to >700	···	···	···

(a) Low strain data. (b) For relatively soft materials: indenter radius of ½ in. and load of 60 kg. (c) Izod test. (d) Range includes some fiber-reinforced material.

present at very low concentrations. These effects relate to the voluminous character of the randomly coiled molecules both in the solid and in the solution.

The viscous properties of polymers are due to the amorphous nature of polymers or to the amorphous regions in partially crystalline polymers. The mechanical behavior of amorphous polymers is thus a special case of the properties of glasses. *Viscosity* is defined by the proportionality between the shearing force per unit area and the gradient of velocity acting on a material, where the area is in a plane perpendicular to the direction of the velocity, as illustrated in Fig. 11.1.

$$\frac{F}{A} = \eta \frac{dv}{dx}$$

where η, the viscosity, is the proportionality constant and is measured in units of poise (1 poise = 1 g/cm · s or 0.1 Pa · s). For example, the rate at which a viscous liquid, such as honey, flows through a tube can be measured and the viscosity determined from the area of the tube and the flow rate. The viscosity of solutions is commonly measured this way in capillary viscometers calibrated using standard liquids of known viscosity. The glasses are normally referred to as having viscous flow deformation above a certain temperature (or temperature range) and

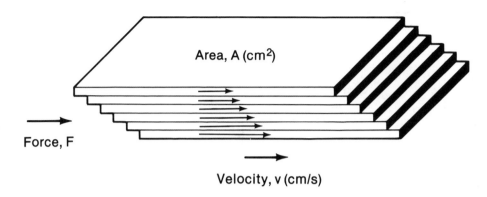

$F/A = \eta\ dv/dx$, where the viscosity, η, is measured in:

1 poise = 1 g/cm • s or 0.1 Pa • s

Fig. 11.1 Viscosity is the proportionality between the rate of change in velocity between layers and the shearing force per unit area

having elastic response below this temperature. This characteristic temperature, the *glass transition temperature*, is also referred to as the *brittle temperature* or the *second-order transition temperature*. It can be determined easily by measurement of a characteristic property—for example, volume or thermal expansion—as a function of temperature (see Fig. 11.2). The glass transition temperatures of some familiar polymers are listed in Table 11.2.

At temperatures significantly above the glass transition temperature, the viscosity follows an Arrhenius equation:

$$\eta = \eta_0\ e^{+Q/RT}$$

where Q is the activation energy, T is the absolute temperature, R is the molar Boltzmann constant, and η_0 is the required proportionality constant which is characteristic of the material. Note that viscosity, η, *decreases* with increasing temperature; in other words, the material becomes *more* fluid at higher temperatures. The average person is probably most familiar with this phenomenon in the behavior of honey or molasses, where the term "slow as molasses" describes the behavior at room temperature even though these materials flow easily when warmed slightly. Another term, *fluidity*, is often used to express this behavior. Fluidity is the reciprocal ($1/\eta$) of viscosity, and fluidity increases with increasing temperature.

The strengths of amorphous polymers in the glassy state are considerably greater than those of the same material above the glass transition temperature. The brittleness and lower impact resistance of glassy materials severely limit their practical application, however, and crystalline and semicrystalline polymers generally have superior combinations of mechanical properties compared with amorphous polymers in the glassy state.

In general, materials having stronger interactions between chain segments will have higher glass transition temperatures. Bulky side groups or highly polar side groups, such as Cl, C=O, or CN raise the glass transition temperature, while polymers with small side groups, long

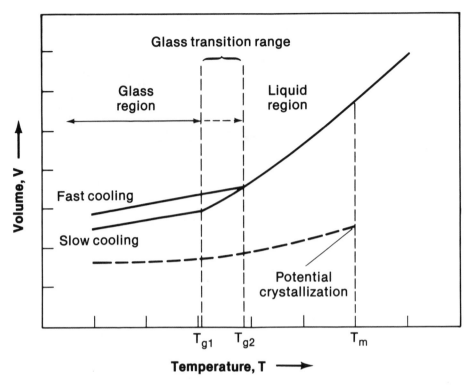

Fig. 11.2   Glass transition temperatures
The rate at which the volume, V, changes during cooling shows an inflection at the glass transition temperature, T_g. T_g depends on the cooling rate. Two transition temperatures are shown. The material exists as a supercooled liquid below the melting temperature, T_m.

Table 11.2   Glass transition temperatures of selected polymers

Polymer	Glass temperature (T_g), °C	Melting point (T_m), °C
Polyethylene (high-density)	−120	140
Polybutadiene	−70 to 90	...
Polypropylene	−13	175
Polymethylacrylate	0	...
Nylon 6/6	50	265
Polyvinyl chloride	75 to 85	210
Polyvinylidine chloride	...	210
Polymethylmethacrylate	90	...
Polyacrylonitrile	...	>250
Polystyrene	100	240
Methacrylonitrile	130	...
Polybutadiene/styrene		
100/0	−70 to 90	...
75/25	−60	...
60/40	−45	...
50/50	−35	...
30/70	0	...
0/100	100	...

flexible branching chain segments, and a highly mobile main chain have lower glass transition temperatures. Often the effects of side groups are proportional to their concentration; for example, in copolymers of acrylonitrile and butadiene, the glass transition temperature varies linearly with the acrylonitrile content over a wide range of concentration.

11.3   Measurement of Viscosity

The measurement of the viscosity of dilute polymer solutions has long been recognized as an important method of characterization. Measuring the viscosity of bulk polymers is very important because the polymers are used in bulk form. Knowledge of the bulk viscosity at various temperatures aids technical operators in the processing and production of polymers, and viscosity measurements are important to design engineers in choosing polymers for use in specific environments.

The ability of a polymer to increase the viscosity of a solution is called the *intrinsic viscosity* and increases with molecular weight. Viscosity measurements are therefore an important way of measuring average molecular weight. However, molecular weight is not simply related to intrinsic viscosity; therefore, the molecular weight, M, must be experimentally correlated with intrinsic velocity, $[\eta]$, before the viscosity measurements can be a useful tool for a specific polymer. Because viscosity is much more easily determined than molecular weight, viscosity is routinely used in polymer science as a measure of molecular weight. Measurements on dilute solutions are best understood on theoretical grounds, but measurements at higher concentrations are more sensitive to molecular weight and are therefore favored by industry. Measurement of the bulk viscosity of polymers of molecular weight <100,000, however, may be used alternatively to those of concentrated solutions.

Viscosities of dilute polymer solutions are measured using capillary viscometers. The measurement itself is the time it takes a given amount of the liquid to flow through the narrow tube at constant temperatures, normally a time of several hundred seconds. The viscosity, η, is calculated as:

$$\eta = \alpha \rho \, (t - \beta/\alpha t)$$

where ρ is the density of the solvent (or solution), t is the flow time, and α and β are calibration constants. After using this calculation, the *relative viscosity*, η_r, is determined by dividing the viscosity of the solution by the viscosity of the solvent. Relative viscosity is then used to evaluate the *specific viscosity*, $\eta_{sp} = \eta_r - 1$. The specific viscosity is an indication of the contribution to the solution viscosity which is attributable to the polymer. The limiting value at infinite dilution (low concentration) of the ratio of specific viscosity to concentration is called the intrinsic viscosity, $[\eta]$:

$$[\eta] = \left(\frac{\eta_r - 1}{c} \right)_{c \to 0} = \left(\frac{\ln \eta_r}{c} \right)_{c \to 0}$$

The procedure for this is relatively simple since the plots of η_{sp}/c versus c are usually linear.

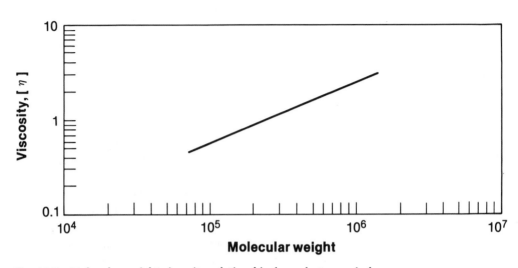

Fig. 11.3 Molecular weight-viscosity relationship for polystyrene in benzene
K' = 1.03 × 10 and a = 0.74

When the logarithms of the intrinsic viscosities of a series of a linear polymer of increasing molecular weight and narrow distribution are plotted against the logarithm of their molecular weights, a linear relationship is obtained (see Fig. 11.3). This relationship between intrinsic viscosity and molecular weight

$$[\eta] = K'M^a$$

involves two empirical constants: K' and a. For polymers with a normal degree of dispersion of the degree of polymerization, an average molecular weight is determined that reflects the molecular weight which would have been determined from measuring the viscosity for each different molecular size at its own concentration. This leads to a viscosity average molecular weight, M_v:

$$M_v = \left(\sum w_i M_i^a \right)^{1/a}$$

which follows an equation analogous to that given in section 10.9 for weight average molecular weights. The measurement of viscosity on bulk polymers is usually an empirical one in which the weight of polymer that is extruded through an orifice for a measured amount of time under a given force is determined.

The intrinsic viscosity is considered to be proportional to the volume of the spherically coiled molecule in solution divided by its molecular weight. This volume—or, more correctly, effective volume—assumes that a linear polymer occupies a roughly spherical space whose radius is proportional to the cube of the root mean square end-to-end distance. The actual root mean square end-to-end distance, $\sqrt{\bar{r^2}}$, is different from the ideal "free space" value by a factor α, which takes into account solvent interactions with the chain segments. This theoretical assumption has led to an expression for the intrinsic viscosity:

$$[\eta] = K \sqrt{M} \, \alpha^3$$

As previously noted (see section 10.12), polymers in dilute solution near the θ point where the onset of separation into two phases occurs are in a condition where solvent-polymer segment interactions are effectively zero and the polymer is in the ideal configuration ($\alpha = 1$). Thus the intrinsic viscosity determined at that point should yield information directly proportional to $\sqrt{M}$. This relationship has been experimentally observed.

11.4   Elastic Response of Polymeric Solids

At low temperatures most polymers deform elastically, whereas at relatively high temperatures they deform viscously. The behavior of polymers as a function of temperature is frequently illustrated by plotting the modulus of elasticity versus temperature, as illustrated in Fig. 11.4. Such a plot exhibits several regions where the slope of modulus versus T has a characteristic behavior. These regions are often referred to as (1) glassy or rigid, (2) leathery, (3) rubbery, and (4) viscous; these terms are descriptive of the type of response to mechanical stress which prevails at that temperature. The degree to which a certain polymer has a given type of response is a characteristic of the polymer type under investigation; for example, an elastomer has a large rubbery region.

The elastic response of a polymer in the rigid zone, which is well below its glass transition point, is very similar to that of metals. Thus a stress-strain curve for a rigid polymer is similar to curves for metals, but most polymers have a modulus of elasticity that is an order of magnitude less than that typical for metals. The bonding *between* molecules in polymers is considerably weaker than the metallic bonding in metals. Nevertheless, the mechanical responses of rigid polymers may be understood using the same principles as for metals.

In the leathery and rubbery region, a different type of response is observed. Instead of the small strain and high modulus typical of metals, we find that polymers can undergo large deformation (large strain) when stressed, and they will regain their original shape (elastic response) when the stress is removed. An important difference is the *time* it takes the material to regain its original dimensions. In the leathery region, which is just above the glass transition temperature, the polymer *slowly* regains its original shape after stress is removed, while in the rubbery region a polymer rapidly returns to its original dimensions upon stress removal. Usually the leathery region occupies a fairly narrow temperature zone, while the rubbery response takes place over a much broader range of temperatures, as illustrated in Fig. 11.5. At some temperature a true melting point occurs for the crystalline regions of the polymer, and the polymer exhibits a more liquidlike viscous response. The exact nature of these responses depends on the individual polymer, and its molecular structure and formulation. For some applications, the melting points and glass transition temperatures are chosen sufficiently low so that the service temperature of the material always occurs above the glass transition point. For other applications—for example, where more rigidity is desired—the structure and formulation of the poly-

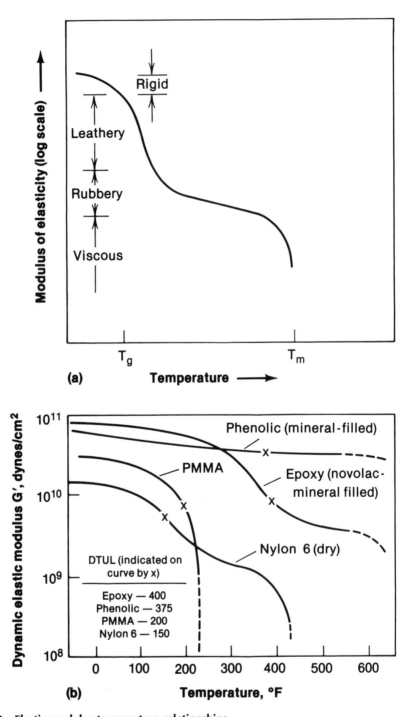

Fig. 11.4   Elastic modulus-temperature relationships
(a) Modulus of elasticity as a function of temperature for a typical thermoplastic polymer with 50% crystallinity. There are four distinct regions of viscoelastic behavior: rigid, leathery, rubbery, and viscous. Source: J.F. Shackelford, *Introduction to Materials Science for Engineers,* Macmillan, 1985, p. 369. (b) Dynamic shear (torsion) modulus versus temperature for representative thermosets and thermoplastics, showing deflection temperature under load (DTUL) at 264 psi. Source: *Modern Plastics Encyclopedia,* McGraw-Hill, 1990, p. 425

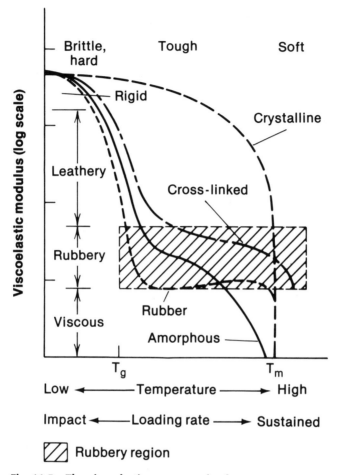

Fig. 11.5   The viscoelastic response of polymers

mer are chosen to ensure that the service temperature falls below the glass transition temperature.

11.5   Viscoelasticity

Perhaps the most fascinating characteristic exhibited by polymers is the transition from rigid to leatherlike then to rubbery, and finally to viscous behavior as the temperature rises. The mechanism by which the stress is resisted by a polymer is not one of relatively uniform bonding (metallic, ionic, or covalent) as in a crystalline solid. In polymers the response to stress is by viscous flow and chain extension by uncoiling. There is no simple, unique response to stress, and the behavior of a polymer is strongly a function of both the temperature and the time during which the stress is applied or the response is observed. Some tests used for polymers are designed to measure response to short-time high stress as in the measurement of impact

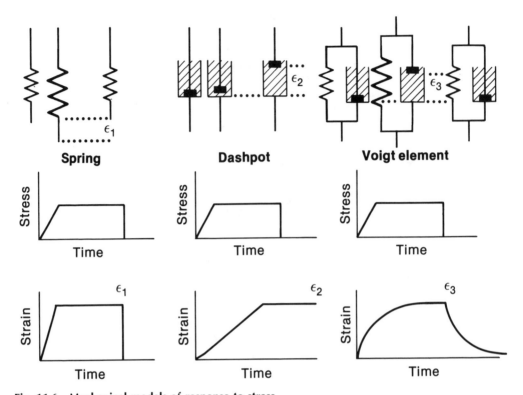

Fig. 11.6 Mechanical models of response to stress
The spring gives instantaneous response to stress, the dashpot responds linearly to stress but with a time delay and without recovery, while the combination (Voight element) has a complex time-dependent response.

strength, while other tests are designed for long-time stress as in tests of stress relaxation or creep. Subjecting a polymer to a cyclic stress tests its short-time constant responses, while lengthening the period of the cyclic stress allows observation of the response of the polymer to a longer-time stress. The response of a polymer to a cyclic stress is subject to both experimental measurement and some theoretical analysis; it also has practical importance.

An understanding of the behavior of polymers may be gained by considering them as *viscoelastic systems*—that is, systems in which the polymer structure is represented by a collection of elastic "Hookian"* springs in parallel with a device exhibiting viscous flow called a dashpot (see Fig. 11.6). The suspension on most automobiles uses exactly this principle: the coil spring, which responds to the shock of the road, is in parallel with a shock absorber, providing the viscous flow of the dashpot. The response of the car to short-term stress (e.g., bumps in the road) is drastically altered as the balance of the suspension system changes when the shock absorber begins to wear out and the viscous element is lessened.

The mathematical analysis of such a system has been known for a long time. When the springs and dashpots of such a system are uniformly distributed along the axis of mechanical

---

*Named for 17th century British physicist Robert Hooke; Hooke's law states that the elastic strain is proportional to the stress for many materials. Materials that obey this law are called "Hookian."

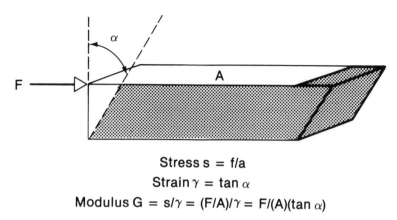

$$\text{Stress s} = f/a$$
$$\text{Strain } \gamma = \tan \alpha$$
$$\text{Modulus G} = s/\gamma = (F/A)/\gamma = F/(A)(\tan \alpha)$$

Fig. 11.7 Shear stress modulus

stress, the mechanical system is directly analogous to an electronic inductive transmission line. The response of the mechanical system to a cyclic stress may be understood in terms of the mathematics for the response of the electrical network to alternating current. Descriptions of polymer response to stress developed from basic considerations of molecular behavior yield the same equations as for the mechanical analogue in those limiting cases where the two approaches can be compared.

The response of a solid to simple shear (Fig. 11.7) is characterized by a *shear modulus*, or *modulus of rigidity*, G:

$$G = \frac{f}{A\gamma} = \frac{f}{A \tan \alpha}$$

where A is the area across which the force, f, is applied to create the strain γ. The strain γ is the tangent of the angle of deformation, α, and is equal to the angle itself for small strains, e.g., small angles of deformation under shear. The reciprocal of the shear modulus is the *shear compliance*, J. Much of the theoretical and experimental work on the response of polymers to cyclic stress has been expressed in terms of G and J.

The response of a solid to simple extension by a normal tensile force f applied to the axis of a material whose ends have area A is characterized by a tensile modulus, E, or Young's modulus of elasticity:

$$E = \frac{\sigma}{\epsilon} = \frac{f}{A\epsilon}$$

where σ, the tensile stress, is f/A and the strain, ϵ, is the fractional increase in length, $\Delta l/l$. The reciprocal of E is the *compliance*, D.

The relationship between G and E is often written in terms of a dimensionless variable, Poisson's ratio: E $= 2G(1 + \mu)$, where μ is Poisson's ratio. The value of Poisson's ratio for a group of representative polymers is given in Table 11.1. A similar relationship exists between the compliances D and J where $D = J/3 + B/9$, where B is the bulk compliance. The bulk compliance is the reciprocal of the bulk modulus K relating the relative change in volume of a solid under compression. For most solids B is the same as the compressibility, but for polymeric solids B is subject to viscoelastic response and is therefore time dependent.

Because many materials must bear stress over quite a long time in their practical applications, the fundamental properties of polymers are important not only over a broad temperature range, but also over a long span of time. The various moduli G, J, B, etc., are all functions of time as well as temperature. Carrying out measurements of physical properties of polymers over long periods of time is frequently impractical. Fortunately, short-term measurements of many of the properties made at higher temperatures can be directly related to long-term behavior at lower temperatures.

11.6   Time-Temperature Relationships of Viscoelastic Properties

11.6.1   Boltzmann's Superposition Principle

If a stress, σ_o, is put on a material at time $t = 0$, the strain as a function of time, $\gamma(t)$, will be $\gamma(t) = \sigma_o J(t)$, where J(t) indicates that the compliance J is time dependent. If a greater stress, σ_1, is added at $t = 0$, the resulting strain will be proportionately larger: $\gamma(t) = (\sigma_o + \sigma_1)J(t)$. However, if the original stress is put on at time 0 and the additional stress is added at time $t = t_1$, the strain observed is a linear superposition of the two strains, each expressed in terms of their individual elapsed times, $\gamma(t) = \sigma_o J(t) + \sigma_1 J(t - t_1)$. This method of combination of effect is often called the *Boltzmann's superposition principle*. The general expression of this effect is:

$$\gamma(t) = \sum{}^i \sigma_i J(t - t_i)$$

A similar relationship holds for the expression of the stress in terms of the superposition of strains:

$$\sigma(t) = \sum{}^i \gamma_i G(t - t_i)$$

The superposition principle is widely accepted as a proper expression of polymeric behavior, and is of use in visualizing the viscoelastic effects in terms of the mechanical models. For example, if a polymer has been subjected to a stress for some time and the stress is removed, the material will gradually return *toward* its original state. The recovery of shape is termed *creep recovery*. Shear creep followed by shear creep recovery in a *cross-linked* polymer is diagrammed in Fig. 11.8. In terms of the mechanical model of Fig. 11.9, this means that none of the springs in the model is missing, so an equilibrium extension will take place. If the original stress has been applied for a sufficient time to reach an equilibrium value of the strain ($\sigma_o J_e$), then the creep recovery will be:

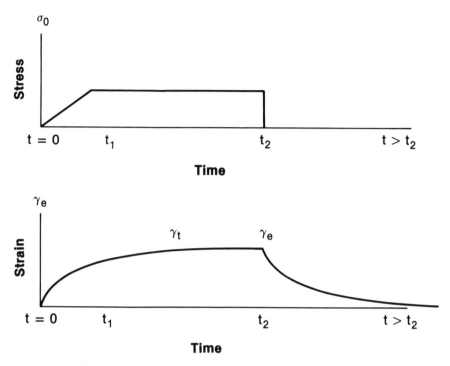

Fig. 11.8 Creep and recovery of a cross-linked polymer
The strain in the cross-linked polymer is the result of extension of the purely elastic portions and the extension of the viscoelastic portions which are constrained by the cross-links. $\gamma_e = \sigma_o J_e$; $\gamma_t = \sigma_o J_t$

$$\gamma_r(t) = \sigma_o \left[J_e - J(t - t_2) \right]$$

This follows from the concept that the removal of stress is equivalent to applying an additional stress of $-\sigma_o$. The shape of the recovery curve would be just the mirror image of the initial creep itself. If the stress were removed before equilibrium had set in and the creep were still changing, the recovery creep would follow the equation:

$$\gamma_r(t) = \sigma_o \left[J(t) - J(t - t_1) \right]$$

This expresses the superposition of the behavior at time t with the behavior at time t_1. Note, however, that these relationships are strictly valid for small extensions only; at very large extensions, nonlinear strains are observed in most polymers.

   If the polymer is *not* cross-linked, there is no equilibrium compliance (see Fig. 11.10). The rate of strain approaches a limiting value and then a steady-state viscous flow sets in, governed by a viscosity, η, whose magnitude may be visualized as the sum of all the individual viscosities of the dashpots in the mechanical model. At the same time, the individual springs in the model are all stretched to an equilibrium value. This elastic portion of the strain (from the springs) is recoverable after the stress is removed. For this material, the shear creep followed

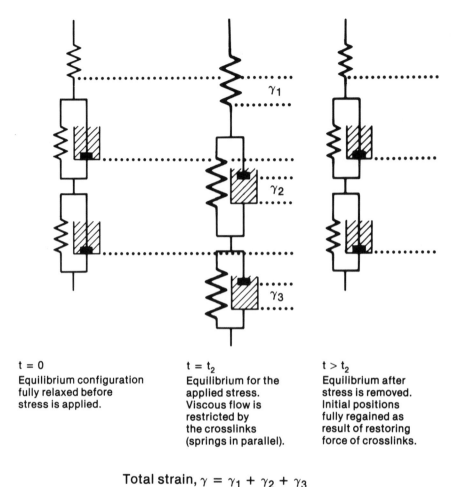

$$\gamma_1$$

$$\gamma_2$$

$$\gamma_3$$

t = 0	t = t$_2$	t > t$_2$

t = 0
Equilibrium configuration
fully relaxed before
stress is applied.

t = t$_2$
Equilibrium for the
applied stress.
Viscous flow is
restricted by
the crosslinks
(springs in parallel).

t > t$_2$
Equilibrium after
stress is removed.
Initial positions
fully regained as
result of restoring
force of crosslinks.

Total strain, $\gamma = \gamma_1 + \gamma_2 + \gamma_3$

Fig. 11.9   Model of a cross-linked polymer

by recovery is shown in Fig. 11.11. After a sufficiently long time, the creep will be $\gamma(t) = \sigma_o(J_e + t/\eta)$. Both J and η could be obtained from a linear plot of values in this region. The recovery process is given by:

$$\gamma_r(t) = \sigma_o \left[J_e + \frac{t}{\eta} - J(t - t_2) \right]$$

In this case, the final value of $\sigma_o t_2/\eta$ provides an alternative measure of η. These examples of the differences in response between cross-linked network polymers and uncross-linked polymers are seen in all of the time-dependent considerations of polymeric behavior.

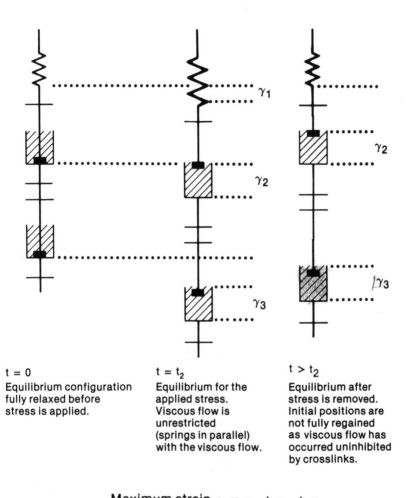

$t = 0$
Equilibrium configuration
fully relaxed before
stress is applied.

$t = t_2$
Equilibrium for the
applied stress.
Viscous flow is
unrestricted
(springs in parallel)
with the viscous flow.

$t > t_2$
Equilibrium after
stress is removed.
Initial positions are
not fully regained
as viscous flow has
occurred uninhibited
by crosslinks.

Maximum strain, $\gamma = \gamma_1 + \gamma_2 + \gamma_3$

Residual strain, $\gamma = \gamma_2 + \gamma_3$

Fig. 11.10   Model of an uncross-linked polymer

11.6.2   Stress Relaxation

When a plastic material is held under constant strain, the magnitude of the associated stress decreases with time due to creep (viscous flow). This phenomenon is known as *stress relaxation*. A familiar example is that of a rubber band holding a pack of letters. When the elastic band is removed after several months, it retains its expanded shape because the restoring stress has relaxed and is no longer present. In many applications where plastics are assembled with a snap-together fit or where plastic bolts or screws are threaded tightly into a part, stress relaxation may destroy the integrity or at least the tightness of fit. A typical stress relaxation curve is shown in Fig. 11.12. Note that the stress decays exponentially with time according to the expression:

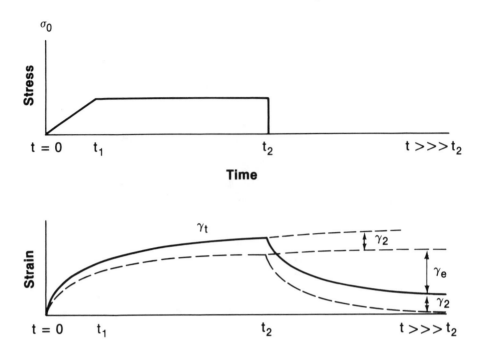

Fig. 11.11   Creep and recovery of a viscoelastic polymer
Strain in the uncross-linked polymer is the sum of the strain from the portion which is elastic, γ_e, and recovers when the strain is removed and the viscous flow which takes place during the time the stress is applied. This viscous flow represents unrecoverable strain, γ_2. $\gamma_2 = \sigma_o t_2/\eta$

$$\sigma = \sigma_o\, e^{-t/\gamma}$$

where σ is the stress, t is the time, σ_o is a constant, and γ is a characteristic value called the *stress relaxation time*. The stress relaxation time is defined as the value of the time at which the stress decays to 1/e of its original value. Note that the stress relaxation phenomenon is a function of the initial rate of straining. This comes from the short time scale viscoelastic effects that are characteristic of the polymer.

Stress relaxation, like other properties of polymers, is also dependent on temperature and follows an equation of the Arrhenius type:

$$\frac{1}{\gamma} = Ce^{-Q/RT}$$

Stress relaxation is often accompanied by creep, a continuing rate of strain under constant stress.

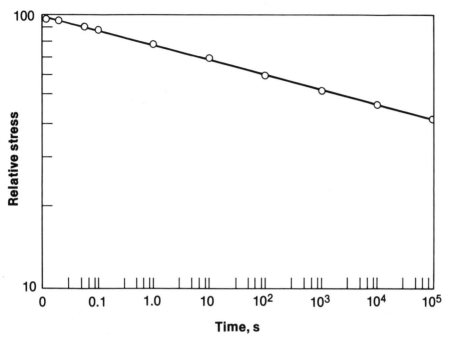

Fig. 11.12   Stress relaxation for a typical polyamide
Initial stress: 2000 psi (140 kg/cm^2)

11.6.3   Creep

When a polymer is kept under a constant stress, it rapidly strains to a position in accord with its modulus, followed by a slower, long-term rate of straining until the stress is removed or rupture occurs. This time-dependent plastic strain is called *creep*, and is also observed in metals at elevated temperatures. A typical creep curve is shown in Fig. 11.13. Three stages are observable: an initial period of rapidly increasing strain, ϵ, followed by a second stage with a slower, relatively constant *rate* of strain (rate = $d\epsilon/dt$), and a third stage of increasing rate of strain ending in rupture.

In the first stage, the rate of strain changes rapidly with time. The strain rate is the amount of incremental strain divided by the time in which it took place; this is the slope of the curve when time is plotted versus strain ($d\epsilon/dt$). There is no fixed modulus, but an instantaneous value called the creep modulus is calculated from the strain rate, $d\epsilon/dt$, at each point. A plot of the creep modulus versus time in logarithmic coordinates is often linear, enabling the extrapolation to points in time where data may not be available.

In the second stage, a constant *rate* of strain is observed. This creep rate of the second stage is the value often given as characteristic of the material, and the level of stress that caused the uniform second-stage creep rate is called the *creep strength*. The creep strength is smaller than the more important value of *creep-rupture strength*, the stress at which the material actually ruptures. The duration of each phase varies for each particular polymer; for example, polymers

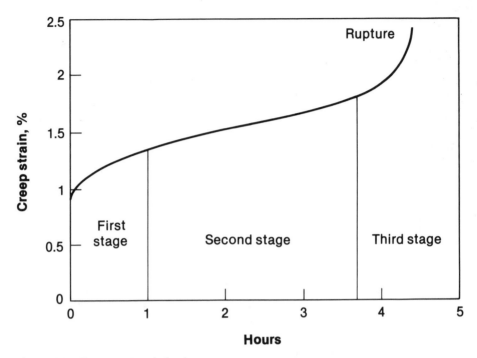

Fig. 11.13   Creep rupture behavior

with low ductility will rupture almost immediately after completion of stage two. The time to rupture for a given polymer sample will be a function of the level of stress applied. At very low stress, the strain rates tend to be small and constant, and the time to rupture becomes immeasurably long.

If one plots a characteristic point of creep such as time to rupture on a log scale versus the stress, a linear relationship is often observed. The data are important in determining the *allowable stress* when designing with plastic materials. For example, in designing a simple beam (Fig. 11.14) of thickness t, width w, and length L, loaded with an applied load of F, one should use the creep-rupture value, S, in the formula:

$$w \cdot t^2 = \frac{3FL}{2S}$$

instead of using the value of the elastic modulus E.

For design purposes, polymers will usually be employed at stresses well below the creep-rupture criterion. Nevertheless, creep is still important. Usually the logarithm of the time-dependent creep modulus (equal to the rate $d\epsilon/dt$ at any particular time) is plotted versus the logarithm of the time. This log-log plot is usually linear within the accuracy of the data available, as shown in Fig. 11.15.

In practice, creep-rupture curves are determined at a series of temperatures covering the range in which the plastic is likely to be used. The time to rupture and the maximum allowable

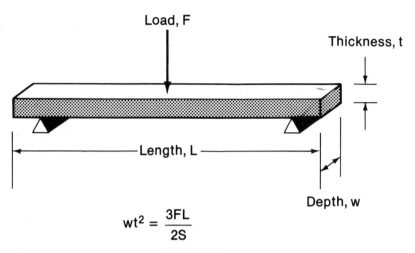

$$wt^2 = \frac{3FL}{2S}$$

Fig. 11.14   Stress on a centrally loaded beam

stress are lower at higher temperatures. In addition, the values of creep modulus are somewhat dependent on the level of stress under which they are determined.

11.6.4   Measurement of Dynamic Mechanical Properties

To measure dynamic properties, a test sample of material is subjected to a cyclic stress and both the stress and strain are measured continuously over time. One common method is to make the polymer specimen the spring in a torsion (twisting) pendulum apparatus (see Fig. 11.16). When the system is allowed to oscillate freely, the elastic shear modulus, G', can be determined:

$$G' = \frac{6.4\,\pi^2\,ILf^2}{\mu\,w \cdot t^3}$$

where I is the moment of inertia in g/cm^2, L is the specimen length, f is the observed frequency in Hz, μ is a shape factor for the specimen cross section, w is the specimen width, and t is the specimen thickness. The free oscillations will decay in amplitude in a manner determined by the *loss shear modulus*, which can be determined by a plot of the decay in amplitudes. The loss shear modulus, G'', is then:

$$G'' = \frac{G'\Delta}{\pi}$$

where the factor Δ, called the *log decrement*, is a dimensionless value given by:

$$\Delta = \ln\left(\frac{A_n}{A_{n+1}}\right)$$

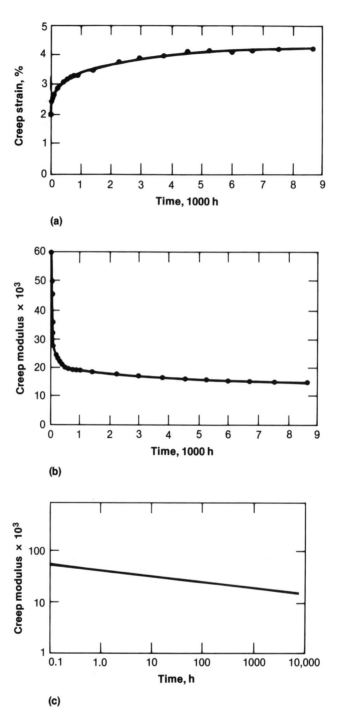

Fig. 11.15   Tensile creep of PTFE at 650 psi, 23 °C, showing derivation of creep modulus
(a) Creep strain versus time on cartesian coordinates. (b) Creep modulus versus time on cartesian coordinates
(creep modulus = applied stress, psi/creep strain, in./in.). (c) Creep modulus versus time on logarithmic coordinates showing linearity. Source: *Modern Plastics Encyclopedia*, McGraw-Hill, 1984-85, p. 410

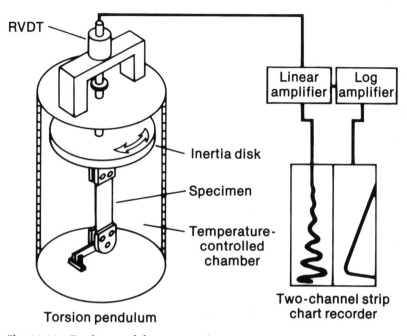

Fig. 11.16   Torsion pendulum test equipment
Source: *Modern Plastics Encyclopedia*, McGraw-Hill, 1984-85, p. 421

where A_n and A_{n+1} are the successive amplitudes of oscillations (refer to ASTM Standard D 2236). Typical torsion pendulum data are given in Fig. 11.17 and 11.18.

Similar experimental data can be obtained using a pendulum or other device that is *driven* with a sinusoidal force. In this case, the frequency is determined by the driving current and the value of the loss tangent is determined instead of the log decrement. Experiments using a driven sinusoidal force allow the time (frequency) dependence of the moduli to be determined and enable the analysis of short-time effects to be separated from the more long-time constant effects.

11.7   Deformation of Network Materials

In a perfect network formed by cross-linking N primary molecules there will be v chains, where a chain is that portion of the structure extending from one cross-link to the next cross-link along the primary molecule. The cross-links are fixed points in the structure in that four chain ends must meet at each cross-link point; however, they are not fixed points in space in the general sense. They have limited mobility in that they may move about from their equilibrium position only by altering the configuration of several of their attached chains. A change in configuration of one of the chains in a network, therefore, requires a change in the configuration of the nearest attached chains. In an infinite network, it is easy to see conceptually how a change in one chain has an effect on all the chains in the network, a sort of "domino" effect. When

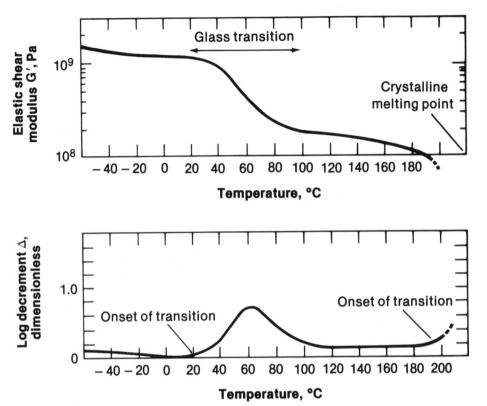

Fig. 11.17 Dynamic mechanical properties by torsion pendulum versus temperature of a typical semicrystalline thermoplastic (nylon 6, dry)
Shown are effects of glass transition and crystalline melting transition and a method of determining their onsets.
Source: *Modern Plastics Encyclopedia*, McGraw-Hill, 1984-85, p. 422

the bulk material is deformed, all of the network branch points assume new equilibrium positions relative to one another.

Any real network must of course contain some flaws, for example, free ends on terminal chains. For every primary molecule there will be two terminal chains, so that for a sample of N primary molecules there are 2N terminal chains, $v - N$ internal chains, and $v + N$ total chains. Thus the fraction of chains subject to a shift in relative position by deformation is given by $x = (v - N)/(v + N)$. This is expressed is terms of macroscopic parameters as:

$$x = \frac{1 - 2M_c}{M + M_c}$$

where M_c is the molecular weight per cross-linked unit, and M is the number average primary molecular weight. Because v equals the total number of units (N_o) times the fraction of units cross-linked, it follows that $M_c = N_o M_o/v$. Under the reasonable assumption that the average

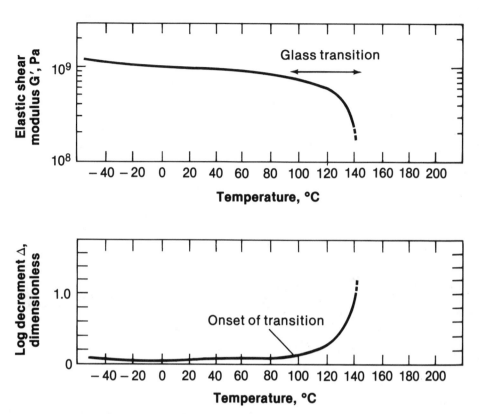

Fig. 11.18   Dynamic mechanical properties by torsion pendulum versus temperature of a typical amorphous thermoplastic (modified polyphenylene oxide, high T_g)
Shown are the effect of glass transition and a method of determining its onset. Source: *Modern Plastics Encyclopedia*, McGraw-Hill, 1984-85, p. 422

length of the terminal chains and the internal chains will be the same (random cross-linking), then x represents the weight fraction of the structure which is constrained in deformation.

Figure 11.19 illustrates the effect of chain end defects in network polymers. Note that in the figure the cross-link point B, which divides the primary chain ABC into two segments AB and BC, does not produce two constrained chain segments; rather, the original segment ABC is still an independent unit with two very long branches, BD and BE.

When N − 1 cross-links are formed, a single giant molecule is produced. Each cross-link beyond this creates one additional internal loop, and a true network quickly forms. Originally these loops will be quite large and will comprise two chain segments of large average molecular weight per segment. If N is the number of cross-links needed for incipient gel (network) formation, then the total number of effective cross-linkages, v_e, is:

$$v_e = v - 2N = v\,(1 - 2N/v)$$

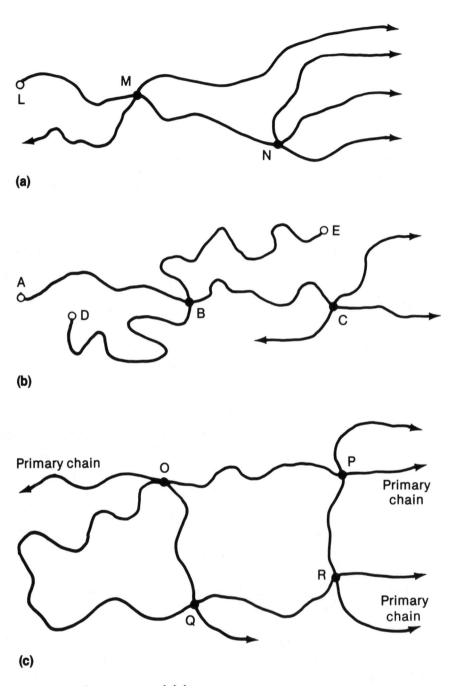

Fig. 11.19   Network structures and defects
(a) Chain LM is the free end of a primary molecule. (b) Cross-link B does not form a constraint. Chain segments AB and BC are not constraining segments since the segment ABC is still free to move independently; it is merely a chain with very long branches, DB and EB. (c) Chain POQR has three internal segments. Chain segments PO, OQ, QR, and RP form a constraint on the primary chain.

Thus, v_e is the number of chain segments effective in elastic response to macrodeformation. Alternatively,

$$\frac{v_e}{v} = \left(\frac{1 - 2M_c}{M}\right)$$

The factor in parentheses differs only slightly from the factor x (given earlier in this section) if the value of $M \gg M_c$. In a more realistic representation, one would have to take into account that these loops will also involve certain chain entanglements, which will also increase the number of elastic units per structure.

11.8   Mechanical Properties of Thermoset Materials

The major mechanical properties of thermoset materials may be understood from the discussion above by noting that the primary chains of a thermoset material are all constrained by these strong covalent network cross-linkages in contrast to the weak associations typical of thermoplastic polymers.

Thermosetting polymers are hard and usually brittle materials with very low creep, high solvent resistance, and a great degree of dimensional stability with temperature. Often they are formed by processing materials which are just below the gel point to the desired final net shape followed by additional cross-links to the thermoset material. Additional processing other than mechanical machining is impossible on a highly cross-linked thermoset. They combine the initial processing advantages of a thermoplastic with the ultimate stability of the thermoset.

Thermoset material properties are the direct result of the degree of cross-linking. When the number of effective cross-links, v_e, becomes quite large, the molecular weight per chain segment decreases significantly. The radius of the equivalent spherically coiled chain segment decreases (end-to-end distance is shorter), and there is less flexibility to extend the chain segments in response to external stress. After a very short strain, the chain segments are fully extended and further strain results in actual fracture of chains, which is, of course, not elastically recoverable. Hence, thermosets are brittle.

11.9   Mechanical Properties of Elastomers

Although linear polymers may undergo a rubbery response to deformation at certain temperatures, some polymers, known as *elastomers*, undergo exceptionally large amounts of elastic strain. The rubbery response is engineered to take place at the service temperature, usually room temperature, and the glass transition temperature is well below the service temperature, as shown in Fig. 11.20. The elastic response is directly related to v_e, the number of effective cross-links, and to M_n, the number average molecular weight.

As Fig. 11.21 shows, the stress-strain curve for an elastomer exhibits a large elongation with relatively low stress followed by a shift to a region of lower strain at higher stress, i.e., a higher modulus. This behavior may be understood in structural terms as resulting from the

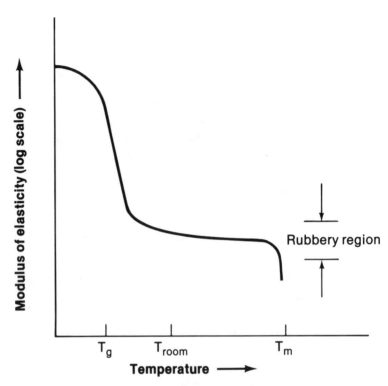

Fig. 11.20   The modulus of elasticity versus temperature plot of an elastomer has a pronounced rubbery region

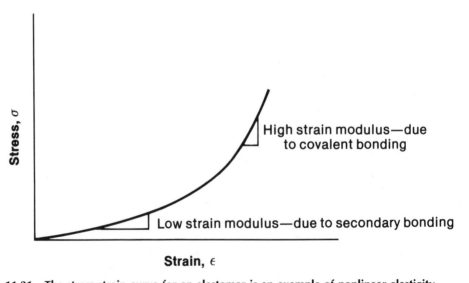

Fig. 11.21   The stress-strain curve for an elastomer is an example of nonlinear elasticity
The initial low-modulus (i.e., low-slope) region corresponds to the uncoiling of molecules (overcoming weak, secondary bonds). The high-modulus region corresponds to elongation of extended molecules (stretching primary, covalent bonds).

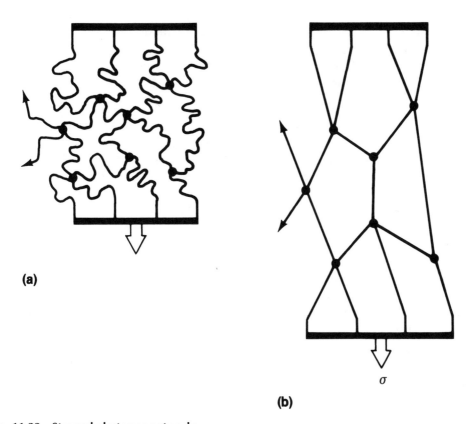

Fig. 11.22   Stressed elastomer networks
(a) Relaxed chains uncoiling; low modulus. (b) Stress on fully extended chains is borne by interatomic bonds; high modulus

coiling and recoiling of the long polymer chain segments. After the molecules have uncoiled as much as possible, additional strain results from stretching the main chain segments. The response to stress is not one of viscous flow accompanied by some chain uncoiling, which would result in a permanent, nonrecoverable change in physical dimensions. The reason is that all elastomers are effectively cross-linked into an infinite network at the service temperature. This means that the low-stress, high-strain region represents uncoiling of the relatively long chain segments between cross-links with a corresponding large increase in the end-to-end distance; and as these sections become more fully extended, the stress is shifted to the covalent bonding along chains (see Fig. 11.22).

11.10   Impact Strength in Polymeric Materials

Polymers are much more likely to fail after receiving a sudden blow or impact than are metals, and the improvement of impact strength is a major goal of polymer science. For a polymeric or other material to survive an impact without failure, it is necessary to:

- Ensure that any internal stresses do not exceed the fracture strength of the material, including stresses localized at stress concentration points
- Provide a mechanism for the material to handle the total kinetic energy of the blow, either by elastic response or absorption or both

The typical test used to measure impact (Izod or tensile impact test) consists of striking a blow against a test specimen of a standard shape. The blow is usually administered by a hammer or pendulum of a specific weight falling under gravity through a measured height. The actual loading and failure of the specimen is complex, as shown in a curve of impact failure (Fig. 11.23). This curve is typical of a filled or reinforced ductile material, and the final stage is primarily due to inclusions in the polymeric material. With other materials, only the initial elastic stage will be observed. Failure has occurred at the end of the elastic stage, regardless of how it proceeds. Therefore, only those properties and mechanisms for elastic response and energy absorption which are active preceding the initial phase of failure are important in preserving the integrity of the material. Tests that measure initial fracture energy (proportional to the energy under the stress-strain curve up to initial fracture) are more useful than those that measure the total energy of the process (Charpy, Izod, etc.). The important mechanical properties of a polymer are the short-time-span properties rather than long-time-span or static properties. Therefore, various moduli obtained from dynamic or cyclic stress measurements may be of particular assistance in ranking polymers for impact. Table 11.1 gives impact test results for several common polymers.

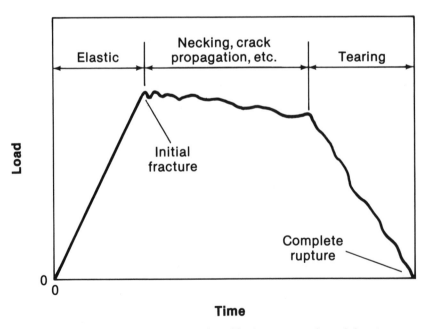

Fig. 11.23 Typical load versus time curve produced by instrumented pendulum impact machines

Steps which may be taken to increase impact resistance of polymers include appropriate choice of polymer, compounding for energy absorption, and designing to eliminate stress concentration points within the material. At the time of impact, the material is highly strained over a short time span. Addition of materials of low modulus and high fracture strain, such as elastomers, may increase impact resistance of a polymer blend. Elastomers are tough because they absorb a large amount of energy on impact and initial deformation. Often the design of the shape of plastic parts themselves enables a controlled response without exceeding allowable stress or buildup of stress concentrations. Designs that allow for deformation over a relatively long span will distribute the stress of a localized blow over a greater amount of the material.

Internal energy absorption over a short time span and conversion to heat (damping) are also effective. Because this absorption is proportional to volume, increasing the wall thicknesses of the shape is effective. Sometimes, however, the increased wall thickness will reduce the impact resistance due to bending deformation, etc., and the expected overall improvement will not be obtained.

Unfortunately, many modifications of materials that improve impact resistance—for example, compounding, plasticization, or introducing copolymers—reduce the primary material properties such as stiffness and long-time-span strength. One must often compromise between static strength and impact strength requirements in a design involving polymers.

11.11   Effect of Fillers, Extenders, and Plasticizers on Mechanical Properties

Fillers, extenders, and plasticizers are materials added to polymers during processing and compounding into commercially useful forms. *Plasticizers* are additives, usually liquid, which are added to soften rigid and brittle polymers and make them flexible. In addition, plasticizers lower processing temperatures. *Fillers* are materials, usually less expensive than the host polymer, that are added in substantial amounts, often over 50 vol%, to reduce cost, improve processibility, or in some cases modify the mechanical or thermal behavior of a polymer. Typical fillers include metallic and mineral powders, glass microspheres, wood flour, and other organic flours and powders. A filler used because it improves mechanical properties is called a *reinforcing filler*; a filler added primarily to reduce cost without sacrificing polymer performance too much is called an *extender*.

All fillers, extenders, and plasticizers modify the mechanical properties of polymers. The way they affect the basic properties of a material can usually be interpreted in terms of the various moduli. However, since a filler may be present at volume fractions larger than 50%, the filler particles themselves may actually carry a large percent of the total load, or they may contribute to the energy absorption of impact. These polymer systems are complex, and estimates of their mechanical properties require advanced calculations such as those used for composite materials.

The primary effect of inert fillers in a polymer matrix is to increase the rigidity in direct proportion to the weight percent of filler (Fig. 11.24). The effect of extenders and plasticizers may be interpreted in terms of the theory of behavior of polymers in concentrated solution. For example, a highly plasticized polymer has greater chain mobility, a lower glass transition

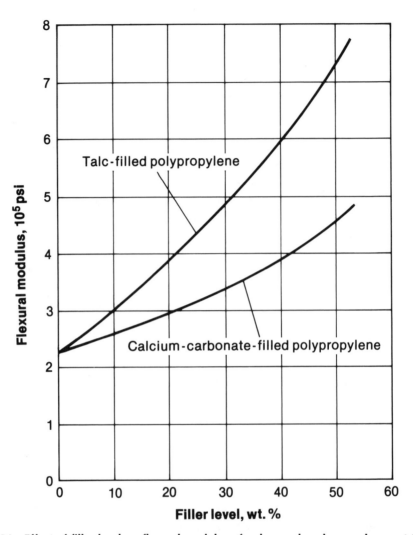

Fig. 11.24   Effect of filler level on flexural modulus of polypropylene homopolymer at 23 °C

temperature, and a greater viscous response to stress than the unmodified material. Plasticizers added to a network polymer form a gel, and the material will be expected to swell in volume and become softer and more compliant.

11.12   Polymer Alloys

Two polymers may be blended to form a material with greatly modified properties—a *polymer alloy*. The effect is similar to that obtained from the inclusion of simple fillers and extenders; however, because the added component is polymeric, some especially useful changes of the basic mechanical properties have been obtained. For example, thermoplastic elastomers have been alloyed (by a special type of blending) with a finely divided dispersion of a highly cross-linked elastomer to produce a material of greatly improved properties which is still easily processed. High-impact polystyrene can be produced by incorporating elastomers such as polybutadiene into a polystyrene blend or alloy. This material is produced either by a special blending and dispersion process or by introducing the elastomer during the styrene polymerization process. In addition to greater impact strength, the resulting material has modified rigidity, clarity, and tensile strength.

11.13   Optical Properties of Polymers

Although optical materials have been traditionally associated with ceramic glasses, crystalline quartz, and sapphire, polymeric materials have been increasingly designed and used for their optical properties. Several plastics (for example, polystyrene, polymethylmethacrylate, and polycarbonate) and their derivatives have important optical applications, perhaps the latest of which is the formation and grinding of optical lenses of soft hydrophilic polymers for use as contact lenses for the eyes. The formation of optical wave guides (fiber optics) is another area where the properties of polymers have been especially useful.

11.13.1   Refractive Index

The most fundamental optical property of a material is its *refractive index*. The refractive index of several polymers used for optical purposes is sufficiently great that they may be employed to form lenses in the same way that optical glass is used. Typical refractive index values for optical glasses vary from 1.5 to 2.5, while those for polymers range from 1.3 to 1.6, as shown in Table 11.3.

Many of the polymers listed in Table 11.3 are translucent due to a combination of partial crystallinity with, in some cases, incipient gel formation. Transparent sheets and optical lenses made from polymers have the advantages of lower weight, shatter resistance, and greater ease of processing compared with similar shapes produced from ceramics. A major disadvantage of polymeric optical devices is their susceptibility to scratching.

Table 11.3   Optical properties of selected polymers

Polymer	Refractive index	Color	Transmission
Cellulose acetate	1.46–1.50	Pale	Good
Nylon	1.53	Pale amber	Clear
Polycarbonate	1.59–1.60	Colorless	88–98%
Polyethylene	1.50–1.54	Colorless	Translucent
Methylmethacrylate	1.48–1.50	Colorless	92–98%
Polypropylene	1.49	Slightly yellow	Translucent
Polystyrene	1.59–1.60	Colorless	Excellent
Polystyrene-acrylonitrile	1.56–1.57	Slightly amber	89%
Polytetrafluoroethylene	1.30–1.40	Slightly gray	Translucent
Polyvinyl chloride	1.50–1.55	Slightly amber	Translucent
Polyacetal	1.48	Colorless	Translucent

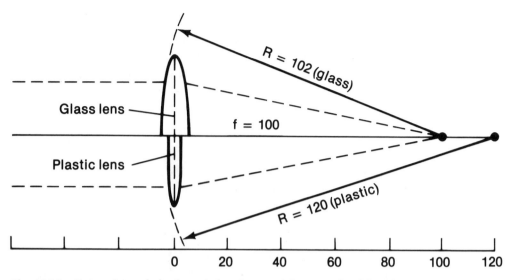

Fig. 11.25   Comparison of plastic and glass lenses of the same focal length

Consider the simple double-convex lens in Fig. 11.25, having a focal length of f = 100 cm, with faces O_1 and O_2 which are portions of a sphere of radius R, and made from optical material of refractive index n. The formula for the principal focus (focal length) is given by:

$$\frac{1}{f} = (n-1) \left(\frac{1}{R_1} - \frac{1}{R_2} \right)$$

For a lens made from an optical polymer of n = 1.6 to be equivalent to a lens of equal focal length made from optical glass of n = 1.51, the radii are 120 cm for the polymeric lens and 102 cm for the optical glass. The polymer lens will be thinner and much lighter in weight than the equivalent glass lens.

11.13.2   Opacity and Scattering

A pigment used for coloring plastic materials appears colored as a result of absorption of selected wavelengths from white light and is effective regardless of the matrix in which it is dispersed. However, most colored objects, paints, and opaque plastic films are bright because the incident light is scattered back to the observer. Ceramic pigments, such as titanium dioxide, are carefully formulated to maximize the light scattered to the observer. In Fig. 11.26, the scattering efficiency is a function of (1) the *relative refractive index*, which is the index of refraction of the pigment divided by the index of refraction of the medium in which it is dispersed, and (2) the particle size of the pigment. For a suspension of particles, an approximation of the magnitude of the scattering often used in the paint and paper industries has the form:

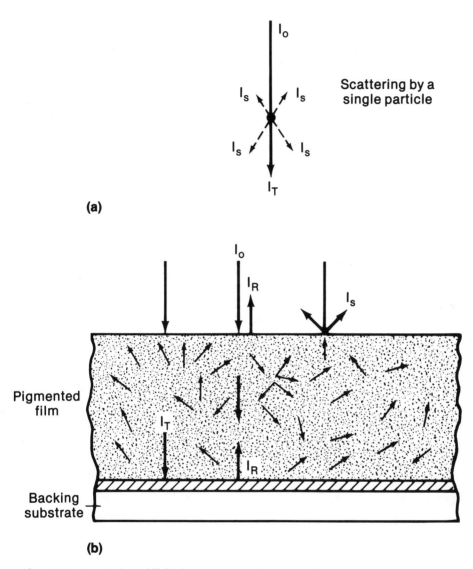

(a)

(b)

Fig. 11.26   Scattering of light from pigmented polymer films

$$R\infty = 1 + \frac{K}{S} - \left(\frac{K^2}{S^2} + \frac{2K}{S}\right)^{1/2}$$

when the thickness, x, is so great that a further increase in thickness does not increase the reflected flux. $R\infty$ is the total reflectivity of the thick section, K is an absorption coefficient, and S is the scattering coefficient, which can be calculated from the particle size and relative refractive index.

A flat wall paint is formulated so that the pigment is only loosely bound by the polymer and is primarily in contact with air. The air-pigment interface has a large value of the relative refractive index; consequently, it is highly reflective (maximum scattering). A gloss wall paint or a pigmented plastic film has the pigment well dispersed in polymers of nearly the same refractive index; this paint has a low value of relative refractive index and produces minimal scattering. The scattering can be increased somewhat by changing the average particle size. Thus, white pigments for polymer film and gloss paints require careful control of pigment particle size. For gloss paints, the preferred particle size is much smaller than that for flat paint formulations.

11.14   Electrical Properties of Polymers

A valuable property of plastic materials is their electrical insulating capability. When a polymer is used as an electrical insulator, several considerations are important, especially if the electrical circuit being insulated carries an alternating current.

Because a polymer is an integral part of the electronic circuit which it insulates, the following must be considered:

- *Resistivity* (both volume resistivity and surface resistivity), which is a measure of the ability to carry a current either through the bulk of the material or across the surface of the material
- *Dielectric constant* or *permittivity*, which determines the magnitude of the capacitive effect between metallic circuit elements or wiring insulated with a polymer; this effective capacitance influences the power loss in a.c. circuits
- *Dissipation factor* or *loss tangent*, which measures the conversion of the alternating electric field into heat in a.c. circuits

Volume resistivity is the property measured using Ohm's law with electrodes placed on either side of a sample of material, usually a thin disk. The volume resistivity is then defined as:

$$\sigma_v = \left(\frac{E}{I}\right)\left(\frac{A}{t}\right)$$

where E and I are the potential and current, A is the area of the sample under the electrodes, and t is the thickness. Volume resistivity is measured in ohm · cm and is of the order of 10^{10} or greater for insulators. In polymeric systems, the resistivity may vary due to heat aging or moisture absorption.

Surface resistivity is the resistance to conduction of a current across the surface of a polymer. The polymer surface is not always identical in composition to the bulk, due to the effects of processing. The most common source of surface conduction comes from the absorption of moisture on the polymer surface. The polymer surface may char slightly when an arc or spark jumps from one electrode to another. Subsequent to such an arc, a track of lower resistivity is formed on the surface and is a source of further breakdown at lower electrical potential.

Dielectric constant is the ratio of the capacitance of a device when the electrode space is filled with the sample material to the capacitance when the electrode space is evacuated. The dielectric constant is a fundamental property of the material and its structure, and it varies with the frequency of the alternating current and temperature. It is a measure of the capacity of the material to store energy under an electric potential. The energy storage comes about through the alignment of dipoles within the material itself. The dipoles are those arising from the different chemical nature of the atoms of the material, especially polar groups attached to the polymer chain.

When a material is subjected to an alternating current, the alignment of the dipoles reverses with each cycle of the current. As a result of this reversal of the alignment of dipoles, a fraction of the energy of the field is converted to heat and a measurable power loss occurs. This power loss is called the *dielectric loss* or *dissipation*, which is usually represented as the tangent of the angle by which the current lags the voltage in a test circuit, thus the term loss angle.

The total power loss in a system using a polymeric insulator in an a.c. circuit is given by:

$$P = \epsilon \tan \alpha \, E^2 \, fv$$

where P is the power loss in watts, ϵ is the dielectric constant, $\tan \alpha$ is the dissipation factor, E is the voltage gradient, f is the frequency, and v is the volume of the insulator in cubic centimeters.

11.15 Problems

1. Describe the differences between a glass transition temperature and a melting point. Some ceramics and polymers exhibit both of these phenomena in the same specimen. Why are both not found in the same metal specimen?

2. Why is viscosity a much more important property of polymers than of metals? Compare the importance of viscosity for glassy versus crystalline ceramics.

3. Why is time generally much more important in affecting properties of polymers compared with metals? What mechanical property of metals is time dependent?

4. Contrast the creep rates of most thermoplastic polymers above their glass transition temperatures and the creep rates of highly cross-linked thermosets. What factors affect the creep rate of thermoplastics?

5. What mechanical properties change when the degree of cross-linking increases in a polymer? Would you expect highly cross-linked polymers to be very soluble in organic solvents? Explain.

6. Explain why elastomers have very great elastic strain. Why are there no elastomeric metals or ceramics?

7. List several examples of fillers, extenders, or plasticizers.

8. What are some advantages and disadvantages of eyeglasses made from polymers?

9. List at least one very unusual property for most polymers; what would happen if some polymers could be developed to change the listed property? Answer this question for an unusual property for ceramics; for metals.

10. Why should an engineer think in terms of properties first and the choice of type of material (polymer, ceramic, or metal) second?

12

Structure & Properties of Composite Materials

12.1 The Nature of Composite Materials

It has long been recognized that combinations of materials having different properties often result in superior performance, each material contributing its own spectrum of properties to achieve a balance of performance in the composite which could not have been obtained separately. Bricks and plaster have been reinforced with straw and other fibrous materials for centuries; it has only been since World War II that the use of other types of composites has become widespread. Much of this growth of composites has come about in an attempt to modify plastic materials for an ever increasing number of applications.

Although composites can be defined as any combination of two or more materials, which would include wood, concrete, brick, and even alloys, this discussion will be confined to two types of composites: (1) those in which a particulate or fibrous material is added to a continuous matrix and (2) sandwiches, or laminates, in which layers of material are bonded together (see Fig. 12.1).

In the first type of composites, the matrices can be polymers, metals, ceramics, or glass, although polymeric matrices are the most widely known. Many of the thermoplastic materials used for consumer goods are molded from plastic materials reinforced with relatively short fiberglass fibers (see Table 12.1). Thermoset polymer composites are often fabricated with continuous fibers that are wound or woven into the structure. Aircraft and automotive production makes extensive use of reinforced plastic composites (Table 12.2).

Metal-matrix composites have not been extensively developed, primarily due to the difficulty of finding inexpensive reinforcing materials that are compatible with the metallic systems at high temperatures. Typical of such metal-matrix composites are two types of aluminum composite: (1) aluminum reinforced with high-strength boron fibers and (2) aluminum reinforced with relatively short SiC fibers. Examples of composite metal laminates and other metal-matrix composites are given in Tables 12.3 and 12.4.

12.2 Properties of Particulate (Nonfiber) Reinforced Composite Materials

The simplest case of a particulate-reinforced composite is that of a uniform dispersion of relatively spherical particulate material in a homogeneous matrix. The behavior of composites

259

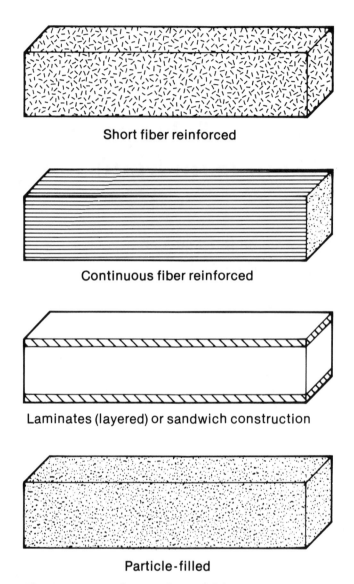

Short fiber reinforced

Continuous fiber reinforced

Laminates (layered) or sandwich construction

Particle-filled

Fig. 12.1   Types of composite materials

in which the reinforcing materials are flakes or fibers can be considered as modifications of this base case. Inexpensive, inert, particulate substances added to increase the volume and lower the cost of the polymer are called *fillers*. Fillers are widely used in commercial polymer materials; typical fillers are calcium carbonate, clay, talc, wood pulp, sand, alumina, calcium silicate, and titanium dioxide. The addition of fillers to a matrix material to lower cost is often called "extending" the material; thus, the added fillers are also referred to as *extenders*. Although fillers are most widely used for their economic benefits, they do influence the properties of the composite (see Table 12.5).

Particulate materials that are added primarily to improve the properties of the composite are usually referred to as *reinforcing agents*, to distinguish them from common fillers. For example, carbon black is added to rubber compounds to improve the wear resistance and other properties of rubber; it is properly called a reinforcing agent, and is essential for the development of the physical properties of rubber. It is used in concentrations as high as 50% of the total composition.

The properties of a composite can be expressed in terms of the properties of the components. Density, heat capacity, and conductivity, for example, can be described in terms of the amount and geometric distribution of each individual material in the composite: the expressions for such properties of composites are termed *mixture rules*. The property, such as density, of a mixture of several materials, e.g., particles dispersed in a matrix, is:

$$p_m = f_1 p_1 + f_2 p_2 + f_3 p_3 + \ldots$$

where p_m refers to the property of the mixture, f_i is the volume or weight fraction of the individual components i, and p_i is the property characteristic of each individual component. If the property being measured is normally expressed in volume units (for example, density), then the f in the expression is a volume fraction. If the property is expressed in terms of mass units, then the f will be a weight fraction. Mixture rules for dispersed phases of spherical particles can be applied even to irregularly shaped or platelike particles, provided a completely random orientation is present and the aspect ratio (length to width) is relatively low.

Example 12.2A. Two different fillers, wood flour and silica flour (SiO_2), are being considered for addition to a phenol-formaldehyde resin. Assume that the density of the resin is 1.3 g/cm^3; for wood flour it is 1.0 g/cm^3 and for SiO_2, 2.2 g/cm^3. What would be the density of each mixture if the mixtures were 50% of each ingredient by weight?

Solution: For the wood flour composite, based on 100 g of composite, we have 50 g of wood flour and 50 g of resin. The volume of wood flour = 50 g/(1.0 g/cm^3) = 50.0 cm^3. The volume of resin = 50 g/(1.3 g/cm^3) = 38.5 cm^3. The total volume of the composite of 100 g = 50.0 cm^3 + 38.5 cm^3 = 88.5 cm^3. The volume fraction for the two ingredients will be: f_1 (wood flour) = 50.0 cm^3/(50.0 cm^3 + 38.5 cm^3) = 0.565; f_2 (resin) = 38.5 cm^3/(50.0 cm^3 + 38.5 cm^3) = 0.435.

For the silica composite, based on 100 g of composite, we have 50 g of silica flour and 50 g of resin. The volume of silica flour = 50 g/(2.2 g/cm^3) = 22.7 cm^3. The volume of resin = 50 g/(1.3 g/cm^3) = 38.5 cm^3. The total volume of the composite of 100 g = 22.7 cm^3 + 38.5 cm^3 = 61.2 cm^3. The volume fraction for these two ingredients will be: f_1 (silica flour) = 22.7 cm^3/(22.7 cm^3 + 38.5 cm^3) = 0.372; f_2 (resin) = 38.5 cm^3/(22.7 cm^3 + 38.5 cm^3) = 0.628.

The density of the wood flour mixture is: $p_m = f_1 p_1 + f_2 p_2 = f_1(1.0) + f_2(1.3) = (0.565 \times 1.0) + (0.435 \times 1.3) = 1.13$ g/cm^3. The density of the silica flour mixture is: $p_m = f_1 p_1 + f_2 p_2 = (0.372 \times 2.2) + (0.628 \times 1.3) = 1.63$ g/cm^3.

Example 12.2B. Such resins as the above are usually *sold by the pound*, but because they are molded into specific shapes, they are *used by volume*. Assuming both materials cost 10 cents per kilogram, what would be the relative cost of parts made from each mixture?

Table 12.1 Properties of glass fiber reinforced resin-matrix composites(a)

Material	Glass fiber content, wt. %	Flexural strength, ksi	Flexural modulus, 10⁵ psi	Tensile strength at yield, ksi	Tensile modulus, 10⁵ psi	Compressive strength, ksi	Ultimate tensile elongation, %	Izod impact strength, ft·lb/in. of notch	Thermal conductivity, Btu·in./ft²·h·°F (K value)	Specific heat, Btu/lb·°F	Flammability (UL94)(b)
Glass fiber reinforced thermosets											
Sheet molding compound (SMC)	15–30	18–30	14–20	8–20	16–25	15–30	0.3–1.5	8–22	1.3–1.7	0.30–0.35	5V
Bulk molding compound (BMC)	15–35	10–20	14–20	4–10	16–25	20–30	0.3–0.5	2–10	1.3–1.7	0.30–0.35	5V
Preform/mat (compression molded)	25–50	17–30	13–18	25–30	9–20	15–30	1–2	10–20	1.3–1.8	0.30–0.33	VO
Cold press molding—polyester	20–30	22–37	13–19	12–20	...	...	1–2	9–12	1.3–1.8	0.30–0.33	VO
Spray-up—polyester	30–50	16–28	10–12	9–18	8–18	15–25	1.0–1.2	4–12	1.2–1.6	0.31–0.34	VO
Filament wound—epoxy	30–80	100–270	50–70	80–250	40–90	45–70	1.6–2.8	40–60	1.92–2.28	0.23–0.25	VO
Rod stock—polyester	40–80	100–180	40–60	60–180	40–60	30–70	1.6–2.5	45–60	1.92–2.28	0.22–0.25	VO
Molding compound—phenolic	5–25	18–24	30	7–17	26–29	14–35	0.25–0.6	1–6	1.1–2.0	0.20–0.30	VO
Glass fiber reinforced thermoplastics											
Acetal	20–40	15–28	8–13	9–18	8–15	11–17	2	0.8–2.8	...	...	HB
Nylon	6–60	7–50	2–26	13–33	2–20	13–24	2–10	0.8–4.5	...	0.30–0.35	VO
Polycarbonate	20–40	17–30	7.5–15	12–25	7.5–17	14–24	2	1.5–3.5	...	...	VO
Polyethylene	10–40	7–12	2.1–6	6.5–11	4–9	4–8	1.5–3.5	1.2–4.0	...	...	VO
Polypropylene	20–40	7–11	3.5–8.2	5.5–10.5	4.5–9	6–8	1–3	1–4	...	...	VO
Polystyrene	20–35	10–17	8–12	10–15	8.4–12.1	13.5–19	1.0–1.4	0.4–4.5	...	0.23–0.35	VO
Polysulfone	20–40	21–27	8–15	13–20	15	21–26	2–3	1.3–2.5	...	...	VO
ABS (acrylonitrile butadiene styrene)	20–40	23–26	9.2–15	11–16	6–10	12–22	3–3.4	1–2.4	...	...	VO
PVC (polyvinyl chloride)	15–35	20–25	9–16	14–18	10–18	13.4–16.8	2–4	0.8–1.6	...	...	VO
Polyphenylene oxide (modified)	20–40	17–31	8–15	15–22	9.5–15	18–20	1.7–5	1.6–2.2	...	...	VO
SAN (styrene acrylonitrile)	20–40	15–21	8.0–18	13–18	9–18.5	12–23	1.1–1.6	0.4–2.4	...	...	VO
Thermoplastic polyester	20–35	19–29	8.7–15	14–19	13–15.5	16–18	1–5	1.0–2.7	1.3	...	VO

Table 12.1 (continued)

Material	Mechanical and physical properties[b]							Chemical resistance[c]				
	Rockwell hardness	Dielectric strength, V/mil	Specific gravity	Density, lb/in.³	Heat distortion at 264 psi, °F	Continuous heat resistance, °F	Coefficient of thermal expansion, 10^{-6} in./in.·°F	Weak acids	Strong acids	Weak alkalis	Strong alkalis	Organic solvents
Glass fiber reinforced thermosets												
Sheet molding compound (SMC)	H50–112	300–450	1.7–2.1	0.061–0.075	400–500	300–400	8–12	G to E	F	F	P	G to E
Bulk molding compound (BMC)	H80–112	300–450	1.8–2.1	0.065–0.075	400–500	300–400	8–12	G to E	F	F	P	G to E
Preform/mat (compression molded)	H40–105	300–600	1.5–1.7	0.054–0.061	350–400	150–400	10–18	G to E	F	F	P	G to E
Cold press molding—polyester	H40–105	300–600	1.5–1.7	0.054–0.061	350–400	150–400	10–18	G to E	F	F	P	G to E
Spray-up—polyester	H40–105	200–400	1.4–1.6	0.050–0.058	350–400	150–350	12–20	G to E	F	F	P	G to E
Filament wound—epoxy	M98–120	300–400	1.7–2.2	0.061–0.079	350–400	500	2–6	E	F	E	G	E
Rod stock—polyester	H80–112	200–400	1.6–2.0	0.058–0.072	325–375	150–500	3–8	G to E	F	F	F	G to E
Molding compound—phenolic	M90–99	150–370	1.7–1.9	0.061–0.069	400–500	325–350	4.5–9	F	P	F	P	F
Glass fiber reinforced thermoplastics												
Acetal	M78–94	500–600	1.55–1.69	...	315–335	185–220	19–35	F	P	F	P	E
Nylon	...	400–500	1.47–1.7	0.049	300–400	300–400	11–21	G	P	E	F	G
Polycarbonate	M75–100	450	1.24–1.52	...	285–300	275	12–18	E	G(d)	G	F	P(f)
Polyethylene	...	450–500	1.16–1.28	...	150–200	280–300	17–27	E	G(d)	E	E	G(h)
Polypropylene	R95–115	500–600	1.04–1.22	...	230–300	300–320	16–24	E	G(d)	E	G	G(h)
Polystyrene	M70–95	350–425	1.20–1.29	0.045–0.048	200–220	180–200	17–22	E	G(d)	G	G	P(f)
Polysulfone	M85–92	...	1.38–1.55	...	333–350	...	12–17	E	E	E	E	G
ABS (acrylonitrile butadiene styrene)	M75–102	...	1.23–1.38	...	215–240	200–230	16–20	E	G(d)	E	E	P(g)
PVC (polyvinyl chloride)	M80–88	500–550	1.45–1.62	...	155–165	...	12	E	G	E	E	P(g)
Polyphenylene oxide (modified)	M95	...	1.20–1.38	...	220–315	240–265	10–20	E	E	E	E	G(j)
SAN (styrene acrylonitrile)	M77–103	...	1.22–1.40	...	210–230	200–220	16–21	G	G(e)	G	G	P(g)
Thermoplastic polyester	R118–M70	560–750	1.45–1.61	...	380–470	275–375	24–33	F	P	P	P	E

(a) Data from Owens-Corning Fiberglas. (b) Classification shown is highest obtainable rating. Less critical applications may permit use of materials with lower classifications. (c) E = excellent (outstanding); G = good (acceptable); F = fair (test before using); P = poor (not recommended). (d) Attacked by oxidizing acids. (e) Disintegrates in sulfuric acid. (f) Soluble in aromatic and chlorinated hydrocarbons. (g) Soluble in ketones and esters, and in aromatic and chlorinated hydrocarbons. (h) Below 176 °F (80 °C). (j) Softens in some aromatic and chlorinated aliphatics. Resistant to alcohol.

Table 12.2   U.S. reinforced plastics shipments by end use (mil lb)

Application	1985	1986	1987	1988	1989	%89/88
Aircraft/aerospace/military	33	37	36	40	42	5.0
Appliance/business equipment	133	136	141	148	158	6.9
Construction	444	456	506	497	497	...
Consumer products	144	149	167	170	181	6.2
Corrosion-resistant equipment	295	291	329	349	361	3.3
Electric/electronic	190	201	214	229	233	1.5
Marine	330	340	413	458	453	−1.0
Land transportation	568	585	656	695	710	2.2
Other, specialties	79	83	75	80	82	3.0
Total(a)	**2216**	**2279**	**2536**	**2665**	**2715**	**2.0**

(a) May differ from actual sums due to rounding
Source: Composites Institute

Solution: Assume the parts molded were 1 cm^3 in volume in order to get a relative cost. Then 1 kg of wood-based material would yield 885 cm^3 of parts (885 parts) for 10 cents, or 0.011 cents per part for material, while the silica-filled material would yield 613 cm^3 (613 parts) for 10 cents, or 0.016 cents per part for material. The wood flour-based material is 0.011/0.016 or 69% of the material cost of the silica flour-based material.

12.3   Effect of Fillers on Physical Properties

The most important polymer properties that are modified by formation of a composite are flexural modulus, tensile strength, and impact strength.

Fillers always increase the *modulus* of a composite. In general, the modulus of the composite is a complex function of the relative moduli of the filler and the matrix, the particle size and size distribution, and whether the polymer is a thermoplastic, thermoset, or elastomeric material. In general, however, the ratio of the composite modulus to the modulus of the matrix polymer increases exponentially as the volume of filler increases. Typically, the modulus of the composite may be as much as 10 to 12 times that of the matrix when maximum filler loading is reached.

The *tensile strength* of filled polymers is influenced by particle size, particle shape, and the degree of bonding at the filler/matrix interface. If the interfacial bonding is strong, then the tensile strength may be greatly enhanced. Conditions under which the tensile stress can be transferred to the filler material can improve the performance of the composite. Fillers with a large *aspect ratio* (ratio of length to cross-sectional diameter), such as fibers, make the greatest improvement and are the most commonly used reinforcing fillers, as discussed in section 12.4.

Fillers are often added to polymers specifically to improve *impact strength*. If the filler has a greater ductility than the matrix, the impact strength of the composite is significantly improved over that of the matrix. Elastomeric materials are often dispersed in more rigid polymers for this purpose (see section 11.10). Because fillers are often rigid, high-modulus materials, the composite will usually be more brittle than the matrix.

Table 12.3 Composite metal laminates

Cladding	Aluminum and alloys	Beryllium copper	Brass	Bronze	Copper	Cupro-nickel	Gold	Gold-nickel alloy	Indium	Fe/Fe-Ni alloys	Magnesium	Molybdenum	Nickel	Ni-Fe alloys	Ni-Cu alloys	Ni-Cr-Fe alloys	Nickel-silver	Platinum	Silver	Stainless steels	Steel, carbon	Tantalum	Titanium	Tungsten	Zinc	Zirconium
																				Base metals						
Aluminum	×	×	×	×	×		×		×	×	×	×	×					×	×		×	×	×	×	×	
Aluminum alloys	×	×	×	×	×				×	×		×		×					×		×	×	×	×	×	
Bismuth alloys	×																									
Brass				×	×																					
Bronze					×																×					
Cadmium	×								×										×							
Calcium																			×							
Copper and alloys	×	×	×	×	×				×	×	×	×	×	×	×	×			×	×	×	×				
Gold and alloys	×				×				×			×	×	×		×	×	×	×	×	×	×				
Indium	×	×	×	×	×				×		×	×	×													
Fe/Fe-Ni low-expansion alloys	×	×	×	×	×				×			×									×	×	×	×		
Lead and alloys	×		×	×	×				×		×	×	×						×		×	×	×			
Magnesium	×										×															
Nickel and alloys		×	×	×	×	×					×	×	×	×	×	×			×	×	×	×	×			
Palladium and alloys	×	×	×	×	×	×			×		×	×	×	×	×	×	×		×	×	×	×				×
Platinum and alloys	×	×	×	×	×	×			×		×	×	×	×	×	×	×	×	×	×	×	×				×
Silver and alloys	×		×	×	×	×	×								×	×	×	×	×	×	×	×			×
Stainless steels	×				×				×											×					
Steel, low-carbon	×														×				×							
Tantalum	×																									
Tin and alloys	×	×	×	×	×	×	×		×		×	×	×				×		×	×	×	×	×	×	×	
Titanium							×				×	×						×	×	×		×			
Zinc	×							×																		

Source: *Stamping/Diemaking*

Table 12.4   Typical mechanical properties of some metal-matrix fiber-reinforced composites

Fiber	Matrix	Reinforce-ment, vol %	Density, g/cm³(a)	Longi-tudinal tensile strength, MPa(b)	Longi-tudinal modulus, GPa(b)	Transverse tensile strength MPa(b)	Transverse modulus, GPa(b)
G T50	201 Al	30	2.380	620	170	50	30
G T50	201 Al	49	...	1120	160	...	...
G GY70	201 Al	34	2.380	660	210	30	30
G GY70	201 Al	30	2.436	550	160	70	40
G HM pitch	6061 Al	41	2.436	620	320	...	...
G HM pitch	AZ31 Mg	38	1.827	510	300	...	...
B on W, 142-μm fiber	6061 Al	50	2.491	1380	230	140	160
Borsic	Ti	45	3.681	1270	220	460	190
G T75	Pb	41	7.474	720	200	...	...
G T75	Cu	39	6.090	290	240	...	...
FP	201 Al	50	3.598	1170	210	(140)	140
SiC	6061 Al	50	2.934	1480	230	(140)	140
SiC	Ti	35	3.931	1210	260	520	210
SiC whisker	Al	20	2.796	340	100	340	100
B₄C on B	Ti	38	3.737	1480	230	>340	>140
G T75	Mg	42	1.799	450	190	...	...
G HM	Pb	35	7.750	500	120	...	...
G T75	Al-7% Zn	38	2.408	870	190	...	...
G T75	Zinc	35	5.287	770	120	...	...
G T50	Nickel	50	5.295	790	240	...	...
G T75	Nickel	50	5.342	828	310	30	40
G (81.3 μm)	2024 Al	50	2.436	760	140	...	...
G (142 μm)	2024 Al	60	2.436	1100	180	...	...
Superhybrid	Graphitic	60	2.048	860	120	220	60
Superhybrid	S-glass	60	2.159	740	80	190	30
Superhybrid	Kevlar	60	1.799	700	80	190	10

(a) To convert g/cm³ to lb/in.³, divide by 27.68. (b) To convert MPa to psi, multiply by 145; to convert GPa to psi, multiply by 145,000.
Source: *Encyclopedia of Composite Materials and Components*, M. Grayson, Ed., John Wiley & Sons, 1983, p. 621

12.4   Continuous Fiber-Reinforced Composite Materials

For the case where the composite comprises a mixture of uniform continuous fibers dispersed evenly in a matrix, the laws of mixture discussed in section 12.2 will apply for properties measured in the direction of the fiber length (see Fig. 12.2). The tensile strength is given by:

$$\sigma_c = \sigma_f V_f + \sigma_m V_m$$

where σ_c, σ_f, and σ_m are the stresses carried by the composite, the fiber, and the matrix, respectively; V_f and V_m are the volume fractions of the fiber and matrix, respectively. An analogous equation holds for ultimate tensile strength, yield strength, stress-rupture strength, compressive strength, and the moduli E and G. For example, the *modulus of elasticity*, E, is:

$$E_c = E_f V_f + E_m V_m$$

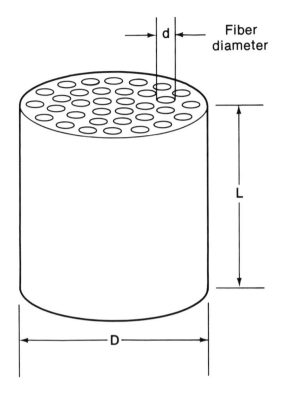

$$\text{Volume of fiber} = (\pi d^2 L)/4$$

$$\text{Volume of matrix} = (\pi D^2 L - \pi d^2 L)/4$$

$$V_f = \frac{\pi d^2 L}{\pi D^2 L} \qquad V_m = 1 - V_f$$

Fig. 12.2 Continuous fiber-reinforced composite
Calculation of the volume fractions of the components

The *flexural modulus*, G, is given as:

$$G_c = G_f V_f + G_m V_m$$

These relationships hold for strains that are strictly elastic. When one component, usually the matrix, undergoes plastic deformation or viscous flow, a modified expression should be used. Figure 12.3 demonstrates this linear relationship between tensile strength of a continuous fiber-reinforced composite with the volume fraction of fiber for tungsten fiber-reinforced copper.

The above expressions make it relatively simple to calculate the expected behavior of any composite, provided the data are available for the individual components. Note that the property improvements occur only in the axial direction. Strengths in the transverse direction can be

Table 12.5   Applications and properties of filled resins(a)

Material formulation(b)	Resin type(b)	Application	Melt flow rate(c), g/10 min	Density(d), g/cm^3	Mold shrinkage(e), in./in.	Tensile strength(f), ksi	Elongation break(g) (2 in. per min), %
Calcium carbonate filled							
10% HP	PP	Sheeting	0.5–2.5	0.975	0.016–0.025	4.8	100
20% HP	PP	Lawn and patio equipment	3.0–7.0	1.04	0.010–0.017	4.2	50
40% HP	PP	Food containers	18.0–25.0	1.28	0.006–0.012	3.8	40
40% HP	PP	Housewares	25.0–35.0	1.25	0.006–0.012	3.5	40
20% CP	PP	Outdoor equipment	12.0–18.0	1.06	0.010–0.016	3.3	>100
40% CP	PP	Toys, housewares	6.0–10.0	1.23	0.006–0.012	3.5	25
Hi 40% CP	PP	Housewares	18.0–25.0	1.23	0.008–0.015	2.4	80
Talc filled							
20% HP	PP	Medical instruments	3.0–7.0	1.04	0.010–0.015	5.3	18
30% HP	PP	Medical instruments	8.0–12.0	1.13	0.005–0.012	4.9	12
40% HP	PP	AC housings, fan shrouds	3.0–7.0	1.24	0.004–0.010	5.0	7.5
20% CP	PP	Cabinets	6.0–10.0	1.03	0.010–0.018	4.1	40
40% CP	PP	Downspouts	3.0–7.0	1.24	0.008–0.014	4.5	20
Mica filled							
30% Unc	PP	Pool equipment	3.0–7.0	1.13	0.005–0.008	4.7	5
40% Unc	PP	Instrument panels	3.0–7.0	1.23	0.003–0.008	5.5	4
12% Cou	PP	Automotive components	6.0–10.0	0.990	0.008–0.015	5.3	10
20% Cou	PP	Automotive components	4.0–10.0	1.03	0.005–0.008	5.8	5
30% Cou	PP	Pool equipment	3.0–7.0	1.13	0.005–0.008	6.3	5
40% Cou	PP	Automotive components	3.0–7.0	1.23	0.003–0.008	6.5	3
50% Cou	PP	Automotive components	6.0–10.0	1.35	0.002–0.006	5.5	3
Glass filled							
10% Unc	PP	Automotive components	...	0.970	0.005–0.008	6.0	4
20% Unc	PP	Appliances	...	1.04	0.003–0.006	6.5	3
30% Unc	PP	Appliances	...	1.13	0.002–0.006	8.0	2
10% Cou	PP	Automotive components	...	0.970	0.005–0.008	7.2	5
20% Cou	PP	Automotive components	...	1.04	0.003–0.006	10.0	3
30% Cou	PP	Automotive components	...	1.13	0.002–0.006	10.5	2

(a) Data from Washington Penn Plastic Co. (b) PP, polypropylene; HP, homopolymers; CP, copolymers; Cou, coupled; Unc, uncoupled. (c) ASTM D 1238. (d) ASTM D 729. (e) ASTM D 953. (f) ASTM D 638. (g) ASTM D 350. (h) ASTM D 570. (j) ASTM D 790. (k) ASTM D 256. (m) ASTM D 648. (n) ASTM D 2240

much lower, usually representing the strength of the matrix. At other angles the critical stress is often shear at the fiber-matrix interface. A number of individual layers or plies oriented in different directions are often used to provide a more isotropic (uniform in all directions) set of physical properties. The analysis of these advanced structures is more complex, however. This anisotropic strength of a continuous fiber-reinforced composite must be considered in practical applications where the stress may not always be exactly along the fiber axis. If the composite is subjected to axial stress but the fibers have a slight angle of alignment in the matrix, the strength can be drastically reduced (see Fig. 12.4). Note that the lowest strength is at a 45° angle to the direction of the applied stress.

Water absorption(h) (24 h at 73 °F), %	Flexural strength(j), ksi	Flexural modulus(j), 10⁶ psi	Izod impact strength(k) at 73 °F, ft · lb/in.				Gardner impact (2 lb) at 73 °F, in. · lb	Heat deflection(m) (⅛ in.), °F		Shore hardness(n)	
			Notched		Unnotched			66 psi	264 psi	A	D
			⅛ in.	¼ in.	⅛ in.	¼ in.					
0.02	6.6	0.250	1.10	0.90	12.80	10.15	30	190	140	95	74
0.02	6.0	0.270	0.78	0.65	13.30	8.20	8	195	145	95	72
0.02	6.5	0.370	0.70	0.60	10.20	5.05	35	220	150	94	74
0.02	6.2	0.350	0.60	0.50	17.60	8.00	40	220	160	96	75
0.02	5.0	0.200	1.30	1.10	22.10	20.00	>130	175	125	96	63
0.02	6.0	0.350	1.20	0.55	9.65	6.50	80	210	145	93	70
0.02	4.0	0.200	1.50	0.90	14.10	9.15	80	190	135	97	67
0.01	7.0	0.400	0.70	0.60	8.05	7.65	10	250	150	87	72
0.02	7.5	0.475	0.65	0.55	6.75	5.80	9	255	158	90	74
0.02	8.5	0.625	0.60	0.50	5.33	4.35	8	268	180	94	75
0.02	5.8	0.280	1.30	1.10	21.00	19.50	60	220	130	85	70
0.02	7.0	0.510	0.75	0.60	5.85	4.15	10	245	150	93	78
0.04	7.5	0.590	0.60	0.53	3.05	2.55	6	263	195	92	75
0.05	8.5	0.900	0.55	0.50	2.20	1.70	4	278	210	97	78
0.03	7.0	0.420	0.80	0.60	10.10	7.70	12	250	170	86	66
0.01	10.0	0.520	0.85	0.55	3.50	3.90	24	275	203	90	77
0.04	10.9	0.730	0.60	0.55	3.25	2.95	7	275	195	93	76
0.05	11.5	1.000	0.55	0.53	3.10	2.15	6	285	220	96	76
0.06	11.0	1.300	0.52	0.45	2.25	1.75	4	310	260	94	76
0.05	8.5	0.350	0.80	0.60	7.00	6.00	6	275	250	98	74
0.05	9.0	0.525	1.1	1.0	7.10	7.35	6	290	260	89	74
0.05	10.0	0.650	1.1	1.0	10.80	7.75	8	295	270	97	78
0.05	10.5	0.470	1.2	0.60	7.45	6.70	7	285	265	96	73
0.05	11.0	0.600	1.3	1.0	7.80	7.45	8	310	275	92	75
0.05	13.0	0.750	1.6	1.4	11.50	7.90	10	320	295	95	77

12.5 Properties of Short Fiber and "Whisker"- Reinforced Plastic Materials

Thermoplastic materials reinforced with glass fibers constitute an important group of composite materials. Nearly all of the common polymeric materials, such as polyethylene, polypropylene, polystyrene, polyamides (nylons), and polycarbonates utilize glass fibers to form a composite or reinforced material with improved properties. To preserve the moldability of the thermoplastic material, the reinforcing fibers are chopped into short lengths of about ½ in.

The property enhancements most often observed are:

- Greater tensile strength
- Greater stiffness (higher modulus in bending)

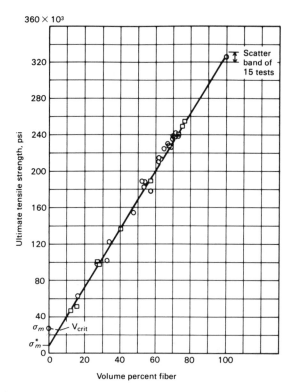

**Fig. 12.3   Tensile strengths of tungsten fiber-reinforced copper composites (for continuous rein-
forcement with 5-mil-diam tungsten fibers)**
Source: D.L. McDanels *et al.*, NASA TN D-1881, 1963; also, D.L. McDanels *et al.*, in *Trans. AIME*, Vol. 233 (No. 4), April 1965

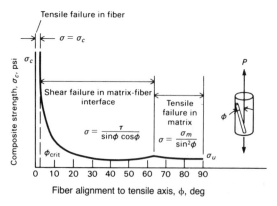

**Fig. 12.4   Failure mode as function of alignment of fibers in discontinuous fiber-reinforced com-
posite**
Source: D.W. Petrasek *et al.*, NASA TN D-3886, March 1967

- Higher impact strength
- Higher fatigue strength
- Higher creep resistance
- Better property retention at high temperatures
- Greater dimensional stability

Table 12.6 compares the properties of some common glass fiber-reinforced polymers to the same properties for the unreinforced material. The fibers used for reinforcement are generally short (½ to 1 in.) and are randomly oriented. The strength of the composite, therefore, cannot be estimated from the simple law of mixtures as in the preceding example.

The mechanism for bearing the load in short fiber-reinforced composites is more complex, because any individual fiber must transfer its share of the load at the end of each individual fiber segment. In the ideal case where the fibers are discontinuous, but aligned parallel to each other, as in Fig. 12.5, the load is transferred in shear over the length, L, by which the fibers overlap each other. This shear strength is equal to or greater than the strength of the fiber so that:

$$\sigma_f = \frac{4 \, \gamma \, L_c}{D}$$

where σ_f is the fiber stress, γ is the shear stress, L_c is the critical fiber overlap, and D is the diameter of the fiber. This leads to the concept of a *critical aspect ratio* (ratio of length to diameter) for reinforcing fibers or other reinforcements. The critical aspect ratio is the length to diameter at which the load transfer is assured and the failure mode under stress becomes fiber fracture rather than failure in shear. The failure in shear can be either through interfacial shear or through shear in the matrix. The necessary aspect ratio is:

$$\frac{L_c}{D} = \frac{1}{2} \left(\frac{\sigma_f}{\gamma} \right)$$

where L_c is the critical length for a fiber of diameter D, σ_f is the fiber tensile strength, and γ is the smaller of the matrix shear strength or the shear strength of the fiber-matrix interface. In most commercial composites, the surfaces of the fibers are modified by special treatments to promote bonding of the fiber to the matrix and increase the interfacial shear strength.

Example 12.5A. A composite is constructed of S glass with a maximum diameter of 20 μm in a matrix of nylon with a shear strength of 9600 psi (66.3 MPa). What is the critical diameter for this fiber reinforcement? The tensile strength of the fiber is 665,000 psi (4590 MPa).

Solution: The critical aspect ratio is: $L_c/D = \frac{1}{2}(\sigma_f/\gamma) = \frac{1}{2}(4590 \text{ MPa}/66.3 \text{ MPa}) = 69.2$. The critical length for the fiber of 20 μm is then: $L_c = 69.2 \times 20 \text{ μm}/10^3 \text{ μm/mm} = 0.14$ mm, or 0.0055 in. Note that the typical ½ in. fiber is approximately 1000 times this critical length. When the fiber length is substantially greater than the critical length, the strength of the composite is closely approximated by that calculated by the rule of mixtures.

Table 12.6 Comparative properties of neat-resin and carbon- and glass-reinforced engineering thermoplastics

Resin type	Tensile strength		Flexural modulus		Impact strength(a), notched/unnotched, J/cm	Heat-deflection temperature, °C
	MPa	ksi	GPa	10^6 psi		
Amorphous resins						
Acrylonitrile-styrene-butadiene (ABS):						
Base resin	55	8	2.8	0.4	1.9/>21	95
30% glass fiber	100	14.5	7.6	1.1	0.75/3.5	105
30% carbon fiber	130	18.8	12.4	1.8	0.59/2.4	105
Nylon:						
Base resin	66	9.5	2.6	0.38	>21/NB	125
30% glass fiber	148	21.5	7.9	1.15	0.64/3.7	140
30% carbon fiber	207	30	15.2	2.2	0.64/4.3	145
Polycarbonate:						
Base resin	62	9	2.3	0.33	1.4/>21	130
30% glass fiber	128	18.5	8.3	1.2	2.0/9.34	150
30% carbon fiber	165	24	13.1	1.9	0.96/5.34	150
Polyetherimide:						
Base resin	105	15.2	2.8	0.4	0.53/13	200
30% glass fiber	197	28.5	8.6	1.25	0.75/5.60	215
30% carbon fiber	234	34	17.2	2.5	0.75/6.67	215
Polyphenylene oxide (PPO):						
Base resin	66	9.5	2.5	0.36	1.4/>21	85–150
30% glass fiber	145	21	9.0	1.3	1.2/5.1	155
30% carbon fiber	159	23	11.7	1.7	0.53/3.0	155
Polysulfone:						
Base resin	69	10	2.8	0.4	0.64/>21	170
30% glass fiber	124	18	8.3	1.2	0.96/7.5	185
30% carbon fiber	159	23	14.5	2.1	0.64/3.5	185
Styrene-maleic-anhydride (SMA):						
Base resin	54	7.8	3.2	0.47	0.2/NB	115
30% glass fiber	103	15	9.0	1.3	0.59/2.4	120
Thermoplastic polyurethane:						
Base resin	14	2	0.1	0.015	>21/NB	32
30% glass fiber	57	8.2	1.3	0.19	5.1/15	170
Crystalline resins						
Acetal:						
Base resin	61	8.8	2.8	0.4	0.69/>21	110
30% glass fiber	134	19.5	9.7	1.4	0.96/4.8	165
20% carbon fiber	81	11.8	9.3	1.35	0.53/1.6	160
Nylon 66:						
Base resin	80	11.6	2.8	0.41	0.48/>21	77
30% glass fiber	179	26	9.0	1.3	1.5/11	255
30% carbon fiber	241	35	20.0	2.9	0.80/6.4	257
Polybutylene terephthalate (PBT):						
Base resin	59	8.5	2.3	0.34	0.48/>21	85
30% glass fiber	134	19.5	9.7	1.4	1.4/9.1	210
30% carbon fiber	152	22	15.9	2.3	0.64/3.5	210
Polyethylene terephthalate (PET):						
30% glass fiber	159	23	9.0	1.3	1.0/ · · ·	225
Polyphenylene sulfide (PPS):						
Base resin	74	10.8	4.1	0.6	0.16/1.9	140
30% glass fiber	138	20	11.0	1.6	0.75/4.5	260
30% carbon fiber	186	27	16.9	2.45	0.59/2.9	265

(a) NB, no break

Source: *Plastics Engineering*, Society of Automotive Engineers, Vol. XLI (No. 4), April 1985, p. 38–41

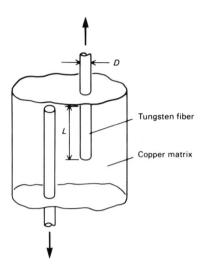

Fig. 12.5 Schematic diagram of shear load transfer mechanism in composites reinforced with discontinuous fibers
Source: D.W. Petrasek *et al.*, NASA TN D-3886, March 1967

Example 12.5B. A sample of nylon resin reinforced with 13 wt.% fiberglass has a tensile strength of 16,000 psi (110 MPa). This rises to 25,000 psi (172 MPa) when compounded with 33% glass fibers, and to 30,000 psi (207 MPa) when reinforced with 43% glass fibers. What is the apparent tensile strength of the polymer and of the fiber? The specific gravity of the composite is 1.22 at 13% fiber content, 1.40 at 33%, and 1.50 at 43%. The specific gravity of the nylon resin is 1.14 and that of the reinforcing fiberglass is 2.6.

Solution: The volume fractions of fiber are calculated as in Example 12.2A. For the 13% composite, based on 100 g of composite, we have 13 g of glass fiber and 87 g of resin. The volume of glass fiber = 13 g/(2.6 g/cm^3) = 5.0 cm^3. The volume of resin = 87 g/(1.14 g/cm^3) = 76.3 cm^3. The total volume of the composite of 100 g = 5.0 cm^3 + 76.3 cm^3 = 81.3 cm^3. The volume fraction for the fiber will be: f_1 (13% fiber) = 5.0 cm^3/(5.0 cm^3 + 76.3 cm^3) = 0.062.

For the 33% composite, based on 100 g of composite, we have 33 g of glass fiber and 67 g of resin. The volume of glass fiber = 33 g/(2.6 g/cm^3) = 12.7 cm^3. The volume of resin = 67 g/(1.14 g/cm^3) = 58.8 cm^3. The total volume of the composite of 100 g = 12.7 cm^3 + 58.8 cm^3 = 71.5 cm^3. The volume fraction for the fiber will be: f_1 (33% fiber) = 12.7 cm^3/(12.7 cm^3 + 58.8 cm^3) = 0.1776.

For the 43% composite, based on 100 g of composite, we have 43 g of glass fiber and 57 g of resin. The volume of glass fiber = 43 g/(2.6 g/cm^3) = 16.5 cm^3. The volume of resin = 57 g/(1.14 g/cm^3) = 50.0 cm^3. The total volume of the composite of 100 g = 16.5 cm^3 + 50.0 cm^3 = 66.5 cm^3. The volume fraction for the fiber will be: f_1 (43% fiber) = 16.5 cm^3/(16.5 cm^3 + 50.0 cm^3) = 0.2481.

A plot of the tensile strengths versus volume fractions yields a straight line, which extrapolates to 12,000 psi (82.8 MPa) tensile strength at 0% fiberglass. The expression for the line is: tensile strength (psi) = 12,000 + f_1 × (30,000/0.2486); tensile strength (MPa) = tensile strength (psi) × 0.0069023. An alternative expression can be written using the law of mixing:

Table 12.7 Properties of fiberglass-reinforced nylon 66(a)

ASTM method	Property	Unreinforced	10% glass	20% glass	30% glass	40% glass	50% glass	60% glass
General								
D792	Specific gravity	1.13–1.15	1.21	1.28	1.37	1.46	1.57	1.70
...	Specific volume, cm^3/kg	...	827	784	730	686	636	589
D570	Water absorption in 24 h, %	1.5	1.1	0.9	0.9	0.6	0.5	0.4
D570	Equil. cont. immersion, %	8.0	7.8	5.6	3.8	3.0	2.6	2.3
D955	Mold shrinkage, mm/mm (in./in.):							
	3.2-mm avg section	0.015	0.0065	0.0050	0.0040	0.0040	0.0030	0.0020
	6.4-mm avg section	0.018	0.0150	0.0060	0.0055	0.0050	0.0040	0.0030
Mechanical								
D638	Tensile strength:							
	MPa	81.4	96.5	131	186	214	221	228
	ksi	11.8	14.0	19.0	27.0	31.0	32.0	33.0
D638	Tensile elongation, %	60	3–4	3–4	3–4	2–3	2–3	1–2
D790	Flexural strength:							
	MPa	103	138	200	262	290	321	345
	ksi	15.0	20.0	29.0	38.0	42.0	46.5	50.0
D790	Flexural modulus:							
	GPa	2.83	4.48	5.86	8.96	11.0	15.2	19.3
	10^6 psi	0.41	0.65	0.85	1.30	1.60	2.20	2.80

Table 12.7   (continued)

ASTM method	Property	Unreinforced	10% glass	20% glass	30% glass	40% glass	50% glass	60% glass
Mechanical (continued)								
D695	Compressive strength:							
	MPa	33.8 (1%)	124	159	165	172	186	207
	ksi	4.9 (1%)	18.0	23.0	24.0	25.0	27.0	30.0
D732	Shear strength:							
	MPa	66.2	68.9	73.1	86.2	87.6	91.7	95.1
	ksi	9.6	10.0	10.6	12.5	12.7	13.3	13.8
D256	Izod impact strength, notched(b)							
	J/m	48.0	42.7	64.1	107	139	139	139
	ft · lb/in.	0.9	0.8	1.2	2.0	2.6	2.6	2.6
D256	Izod impact strength, unnotched(b)							
	J/m	…	267–320	427–480	907	1015	1070	1070
	ft · lb/in.	…	5–6	8–9	17	19	20	20
D785	Rockwell hardness	R118	R121 M92	R121 M93	R121 M96	R121 M96	R121 M100	R121 M104
Thermal								
D621	Deformation under load(c), %	1.4(d)	0.9	0.7	0.6	0.4	0.3	0.2
D648	Deflection temperature at 0.45 MPa (66 psi), °C	182	260	260	260	260	260	260
D648	Deflection temperature at 1.82 MPa (264 psi), °C	66	252	252	254	260	260	260
Cenco	Thermal conductivity, W/m · °K	0.25	0.39	0.42	0.49	0.52	0.55	0.58
D696	Coefficient of linear thermal expansion, 10^{-5} m/m · °K	8.1	4.9	4.1	3.2	2.5	1.8	1.6

(a) Data from LNP Corp. Test specimens dry, as molded. (b) Izod impact test bars 6.35 × 12.7 mm. (c) In 24 h at 28 MPa and 50 °C. (d) At 14 MPa.

tensile strength (psi) = $f_2 \times 12,000 + f_1 \times$ (tensile strength of fiber), where f_2 is the volume fraction of resin, and is $1 - f_1$. Choosing $f_1 = 0.2486$ (43% fiber), the equation simplifies to: fiber tensile strength = [tensile strength (psi) $- (f_2 \times 12,000)]/f_1$; fiber tensile strength = $[30,000 - (0.7514 \times 12,000)]/0.2486 = (30,000 - 9017)/0.2486 = 84,400$ psi, or 583 MPa.

The tensile strength of a fiberglass filament is between 200,000 and 300,000 psi (1380 to 2070 MPa). The low apparent value of fiberglass strength calculated from the law of mixing (84,400 psi, or 583 MPa) results from the random orientation of the glass strands. Many strands bear little or no stress, because of the random orientation and because the strands are discontinuous in the axis of stress. However, the composite is isotropic, with equal strength in any direction. Composite materials fabricated with long continuous filaments aligned in the axis of maximum stress are much stronger in the maximum stress direction, but less strong in the transverse direction, as previously noted (see Fig. 12.4). Reinforcement with different values of fiber content allows the designer a wide choice in the balance of properties using a single polymer type. For example, Table 12.7 lists the properties developed in a typical polyamide (nylon) resin at several levels of fiber content and compares them with the properties of the unreinforced material.

Reinforcement fibers can be chosen from a variety of materials; however, fiberglass is most often chosen because it has the best combination of strength, price, and availability. Other

Table 12.8   Some fibers used in composites

Fiber(a)	Fiber diameter, μm	Density, Mg/m³	Tensile strength MPa	ksi	Modulus GPa	10⁶ psi	Use limit, °C	Price/lb (1985)
Plastic matrix								
E glass (C)	3–20	2.49	3450	500	72.4	10.5	425	$0.80–$1.20
S glass (C)	10–20	2.49	4590	665	86.9	12.6	425	$4
Kevlar (C)	12	1.44	2760	400	125	18	425	$16
Carbon/graphite- PAN (C)	7.0	1.72–1.80	2410–4830	350–700	230–395	33–57	>1650(b)	$17–$450
Carbon/graphite- pitch (C)	5.1–12.7	1.99–2.16	2070	300	380–690	55–100	>1650(b)	$26–$1250
Processed mineral fiber (DC)	4–6	2.68	830	120	105	15	760	$0.30–$0.50
Fiberfrax (DC)	2–5	2.60	1030	150	105	15	1150	$1
Fibermax (DC)	3–6	2.99	860	125	150	22	1760	$16.65
Metal matrix								
Boron (C)	102–103	· · ·	3450	500	400	58	540	$262
Carbon/graphite- PAN (C)	7.0	· · ·	2410–4830	350–700	230–395	33–57	>1650(b)	$17–$450
Carbon/graphite- pitch (C)	5.1–12.7	· · ·	2070	300	380–690	55–100	>1650(b)	$26–$1250
SiC monofilament (C) ..	140	· · ·	4140	600	425	62	930	$800
SiC (W)	6	· · ·	3340	485	485–825	70–120	930	$95
FP alumina (C)	20	· · ·	1380	200	380	55	>1650	$200
Fiberfrax (DC)	2–5	· · ·	1030	150	105	15	1150	$1
Fibermax (DC)	3–6	· · ·	830	120	150	22	1760	$16.65

(a) C, continuous; W, whisker; DC, discontinuous. (b) Oxidation begins at lower temperatures.

fibers in use are given in Table 12.8. For aerospace applications there is a premium on maximum strength at minimum weight. In these applications, the cost of transporting one extra pound is well known to aircraft and missile manufacturers. Therefore, a reinforced material with a high strength-to-weight ratio can command a much higher price for aerospace applications than for automotive applications, for example.

12.6 Layer-Type Composite Materials

A composite material can be fabricated by bonding together two or more layers of dissimilar materials to form a "sandwich"-type structure. Plywood is a typical example of such a material. Lightweight "honeycomb" sandwich panels have been heavily used in aircraft construction (see Fig. 12.6). Sandwich panels of lightweight plastic (often foamed) cores with metal faces, or skins, are widely used in building construction.

The basic mechanics of a sandwich panel bearing a load on its faces are similar to those of an "I" beam subjected to bending loads. One face is in tension, one face is in compression, and the core is subject to shear. In designing the composite, therefore, each component can utilize a material chosen to maximize the specific property required for that component, while minimizing the overall weight and use of expensive materials. When evaluating a composite panel for a specific application, however, the type of loading of the panel becomes very important. Stress patterns in the components of the composite vary significantly with different loading arrangements (see Fig. 12.7). Usually the applications can be designed using well-

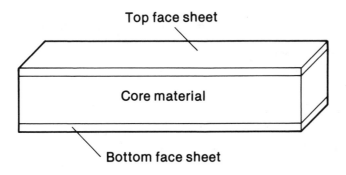

Relative mechanical properties	Solid face material of unit thickness	Sandwich of 2 × unit thickness	Sandwich of 4 × unit thickness
Strength	1	3.5 ×	9.3 ×
Stiffness	1	7 ×	37 ×

Fig. 12.6 "Sandwich"-type composite

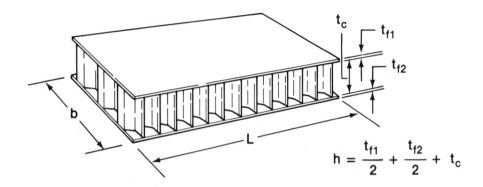

$$h = \frac{t_{f1}}{2} + \frac{t_{f2}}{2} + t_c$$

(a)

Bending stress in facings

$$\sigma_{fi} = \frac{M}{t_{fi}hb}$$

where M is determined from
Fig. 12.7(c) and i = 1 or 2

Core shear stress

$$\tau_c = \frac{V}{hb}$$

where V is determined
from Fig. 12.7(c)

Face dimpling

$$\sigma_{CR} = \frac{2E_f}{\lambda}\left[\frac{t_f}{s}\right]^2$$

Face wrinkling

$$\sigma_{CR} = 0.82\, E_f\left[\frac{E_c t_f}{E_t t_c}\right]^{1/2}$$

Deflection

$$\Delta = \frac{2K_b PL^3 \lambda}{E_f t_f h^2 b} + \frac{K_S PL}{hG_c b} \qquad \text{(for same skin materials)}$$

or

$$\Delta = \frac{K_b PL^3}{D} + \frac{K_S PL}{hG_c b}$$

(for most long beams, the second term is
relatively small, but should be checked if
deflection is critical.)

K_b and K_S are from
Fig. 12.7(c) and

$$D = \frac{E_1 t_1 E_2 t_2 h_2 b}{E_1 t_1 \lambda_2 + E_2 t_2 \lambda_1} \quad \text{or}$$

$$\frac{E t_1 t_2 h^2 b}{(t_1 + t_2)\lambda} \quad \text{or} \quad \frac{E t h^2 b}{2\lambda}$$

(b)

Fig. 12.7   Analysis of flat rectangular sandwich beams
Source: Hexcel Corporation

Beam type	Maximum shear force, V	Maximum bending moment, M	Bending deflection constant, K_b	Shear deflection constant, K_S
Simple support P = ql Uniform load 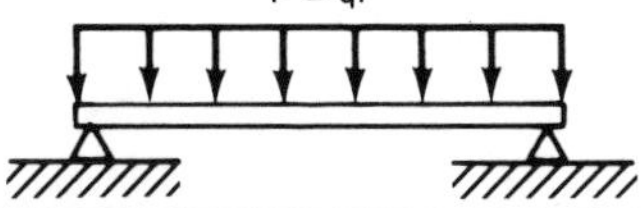	0.5P	0.125PL	0.01302	0.125
Both ends fixed P = ql Uniform load 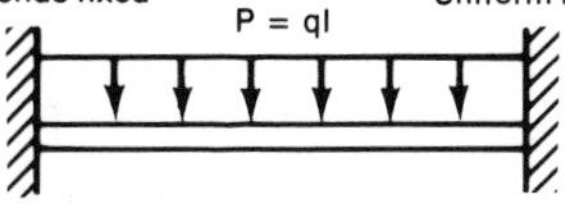	0.5P	0.08333PL	0.002604	0.125
Simple support P Center load 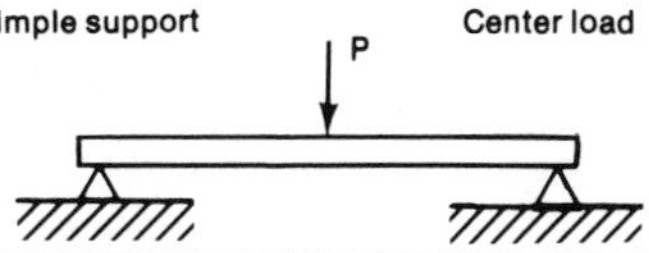	0.5P	0.25PL	0.02083	0.25
Both ends fixed P Center load 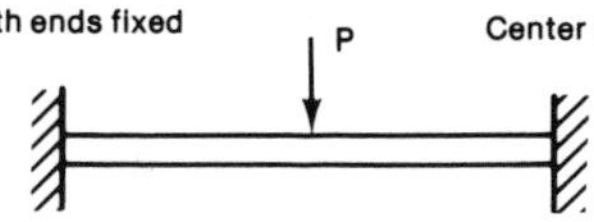	0.5P	0.125PL	0.00521	0.25
Cantilever P = ql Uniform load 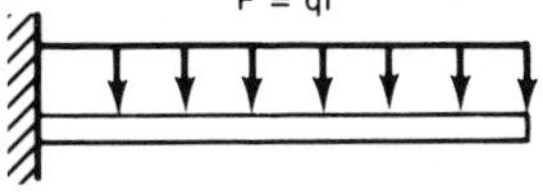	P	0.5PL	0.125	0.5
Cantilever End load P 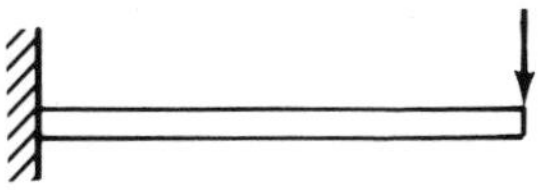	P	PL	0.3333	1
Cantilever Triangular load q P = ½ ql 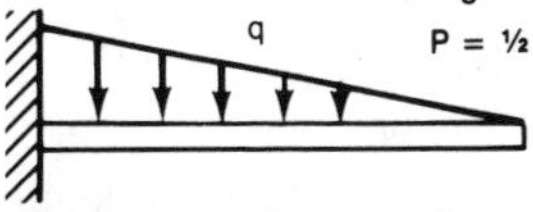	P	0.3333PL	0.06666	0.3333
One end simply supported One end fixed q Uniform load P = ql 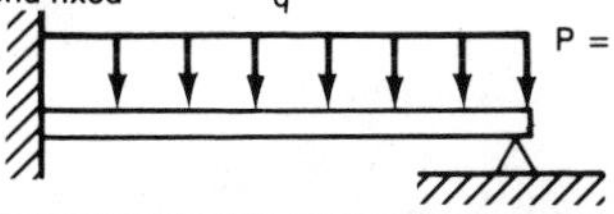	0.625P	0.125PL	0.005405	0.07042

(c)

Fig. 12.7 (continued)   Analysis of flat rectangular sandwich beams
Source: Hexcel Corporation

known formulas developed for beams. The properties of typical materials used in faces are given in Table 12.9, and examples of typical core materials are given in Table 12.10.

Example 12.6A. A standard 4 × 8 ft (48 × 96 in.) sandwich panel is fabricated from aluminum honeycomb material and 5042-H34 aluminum facings. Assume that the facings are 0.020 in. thick and the core is 1 in. thick. If the composite panel is to be supported at each end (across the full width) and will bear a uniformly distributed load of 20 lb/ft^2, what is the deflection of the beam and the shear stress on the core? For the faces, E = 10.1×10^6 psi (69.6 GPa) $\lambda = (1 - \mu) = 0.89$, where μ is Poisson's ratio. For the core, the shear strength (γ) is 85 psi (12.3 mPa).

Solution: Using the equations for shear stress (γ) and deflection (Δ) in Fig. 12.7, P = 20 lb/ft^2 × 4 ft × 8 ft = 640 lb. For a uniformly distributed load, V = 0.5 × P = 320 lb, K_b = 0.01302, and K_s = 0.125. Calculate deflection, Δ: L = 8 ft × 12 in./ft = 96 in.; b = 4 ft × 12 in./ft = 48 in.; Δ = $K_b(PL^32\lambda)/(Eth^2b) + K_s(PL)/(hG_cb)$. The panel thickness is the total of both faces and the core. For most beams the second term in the equation for deflection is usually small and can be neglected unless the deflection is a critical specification. Neglecting the second term, Δ = $(2)(0.0130)(640$ lb$)(96$ in.$^3)(0.89)/(10.1 \times 10^6$ lb/in.$^2)(0.020$ in.$)(1.020$ in.$^2)(48$ in.$)$ = 1.30 in. deflection. Evaluation of the second term shows that it will add 0.01 in. for a total of 1.31 in. Evaluating the shear stress: γ = V/(hb) = 320 lb/(1.020 in.)(48 in.) = 6.54 lb/in.^{2}, or 0.94 mPa. The calculated shear stress is less than one-twelfth the stress (12.3 mPa) that the core can withstand.

Example 12.6B. What thickness of polyester-fiberglass facing would be needed to give the same deflection?

Solution: Using woven roving polyester combination with E = 1.85×10^6 psi (268.25 Pa) and λ_s = 0.98, the deflection formula above will have these values substituted for the values for aluminum faces. Δ = 1.30 × (correction factor derived from Example 12.6A) = 1.30 × (0.98/0.89) × (10.1 lb/in.^{2}/1.85 lb/in.^{2}) × (0.020 in./t in.) × [(1.020 in./(1.0 in. + t in.)]^2. For the deflection to remain at 1.30, the correction factor must equal 1. Therefore, (0.98/0.89) × (10.1 lb/in.^{2}/1.85 lb/in.^{2}) × (0.020 in./t in.) × [(1.020 in./(1.0 in. + t in.)]^2 = 1. Thus, 1.101 × 5.459 × (0.020 in.)(1.020 in.)^2/(t in.) × (1.0 + t in.)^2 = 1, and 1.101 × 5.459 × 0.020808 in.^{3} = (t) × (t^2 + 2t + 1) = t^3 + 2t^2 + t. An approximate solution can be obtained by neglecting the t^3 term. Then, 0.125 in.^{3} = 2t^2 + t or t^2 + 0.5t − 0.0625 in.^{3} = 0. Therefore, t = 0.1035 in., or about five times the thickness of the aluminum skins. The total thickness is now 1.207 in. This is only [(1.207/1.04) − 1] × 100, or 16% thicker.

Example 12.6C. From Table 12.10, what alternative materials can be used as core material for a panel of the same core thickness?

Solution: The calculated shear stress is 6.54 psi (0.95 mPa). Any material capable of sustaining this shear stress can be used, although a margin of safety would normally be assumed. From Table 12.10, one could use a core of foamed polystyrene, or of nylon, polypropylene, etc.

Table 12.9 Properties of typical sandwich facing materials(a)

Facing material	Yield strength(b) MPa	ksi	Modulus of elasticity GPa	10^6 psi	λ_f (h), $1 - \mu^2$	Weight per mil thickness, kg/m^2	Comments
Aluminum:							
1100-H14	89.6C	13C	68.9	10.0	0.89	0.068	Moderate cost, workable, excellent chemical resistance, scars easily
3003-H16	124C	18C	68.9	10.0	0.89	0.068	Fair strength, good weather resistance, moderate cost
5052-H34	165C	24C	69.6	10.1	0.89	0.068	Better strength, good weather resistance, moderate cost
6061-T6	241C,T	35C,T	68.9	10.0	0.89	0.068	Good strength, workable, only heat treatable alloy easily welded
2024-T3	290T	42T	72.4	10.5	0.89	0.068	Excellent strength, heat treatable, soft stage for working, fair corrosion resistance
7075-T6	455T	66T	71.0	10.3	0.89	0.068	High strength, fair corrosion resistance
Mild carbon steel	345	50	200	29	0.91	0.20	Low cost, high weight, good availability
Stainless steel:							
316	414	60	193	28	0.94	0.20	High cost, corrosion resistant
17-7 PH	1380	200	200	29	0.94	0.20	High strength, heat treatable
Graphite, woven	586	85	55.8	8.1	0.99	0.039(c)	High cost, strength, and modulus
Graphite, unidirectional	1290	187	126	18.3	0.99	(d)	High cost, strength, and modulus
Fiberglass prepreg:							
Epoxy F155(e)	427	62	22.8	3.3	0.98	0.046	Excellent strength, low-temperature cure
Epoxy F161(e)	399	57.9	25.5	3.7	0.98	0.044	Heat resistant, good strength
Phenolic F120(e)	331	48	24.1	3.5	0.98	0.043	High temperature, good strength
Polyester F141(e)	331	48	24.1	3.5	0.98	0.044	Good strength, low cost
Polyimide F174(e)	414	60	24.1	3.5	0.98	0.042	High temperature resistant
Epoxy unidirectional F155	690	100	44.8	6.5	0.98	(d)	Highest strength
Kevlar F155(f)	221C	32C	28.3C	4.1C	1.00	0.034	High cost, high tensile
Kevlar F155(f)	483T	70T	28.3T	4.1T	...	...	Low weight, tough, low compression
Fiberglass mat, polyester resin ...	96.5	14	6.3	0.92	0.98	0.034	Very low cost
Woven roving, polyester resin ...	262	38	12.8	1.85	0.98	0.034	Very low cost
Ext fir plywood(g)	18.3	2.65	12.4	1.8	0.99	0.015	...
Pine plywood(g)	20.7	3.0	12.4	1.8	0.99	0.015	...
Luan plywood(g)	15.5	2.25	12.4	1.8	0.99	0.010	...
Tempered hardwood, 1.12 Mg/m^3 (70 lb/ft^3)	24.8	3.6	4.5	0.65	0.99	0.024	Good hardwood, low cost, smooth surface
Gypsum board	0.83	0.12	2.1	0.30	0.98	0.020	Fire resistant, low cost

(a) Data from Hexcel-Trevarno. (b) Yield strength is the lower of tensile (T) or compressive (C). (c) 3K-70-PU (12½ × 12½ Thornel 300/3000 Tow 7 mil/ply). (d) Dependent on fiber areal weight of 145 g/m^2 (4.28 oz/yd^2). (e) All values are for 1581 fabric and autoclave cure. (f) Both compressive (C) and tensile (T) shown for 285 styles. (g) For calculations involving plywood, "effective thickness" should be used except for locating centroid: actual thicknesses of 6.35, 9.52, and 12.7 mm (¼ or 0.250, ⅜ or 0.375, and ½ or 0.500 in.) correspond to effective thicknesses of 3.56, 3.56, and 6.60 mm (0.14, 0.14, and 0.26 in.), respectively. (h) μ = Poisson's ratio of facing: $\lambda = 1 - \mu^2$

Laminates have been fabricated from alternating layers of dissimilar metals. A common example is found in U.S. coinage, where silver coins have been substituted by a composite with a copper core and copper-nickel alloy faces. In this case, the primary motivation appears to be economic, substituting the copper for the more expensive silver in the bulk of the coin. Another

Table 12.10 Honeycomb property data

Honeycomb type(a)	Nominal density, lb/ft³	Compressive						Plate shear					
		Bare Strength, psi		Stabilized			Crush strength (typ), psi	"L" direction			"W" direction		
				Strength, psi		Modulus (typ), ksi		Strength, psi		Modulus (typ), ksi	Strength, psi		Modulus (typ), ksi
		Typ	Min	Typ	Min			Typ	Min		Typ	Min	
Aluminum alloy honeycombs													
5052-⅛-0.0007	3.1	270	200	290	215	75	130	210	155	45.0	130	90	22.0
5052-³⁄₁₆-0.0015	4.4	500	360	525	385	145	250	330	280	68.0	215	160	30.0
5052-¼-0.001	2.3	165	120	175	130	45	75	140	100	32.0	85	57	16.2
5052-⅜-0.0007	1.0	30	20	45	20	10	25	45	32	12.0	30	20	7.0
5056-⅛-0.0007	3.1	340	250	360	260	97	170	250	200	45.0	155	110	20.0
5056-³⁄₁₆-0.0015	4.4	600	460	650	490	180	310	410	340	68.0	245	198	27.5
5056-¼-0.001	2.3	205	145	210	155	58	100	170	130	32.0	105	62	15.0
5056-⅜-0.0007	1.0	35	25	50	35	15	35	60	45	15.0	35	25	9.0
2024-⅛-0.0015	5.0	700	525	780	620	200	425	500	400	82.0	315	250	33.0
2024-³⁄₁₆-0.0015	3.5	330	250	370	290	86	200	290	230	55.0	180	143	23.0
Glass fabric reinforced plastic resin honeycombs													
Phenolic-³⁄₁₆	4.0	500	350	600	480	57	...	260	210	11.5	140	110	5.0
Phenolic-¼	3.5	350	260	500	400	46	...	230	170	9.0	120	110	3.5
Phenolic-⅜	3.2	320	245	440	350	38	...	200	160	8.0	105	85	3.0
NP(b)/polyester-³⁄₁₆	4.5	520	365	670	470	80	...	280	195	13.5	130	90	5.2
NP(b)/polyester-¼	4.0	420	295	560	390	68	...	260	180	13.0	120	85	5.0
Aramid fiber paper honeycombs(c)													
⅛ (1.5 gauge)	1.8	110	70	130	85	...	...	90	65	3.7	50	36	2.0
⅛ (2 gauge)	3.0	300	180	330	270	20	...	180	162	7.0	95	85	3.5
³⁄₁₆ (2 gauge)	2.0	150	90	170	105	11	...	110	72	4.2	55	40	2.2
³⁄₁₆ (5 gauge)	6.0	650	580	700	650	...	...	390	330	14.5	185	150	6.0
¼ (2 gauge)	1.5	90	45	95	55	6	...	75	45	3.0	35	23	1.5
¼ (5 gauge)	3.1	275	180	285	240	...	...	170	135	7.0	85	60	3.0
⅜ (2 gauge)	1.5	90	45	95	55	6	...	75	45	3.0	35	23	1.5
⅜ (5 gauge)(d)	3.0	285	...	300	...	17	...	170	...	5.6	95	...	3.0

Table 12.10 (continued)

Material	Flexural		Compressive		Shear	
	Strength, psi	Modulus, ksi	Strength, psi	Modulus, ksi	Strength, psi	Modulus, ksi
Other core materials						
Polystyrene foam						
2.2 lb/ft³	60	2	40	1.2	40	0.55
2.4 lb/ft³	75	2.5	60	1.6	45	0.60
3.5 lb/ft³	100	2.5	115	2	70	0.70
Isocyanate foam						
2 lb/ft³	...	...	...	...	25	...
3 lb/ft³	...	...	...	...	35	...
4 lb/ft³	...	...	...	...	48	...

(a) Materials designated as follows: Material/alloy–cell size–foil thickness for aluminum alloy honeycombs. All cell configurations are hexagonal. (b) NP = initial impregnation of fabric is nylon-modified phenolic resin; final dip coats are polyester. (c) Du Pont's Nomex, an aramid-fiber paper treated with phenolic resin. (d) Preliminary data

reason, however, is that silver is much more valuable in applications such as photography, where it is difficult to find a substitute for silver.

12.7 Problems

1. How much wood filler, having a density of 0.68 g/cm^3, should be added to 100 g of polymer, having a density of 1.40 g/cm^3, to obtain a composite having an overall density of 1.10 g/cm^3?

2. If one wanted to decrease the coefficient of thermal expansion of a given polymer, what sort of filler would you recommend?

3. Calculate the tensile strength of a composite comprising 50 g of glass having a density of 2.24 g/cm^3 and 50 g of polymer having a density of 1.32 g/cm^3. The tensile strength of the polymer is 6000 psi, and the tensile strength of the glass fibers is 200,000 psi.

4. List one or two examples of metallic sandwich composites. Explain the advantage gained by the use of a composite as opposed to a pure metal.

5. What is the function of straw in mud bricks?

13

Electrical & Electronic Properties of Materials

The electron structure of materials is important primarily for understanding and describing electrical, magnetic, thermal, and optical properties. Ultimately, nearly all properties depend on electron structure because chemical bonding is a reflection of the electron structures of atoms and molecules. The electron structure of isolated atoms is described in most introductory chemistry texts and will not be repeated here. The interactions of electrons in materials introduce new aspects not dealt with in the electron structures of isolated atoms. In order to develop the modern electron theory of materials, wave mechanics must be employed. Although wave mechanics are beyond the scope of this text, the results of the use of wave mechanics will be discussed.

13.1   The Band Model

An *electron orbital* is the probability of finding a given electron at a certain point in space. Generally, electron orbitals are considered to be the volume in space in which the electron may be found with about 95% certainty. An orbital may be viewed as an electron cloud with regions in the cloud where there is a high probability of finding the electron at any time. Each orbital may contain two electrons having opposite spins. Orbitals are described mathematically by wave functions; we shall omit discussion of wave functions and the derivation of the band model for electrons in materials. The following is derived from the theory of quantum mechanics, and the results agree qualitatively with experiments.

A metal such as sodium has one valence electron in its outermost orbital. Two sodium atoms isolated in space have the same energy for the valence electron: the energy levels are *degenerate*. When the two sodium atoms are brought closer together, at some distance the degeneracy of the levels is split, or the energy levels divide into two levels, as shown in Fig. 13.1. If six isolated sodium atoms were brought together, then the energy level of the six valence electrons would be split into six energy levels, as shown in Fig. 13.2. Likewise, if 1 mol of isolated sodium atoms were brought together to form 23.0 g of solid sodium metal, the electron energy level would be split into 6.022×10^{23} energy levels, very closely spaced. In general, N atoms give rise to N closely spaced energy levels. Each energy level may accommodate two electrons provided that they have opposite spins.

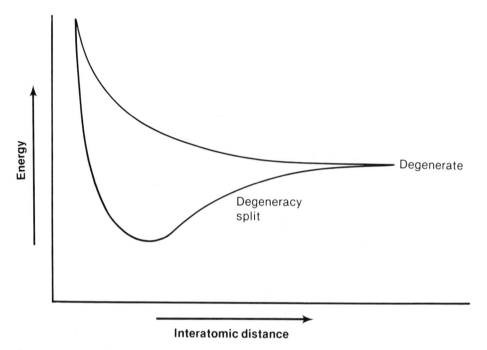

Fig. 13.1   Energy levels of separated and combined sodium atoms

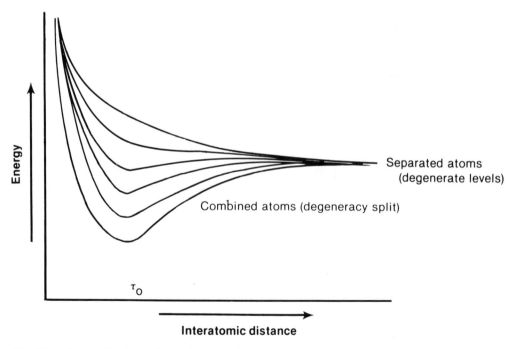

Fig. 13.2   Energy levels of six combined sodium atoms

In Figure 13.2, τ_o is the equilibrium interatomic distance, and corresponding to this distance there are six energy levels for six sodium atoms, and N energy levels for N atoms. A *band* is a group of very closely spaced energy levels in a solid material. Even for a minute quantity of material, the energy levels in a band are extremely close together, and at nearly all temperatures the electrons have sufficient energy to move from energy level to energy level: the band is practically a continuum of energy.

In 1 mol of sodium metal, there are 6.022×10^{23} energy levels for the valence electrons, and each level may accommodate two electrons. Therefore, only one-half of the energy levels, namely, the lowest energy levels, will be fully occupied (actually, this is true only at 0 °K).

The above band is the 3s band ("3" refers to the fact that the valence electron has a principal quantum number, n = 3, and "s" refers to the fact that the second quantum number is zero, meaning that the shape of the atomic orbital is spherical). Another band, the 3p, overlaps the 3s band, as shown in Fig. 13.3. Incidentally, three times as many electrons may be accommodated in a p band as in an s band, because a p band contains three orbitals. Because the s band can hold 2N electrons and the p band can accommodate 6N electrons, for N sodium atoms the entire band is only one-eighth full of electrons.

Conduction of electricity is possible because there are so many vacant energy levels very close in energy to the occupied levels. In order for an electron to move toward a positive pole, the electron must be given a somewhat higher energy. If the band of orbitals is filled with electrons, then conduction cannot easily occur, and the material is an insulator.

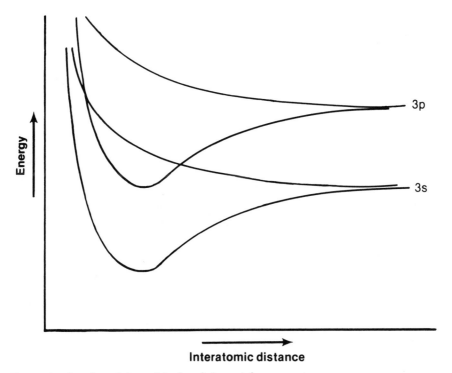

Fig. 13.3   Overlap of 3s and 3p bands in metals

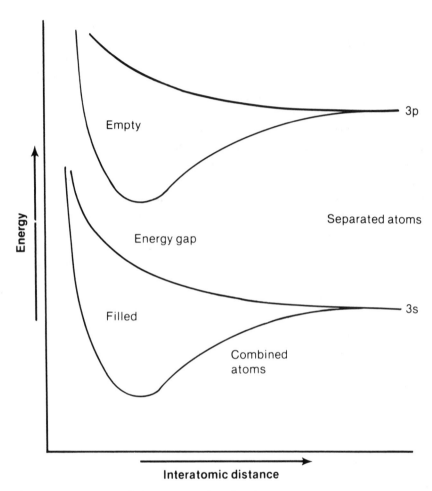

Fig. 13.4   Energy gap between 3s and 3p bands in an insulator
Note: the 3s band is filled with electrons; the 3p level is empty.

The bands due to the inner electrons of sodium are narrow, and they are separated by a large energy gap. Furthermore, these inner bands are completely filled with electrons. Therefore, the inner electrons do not participate in the conduction of electricity. The energy between bands is called the *energy gap*. The energy gap between the 1s and the 2s bands is large, and electrons cannot have an energy in the energy gap—this is a "forbidden" energy region.

Because metals have only partially filled bands of valence electron orbitals, and because the energies separating these energy levels are minute, electrons in metals readily absorb light of a wide range of wavelengths: an electron moves to a higher energy level when it absorbs a photon of light. When electrons fall from a given energy level to a lower one, they emit a photon. Metals appear lustrous because light is continually being absorbed and emitted. Re-emission of light occurs in random directions and is different from simple reflection, so that the appearance of luster is different from reflected light from a glossy white painted surface, for example.

Intrinsic semiconductors are materials such as silicon or germanium that have a relatively small energy gap between the filled valence band and the next higher band, which is empty (see Fig. 13.3). Electrical conduction occurs when an electron is excited in energy so that it moves to the higher, empty levels. Thus, the electrical conductivity of an intrinsic semiconductor increases markedly with temperature. Higher temperatures provide greater thermal energy to promote electrons into the high energy level band, which is largely empty. Furthermore, when an electron is excited to a higher energy level, it leaves a vacant energy level that may also be used for electrical conduction of another electron.

Insulators are poor conductors of electricity because a given band has all of its energy levels occupied by electrons, and a relatively large energy gap exists between the given band and the next higher band. As shown in Fig. 13.4, the 3s band is completely filled with electrons, but the 3p band is empty and situated at a much higher energy so that it is nearly impossible to excite electrons to the conducting (empty, or partly empty) band.

When a metal is heated, the electrons move faster, and this motion is quickly transmitted throughout the metal because the electrons in a metal behave similar to molecules in a gas. Heat conductivity, therefore, is high in metals and, furthermore, the mobility of electrons is confirmed.

13.2   Density of States

The energy levels within a given band are not distributed evenly. According to quantum mechanics, the relative density of states, N(E), for a given small energy range, is greatest in the center of the band. In the lower part of the band, there are many fewer energy states. Figure 13.5 shows a plot of the relative density of states, N(E), plotted against energy. The total number of states, or levels, equals N, the number of atoms, for a metal having one valence electron. Of course, the overlapping p levels add an additional 3N of levels.

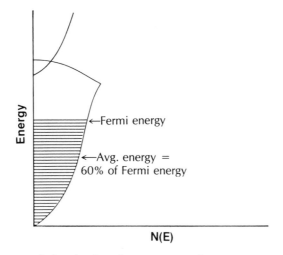

Fig. 13.5   Relative density of states versus electron energy at 0 °K

The unusual shape of the plot in Fig. 13.5 near the maximum energy in a given band is a consequence of diffraction of electrons moving in certain directions and having certain energies or wavelengths. When diffraction occurs from a given family of planes, it means that the electron can no longer move in its original direction; therefore, its movement is confined to certain directions. Near the top of the first band, the electrons are very restricted in their movement, or the number of states becomes fewer and fewer.

13.3   The Fermi Energy

Electrons occupy the lowest energy states available unless they are excited or activated—by elevated temperatures, for example. At 0 °K, all the lowest states of a band are filled. This is shown in Fig. 13.5. When the temperature is increased, say to room temperature, some electrons are excited to higher energy states, leaving unoccupied the states from which the electrons originated. This effect is shown in Fig. 13.6.

The *Fermi energy* is the maximum energy of the occupied states at 0 °K; it is more generally defined as the energy at which half the energy states are occupied (see Fig. 13.6). Note that the distribution is symmetrical on both sides of the Fermi energy. Hence, the Fermi energy does *not* change with temperature. As temperature increases, the distribution of electrons in the various states changes, but the Fermi energy remains fixed.

Figure 13.6 shows that only a relatively small fraction of valence electrons is excited when the temperature is increased. Because only a few valence electrons absorb thermal energy, the result is that electrons add very little to the heat capacity of materials. The heat capacities of many materials are 6 cal/°K·mol for each atom present (see Chapter 3): chromium metal has a

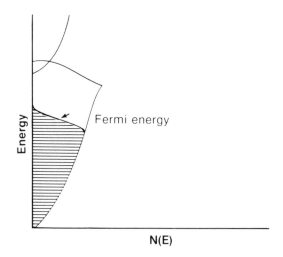

Fermi energy

N(E)

Fig. 13.6   Relative density of states versus electron energy at room temperature

heat capacity of about 6 cal/°K·mol; NaCl is about 12 cal/°K·mol, and $CaCl_2$ is about 18 cal/°K·mol. In an ideal solid, each atom vibrates essentially independently about its equilibrium position, leading to a theoretical heat capacity of 3R, or about 6 cal/°K·mol; R is the molar Boltzmann constant (i.e., the "gas law constant" of 1.987 cal/°K·mol). If electrons contributed to heat capacity, instead of 6 cal/°K·mol, materials would have an extra 3R/2 cal/mol for one valence electron, or a total of about 9 cal/°K·mol. Experiments have shown that this is not so, and therefore valence electrons add very little to heat capacities. This is in agreement with the distribution of electrons around the Fermi energy, as shown in Fig. 13.6. Heat capacity due to electrons is only about 0.03 cal/°K·mol; therefore, only about 1% of the valence electrons are thermally excited.

The Fermi energy of silver is 130,000 cal/mol. This is a very high energy, particularly when compared with thermal energy. At room temperature, the thermal energy is RT, or about 575 cal/mol. Thus, thermal vibrations are able to change the energy of electrons from 130,000 cal/mol to 130,600 cal/mol (rounding off). This confirms the fact that thermal energy excites only a small percentage of valence electrons and adds little to the heat capacity of the material. Heat capacity is due almost entirely to vibrational energy absorbed by atoms.

The average electron in a metal at absolute zero has an energy that is 60% of the Fermi value. For instance, in silver, $E_{avg} = 0.6 \times 130,000$ cal/mol $= 78,000$ cal/mol.

13.4   The Work Function

Electrons in a metal are in a potential energy "well": it requires energy to lift an electron from the Fermi energy and remove it from the metal. The energy required is $e\phi$, where e is the electron charge and ϕ is a voltage. The product, $e\phi$, is called the *work function* and is commonly measured in electron volts (1 eV = 23,061 cal/mol). The work function is analogous to the ionization energy required to remove an electron from an isolated atom except that the work function involves an aggregate of many metal atoms. Contrary to ionization potential, the second, third, and subsequent electrons have the same work function within experimental error because the energy levels of electrons in a given band of a metal are extremely close together. The work function of silver is 4.7 eV, or 108,000 cal/mol. For other values of the work function, see Table 13.1.

Contact potential is the difference in voltage created when two different metals are in electrical contact. The first metal has a given work function, $e\phi_1$, which is different from $e\phi_2$ for the second metal because the two metals have different Fermi energies. Electrical contact enables the electrons having the higher Fermi energy to spill over into the other metal until the electrons in both metals have the same maximum energies. Thus, the Fermi energy of one metal decreases, and it increases in the other metal until the two Fermi energies are the same (see Fig. 13.7). Therefore, an excess negative charge accumulates on one metal and a positive charge forms on the other. This process is analogous to having two wells with different levels of water in them. If the wells are connected by a pipe beneath both levels of water, the water levels will equalize.

Table 13.1   Work function and magnetic susceptibilities of metals

Metal	Work function $e\phi$, eV	Magnetic susceptibility, $10^6 \chi$	Metal	Work function $e\phi$, eV	Magnetic susceptibility, $10^6 \chi$
Aluminum	4.2	+0.65	Molybdenum	4.2	+0.04
Antimony	4.1	−0.87	Nickel	4.9	Ferromagnetic
Arsenic	5.1	−0.31	Niobium	4.0	+1.5
Barium	2.5	+0.9	Palladium	5.0	+5.4
Beryllium	3.4	−1.00	Platinum	5.3	+1.10
Bismuth	4.4	−1.35	Potassium	2.2	+0.52
Cadmium	4.0	−0.18	Rhenium	5.0	· · ·
Calcium	2.9	+1.10	Rhodium	4.6	+1.11
Chromium	4.4	+3.08	Rubidium	2.1	+0.21
Cobalt	4.0	Ferromagnetic	Silicon	4.2	−0.13
Cesium	1.9	−0.22	Silver	4.7	−0.20
Copper	4.5	−0.086	Sodium	2.25	+0.51
Gallium	3.9	−0.24	Strontium	2.7	−0.20
Germanium	4.8	−0.12	Tantalum	4.1	+0.93
Gold	4.8	−0.15	Tellurium	4.8	−0.31
Hafnium	3.6	· · ·	Thallium	3.8	−0.24
Iridium	4.6	+0.15	Thorium	3.5	+0.11
Iron (BCC)	4.48	Ferromagnetic	Tin	4.3	−0.25
Iron (FCC)	4.21	· · ·	Titanium	4.1	+1.25
Lead	4.0	−0.12	Tungsten	4.5	+0.28
Lithium	2.4	+0.50	Uranium	3.6	· · ·
Magnesium	3.7	+0.55	Zinc	4.3	−0.157
Manganese	3.8	+11.8	Zirconium	3.8	−0.45
Mercury	4.5	−0.168			

Note: Gadolinium and a few other rare earth metals are ferromagnetic at low temperatures.

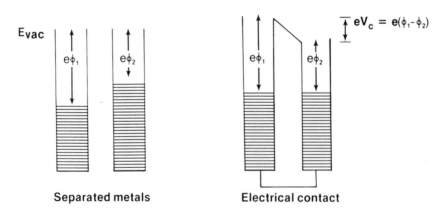

Fig. 13.7   Contact potential between two metals

13.5   Electrical Resistivity

A prerequisite for electrical conduction is the ability of electrons to move to a vacant higher energy level. Such levels abound in metals, and in general, metals are excellent conductors. Tables 3.5 and 3.6 contain the electrical resistivities of metals at their melting points and at

20 °C, as well as the temperature coefficient of resistivity. For an electron to exhibit net motion in one direction, it must gain velocity and momentum as it moves toward the positive electrode.

Only the electrons near the Fermi energy level act as charge carriers, because other electrons deeper in the potential well would require too much energy to excite them to a higher unoccupied level. The upper energy levels contain sufficient electrons for electrical conductivity to occur easily. The limiting factor for conduction is the *mobility* of electrons rather than the number available for conduction.

The mobility of an electron is defined by its *mean free path,* which is the average distance between scattering points. Electrons are diffracted and scattered by irregularities in the crystal structures of metals. When an electron moves between two scattering points, it accelerates if it is moving toward the positive electrode or decelerates if it is moving away from the positive electrode (or has a component of motion away from the positive electrode and toward the negative electrode). A larger mean free path results in a greater drift velocity because the electrons have a longer distance for acceleration. The overall drift velocity is much less than the velocity of the electrons because of the frequent collisions at scattering points, backtracking, and other detours taken by the electron in its net motion toward an electrode.

Higher temperatures in a metal result in greater atomic vibrations, which has the same effect as more structural disorder to diffract and scatter electrons. Consequently, the mean free path decreases with increasing temperature. *Resistivity,* the inverse of conductivity, increases with increasing temperature. The contribution to resistivity by imperfections other than thermal is independent of temperature. Figure 13.8 shows the typical dependence of resistivity on temperature: at elevated temperatures, resistivity is proportional to the temperature; at very low temperatures, the resistivity is proportional to T^5.

At very low temperatures, resistivity is due primarily to lattice imperfections (impurities, etc.) rather than to thermal vibrations. In a very pure metal, the resistivity ratio of ρ_{273}/ρ_4 is about 10^4 to 10^5, whereas in impure metals the resistivity ratio ρ_{273}/ρ_4 may be only about 10.

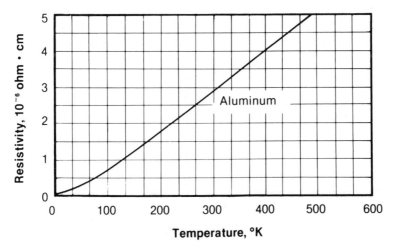

Fig. 13.8   Electrical resistivity as a function of temperature

Measuring the resistivity ratio at room temperature versus 4 °K is a good practical check for the purity and perfection of a metal.

Alloying reduces conductivity because the crystal structure is strained around each solute atom; these structural irregularities scatter electrons, thereby reducing conductivity (see Fig. 13.9). As one would expect, conductivity decreases more drastically in proportion to the difference in size and valence between the solvent and solute atoms (i.e., the greater the difference, the lower the conductivity).

Structural imperfections of any sort, such as vacancies, dislocations, solute atoms, or grain boundaries, reduce the electrical conductivity. Cold working, or strain hardening, of metals introduces imperfections and, as one would expect, conductivity is lower in cold-worked metals than in annealed metals.

Metals having only one valence electron or three valence electrons, such as sodium, copper, or silver (one valence electron) or aluminum (three valence electrons), usually have good to excellent conductivity compared with other metals. These metals have a half-filled band, and there are many electrons near the Fermi energy because N(E) is large there (see Fig. 13.5). One- or three-electron metals have many conducting electrons compared with divalent metals (e.g., calcium), which have lower conductivities. In divalent metals, the first band is nearly full and the second band is only very slightly filled; the total number of energy levels in this region is much lower than for one- or three-electron metals. Therefore, relatively few electrons conduct electricity, leading to the lower conductivities determined experimentally.

Resistivity is isotropic (the same in the directions [100], [010], and [001]) in cubic structures, but it is anisotropic in noncubic crystals, such as HCP crystals.

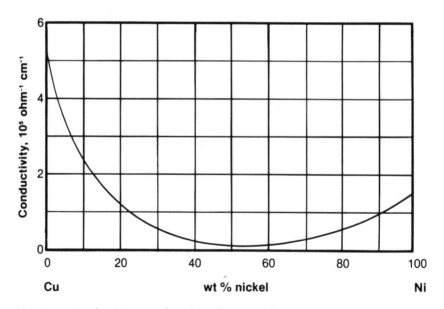

Fig. 13.9 Conductivity as a function of composition

13.6 Thermal Conductivity

The heat capacity of metals is only slightly affected by electrons because relatively few electrons are thermally excited to higher energy levels at room temperature or even above. In metals, thermal conductivity depends primarily on the motion of thermally excited electrons. These electrons rapidly transmit their thermal energy to atoms and other electrons because of the relatively high mobility of electrons in metals. Consequently, metals in general are excellent conductors of heat: thermal conductivities of metals are given in Tables 3.6 and 3.7. Insulators conduct heat slowly by means of elastic waves, or *phonons,* and this mechanism also occurs in metals, but is overshadowed by the heat conductivity due to mobile electrons.

Long ago it was noted that thermal conductivity and electrical conductivity in metals are proportional to one another. This is due to the fact that both phenomena depend on the motion of valence electrons from one atom to another and to another and another, and so on.

13.7 Thermoelectric Effects

Certain thermal and electric effects are related in metals. The four common thermoelectric effects are (1) resistance heating, (2) the Peltier effect, (3) the Thomson effect, and (4) the Seebeck effect, which enable the use of thermocouples to measure temperatures.

Resistance heating is the heat evolved due to the passage of a current in a conductor. The heat generated per second equals I^2R, where I is the electric current in amperes and R is the metal resistance in ohms. Some of the extra energy of the electrons is transferred to the atoms during conduction of current, and this amount of energy equals I^2R.

The *Peltier effect* occurs when current flows through a junction between two different metals: either heating or cooling may occur, depending on the direction of the current flow. The magnitude of heating or cooling depends on the nature of the two metals, the temperature, and the current. Note that this is different from resistance heating. As a matter of fact, the Peltier effect could be used for cooling without compressors or the usual air-conditioning equipment. However, the process simultaneously produces unwanted resistance-type (I^2R) heating.

The *Thomson effect* is the production of a potential gradient in a homogeneous metal when a temperature gradient exists along it. This is due to the fact that there are more energetic electrons at the hot end than at the cold end. The higher energy electrons move more readily toward the cool end than vice versa, producing an electromotive force (emf) difference.

The *Seebeck effect* is the production of a current when two wires of different metals are joined at two ends and these ends are placed in a hot zone and a cool zone. The emf in the circuit depends on the two metals employed and the temperatures of the junctures. *Thermocouples* are devices made of two metals having one junction at room temperature or at 0 °C and the other juncture, which is welded, at the temperature that one desires to measure. Thermocouples are most useful to measure temperatures beyond the range of thermometers. Because the current depends on the resistance of the wires, it is common practice to measure the emf in thermocouples, usually with a potentiometer. Then the difference in temperature between the hot end and the cooler end is determined by referring to appropriate emf tables in handbooks for the given thermocouple metals. Modern instruments are calibrated to give a direct temperature readout when the temperature of the cool end is compensated for.

13.8 Superconductivity

Superconductivity is the disappearance of all electrical resistivity by certain materials when they are cooled below a *critical temperature,* T_c, and placed in a magnetic field below a critical field strength, H_o. Over 1000 alloys and intermetallic compounds and oxides, as well as many metallic elements, are superconducting; the primary exceptions are monovalent and ferromagnetic metals.

Superconductivity would be of tremendous practical value if a material could be found for which T_c is greater than room temperature. Advances in superconducting materials have elevated the maximum known value of T_c from about 23 °K around 1985 to about 125 °K two or three years later. This breakthrough occurred in the discovery of a new class of materials, complex oxides having certain structural features. This achievement should enable superconductivity to be much less costly, because liquid nitrogen can be used to obtain temperatures less than T_c compared with liquid hydrogen in the past.

Superconductivity is extremely important in achieving strong magnetic fields in masers, lasers, infrared detectors, and particle accelerators, such as the superconducting supercollider. Table 13.2 presents the critical temperatures, T_c, and critical fields, H_o, for a number of

Table 13.2   Superconductivity in metals

Metal	Critical temperature, T_c, °K	Critical field, H_o, Oe	Metal	Critical temperature, T_c, °K	Critical field H_o, Oe
Al	1.20	99	Nb_3Al	18.0	· · ·
Al_2CMo_3	10.0	1700	Nb_3Ga	14.5	· · ·
$Al_{0.8}Ge_{0.2}Nb_3$	20.7	· · ·	Nb_3Pt	10.9	· · ·
AsPb	8.4	· · ·	Nb_3Sn	18.1	· · ·
$AuNb_3$	11.5	· · ·	Nb_3Sn_2	16.6	· · ·
Cd	0.56	30	Os	0.66	65
CaPb	7.0	· · ·	Pb	7.19	803
Ga	1.09	51	Re	1.70	198
Ga_2Mo	9.5	· · ·	$RhZr_2$	10.8	· · ·
GaV_3	16.8	· · ·	Ru	0.49	66
$Hg(\alpha)$	4.15	411	Sn	3.72	305
In	3.40	293	$SnTa_3$	8.35	· · ·
$InLa_3$	10.4	· · ·	Ta	4.48	830
Ir	0.14	19	TaN	13	· · ·
$La(\beta)$	6.06	1600	Th	1.37	162
$LaOs_2$	6.5	· · ·	Ti	0.39	100
LiPb	7.2	· · ·	Tl	2.39	171
Mo	0.92	98	V	5.30	1020
Mo_2C	12.2	· · ·	V_3Si	17.1	· · ·
MoN	12	· · ·	W	0.012	1.07
Mo_3Re	10.0	· · ·	Zn	0.88	53
$Mo_{0.57}Re_{0.43}$	14.0	· · ·	Zr	0.55	47
Nb	9.26	1980			

materials. In metals, superconductivity is caused by the motion of pairs of conduction electrons. When the electron pair encounters an obstacle, what one electron loses in momentum the other one gains, and thus the electron pair continues with no net loss of momentum. Temperatures above T_c break up the electron pairs, as do magnetic fields above H_o. The critical value of H is a function of temperature, as shown schematically in Fig. 13.10. H_o decreases with increasing temperature because the effects of magnetic field and temperature on superconductivity are additive. Notice that the transition from the normal state to the superconducting state is abrupt. Generally, superconductivity occurs most readily among metals having (1) low conductivities of the normal type or (2) three, five, or seven valence electrons.

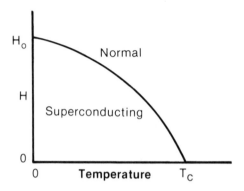

Fig. 13.10   Superconductivity as a function of temperature and magnetic field

13.9   Extrinsic Semiconductors

Intrinsic semiconductors are materials having some conductivity due to a relatively small energy gap between the filled band of valence electrons and the next higher band of empty energy levels. Thermal energy is sufficient to promote some electrons to the energy levels in the empty band, thereby leaving some levels empty in the lower, previously filled band. Diamond is an insulator which has the relatively large energy gap between the filled and empty bands of 5.30 eV or 122,000 cal. Silicon, a semiconductor, has a gap of only 1.06 eV or 24,400 cal, and germanium has a gap of 0.67 eV or 15,500 cal. Tin, a metal which is also located in Group IV, has an energy gap of only 0.06 eV or 1400 cal. This small gap makes tin a metallic conductor because of the relatively large number of electrons that are able to jump into the conducting band.

Silicon has four valence electrons. If an atom, such as phosphorus, having five valence electrons is placed into the silicon crystal lattice, then the extra electron, called a *donor electron*, has an energy just below the vacant conduction band of energy levels. It is then very easy for this donor electron to obtain enough additional energy to jump into the conduction band. The phosphorus, which is added in low concentrations to very pure silicon, is called a *dopant*, meaning that it is a deliberate, small impurity. When an extrinsic semiconductor has an excess of valence electrons due to a dopant, it is called an *n-type semiconductor* (the "n" signifies "negative"). By doping silicon with phosphorus, the conductivity may be increased by a factor of 10,000! n-type semiconductors are *extrinsic semiconductors;* that is, they are semiconductors because an impurity was deliberately added to the material to make it semiconducting.

If the dopant atom has only three valence electrons, e.g., aluminum, instead of four as in silicon, then an *electron hole* is created. The effect of the electron hole is to create a vacant energy level just above the filled band. Again, it is easy through thermal energy for an electron to obtain the necessary additional energy to jump into the higher energy level and become a conducting electron. Even though there is a deficiency of valence electrons, the charge is neutral: each aluminum atom supplies only three valence electrons, but the nucleus of each aluminum atom has an excess of only three protons over the number of core electrons; i.e., Al is neutral, so the entire crystal remains neutral. The creation of electron holes forms what is known as a *p-type semiconductor,* which is also an extrinsic semiconductor. Extrinsic semiconductors are either n-type or p-type.

13.10   Electronic Devices

The joining of p- and n-type semiconductors produces many useful electronic devices, including transistors, rectifiers, integrated circuits, and solar cells. A p-n junction creates an interesting situation: in a thin zone (about 0.5 μm thick), the n-type material will contribute an electron which will then reside in the p-type material. This short diffusion of electrons creates a thin zone that is positively charged on the side where the electrons left and negatively charged on the side where they arrived. In effect, a barrier is set up in which electron holes are repelled by the positive side of the n-type part of the barrier; electrons are repelled by the negative charges on the p-side of the barrier.

A current flows through the material when an external potential is applied so that the positive connection is to the p-type material, and the negative connection is to the n-type material. This connection causes the charges in the junction interface to be neutralized, which enables electrons to move in one direction by jumping into holes. Holes "move" in the opposite direction whenever electrons jump into them (actually, one hole is obliterated while another hole is created where the electron had been).

When the external potential is applied in the opposite manner, then practically no current flows. The negative-positive zone is accentuated. If an alternating current is applied to the device, a direct current output is obtained because the electrons move in only one direction rather than back and forth as in alternating current.

A *transistor* is a device in which two p-n junctions are arranged either as p-n-p or n-p-n. Three connections are made, one to each type material. Depending on the direction of the flow of electrons, it is possible to amplify the current. For details on how these most valuable devices operate and how they may be combined to produce useful objects, refer to a standard text on electrical engineering.

13.11   Dielectric Materials

Many ceramics and polymers have electrical resistivities about 10^{20} times higher than the resistivities of metals. An *insulator* is a material with a very high resistivity. Ceramic insulators include mica, porcelain, alumina (Al_2O_3), and glass, whereas polymeric insulators include paper, textiles, elastomers, and waxes. A *dipole* is a separation of a negative from a positive charge. A *capacitor* is a pair or series of surfaces or "plates" on which relatively large amounts of positive or negative charges are stored. A *dielectric* is an insulator composed of ions, polar molecules, or materials in which a dipole may be induced by an electric field. Dielectrics are often used in capacitors.

Consider a parallel plate capacitor in a vacuum. When subjected to a voltage, one plate is charged positively and the other plate is charged negatively. The electric field, E, is equal to the voltage difference divided by the distance between the plates: $E = V/l$. The electric charge density on the plates is D, where $D = \epsilon_o E$. ϵ_o is a constant equal to 8.854×10^{-12} C/V·m. Rearranging this equation, $\epsilon_o = D/E$. Units are C/V·m = $(C/m^2)/(V/m)$.

Polarization is the formation of induced dipoles or the increase in the strength of dipoles in the presence of an electric field. An electric field tends to align permanent dipoles so that the positive end of the dipole points toward the negatively charged plate, and the negative end of the dipole points toward the positively charged plate. An electric field may induce dipoles in a material. Polarization may be due to (1) electron clouds which deform in the electric field, (2) ions (positive ions move slightly closer to the negative plate and negative ions move slightly closer to the positive plate), or (3) orientation of polar molecules, in which the molecules may partially or completely align themselves so that the positive end points toward the negative plate, etc. Another rarer type of polarization occurs in polyphase materials in which charge accumulates at phase boundaries.

When a polarizable insulator is placed between plates in an electric field, the material acts as a dielectric, and the surface charge density on the plates, D, increases. The increase is due to the following: the voltage source imposes a given voltage between the plates, and the dipoles act as charge carriers; the magnitude of the *charge density* on the plates, D, increases until E in the capacitor reaches the same value as for a vacuum. The net effect is that $D = \epsilon E$, where ϵ is greater than ϵ_o. ϵ is called the *permittivity,* and the relative permittivity is the *dielectric constant,* ϵ_r: $\epsilon_r = \epsilon/\epsilon_o$. The dielectric constants of various materials are given in Table 13.3.

When a dielectric material is present in an electric field which has a change in polarity, the dipoles will reverse themselves at a certain rate. Electronic dipoles and ion dipoles can reverse extremely rapidly because of the short distances involved. Also, no rotation is involved: the electrons merely shift from being predominantly on one side of an atom to the other side, or ions which were attracted to a given direction merely move slightly in the opposite direction when the polarity of the electric field is reversed.

Polarization that depends on molecular orientation, however, may take much longer to reverse itself: consider a water molecule (remember that the water molecule is *not* linear—the oxygen end has a slight negative charge and the hydrogen end has a slight positive charge). In an electric field the water molecule is aligned so that the oxygen atom points toward the positively charged plate. When the electric field is reversed, the water molecule must rotate if it is to be properly aligned. If liquid water is present, this is not very difficult, although it takes time. In ice, however, the water molecule may not be free to rotate, or it would take a relatively long time for this to occur because of constraints on the movement of the water molecule.

Table 13.3 Dielectric constants of some materials

Material	Dielectric constant	Special conditions(a)
Organic		
Asphalt	2.7	
Benzene	2.3	Liquid
Glycerol	42.5	Liquid
Natural rubber	3.0	
Nylon	3.6	
Phenol-formaldehyde	4.8	
Polyester	4–6	Glass fiber reinforced
Polyethylene	2.2	
Polypropylene	2.0	
Polystyrene	2.6	
Polytetrafluoroethylene	2.1	
Polyvinylchloride	3.8	
Polyurethane	~6.6	Glass fiber reinforced
Ceramic		
Calcium carbonate	6.1	
Gypsum	5.7	10^4 Hz
Diamond	5.5	10^8 Hz
Glass (soda-lime)	6.9	
Mica	5.4–8.7	
Porcelain	6.0	
Rutile	86	$\perp$ Optic axis; 10^8 Hz
Rutile	170	$\parallel$ Optic axis; 10^8 Hz
Silica (fused)	3.8	
Steatite ($MgO \cdot SiO_2$)	~6.5	
Titanates	15	
Zircon	12	10^8 Hz
Water	78.5	Liquid

(a) Most measurements were made at 10^6 Hz.

Relaxation time is the time required for dipole reversal, and the reciprocal of relaxation time is the *relaxation frequency*. If the relaxation frequency is less than the field reversal frequency, then the given dipole does not contribute to the total polarization of the material.

Both electronic and ionic polarization are insensitive to temperature. An increase in temperature does not affect the magnitude of the dipole or the relaxation time for these two types of polarization. Frequently, however, the orientation type is strongly affected by temperature. For instance, when ice melts, water obviously can then respond much more rapidly to a reversal of the polarity of the electric field. Consequently, there is a strong dependence of the relaxation time in water on temperature.

If the frequency of reversal of an electric field is much *less* than the relaxation frequency, then the dipole can respond easily and quickly to the reversal: polarization varies instantly with the electric field. If the relaxation time is large compared with the time to reverse the electric field, then that particular polarization process will be inoperable: no effect will be noticed. As the frequency of field reversal approaches that of the relaxation frequency, then the reversal of the dipoles will lag behind the reversal of the electric field: this lag is called *hysteresis*. When D, the plate charge density, is plotted versus E, the electric field, then as the field polarity is

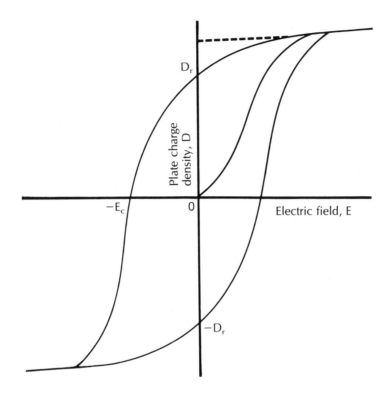

Plate charge density, D

D_r

$-E_c$

0

Electric field, E

$-D_r$

Fig. 13.11   Hysteresis curve for a dielectric material

changed back and forth, a hysteresis loop is experienced, as shown in Fig. 13.11. On repeated reversal of the electric field polarity, the charge density, D, always lags behind, and the loop is traversed over and over.

The area inside the hysteresis loop represents the energy loss in reversing the electric field twice. When this energy loss per cycle is multiplied by the frequency of field reversal in cycles per second, then the *power loss* is obtained. The power loss is a function of the frequency of polarity reversal of the electric field, and hence certain dielectrics are undesirable for certain frequency ranges. Obviously, this is an important design criterion.

At some value of voltage gradient, each dielectric material will eventually break down. Breakdown probably occurs by the excitation of some electrons into the conducting band above the energy gap. The high field then accelerates these electrons sufficiently to knock other electrons into the conducting band, and an arc may then occur. Excessive heating occurs, which can lead to decomposition, burning, or vaporization of the dielectric. Consequently, many dielectrics, such as oil in a high-voltage transformer, may need to be changed regularly.

Ceramics generally exhibit the best performance as high-quality dielectric materials. Many ceramic dielectrics are comprised of varying proportions of silica (SiO_2), alumina (Al_2O_3), and magnesia (MgO). Porcelain dielectrics also contain feldspar ($K_2O \cdot Al_2O_3 \cdot 6SiO_2$). These are used in power-line insulation.

13.12   Ferroelectricity

Barium titanate, $BaTiO_3$, is an important ferroelectric material. A *ferroelectric* material is one that contains permanent dipoles, all of which are aligned in the same direction within a small volume called a *domain*. In an electric field all the domains may be lined up so that the material exhibits a high dielectric constant.

In $BaTiO_3$, the unit cell is not quite cubic, because the single titanium atom in each unit cell is located just a bit off center. The oxygen atoms are located in face-centered sites, and they, too, are slightly out of position opposite to the titanium atom. The barium atoms occupy corner sites. Because of the off-centered positioning of titanium and oxygen atoms, the material has a permanent dipole (titanium is positive and the oxygens are negative). In a given domain, which is much larger than a unit cell, but usually smaller than a crystal, the titaniums are "up" in the same direction. Thus, each domain has a dipole moment. At room temperature in the absence of an electric field, the domains are randomly oriented so that the material has no net dipole moment. When an electric field is applied, each domain tends to align with the titanium atoms closest to the negative plate and the oxygen atoms closest to the positive plate.

Ferroelectric materials may be used for information storage. When the electric field is removed, the polarization remains for a limited time. Eventually, the material depolarizes.

Ferroelectrics, such as $BaTiO_3$, and other dielectrics, such as quartz, are *piezoelectric* materials, which expand or contract when in an electric field. Likewise, piezoelectric materials will develop a potential difference when they are stressed. Therefore, they are useful as *transducers,* which convert electrical energy into mechanical energy and vice versa. Piezoelectrics are used in microphones, phonographs, and equipment for measuring mechanical strain.

13.13   Problems

1. Why do very few electrons contribute to the heat capacity of a metal? What is the evidence for this phenomenon?

2. Define conductors, semiconductors, and insulators in terms of the energy gap between s and p orbitals in materials.

3. Why is the ionization energy constant for removing the second, third, fourth, and so on, electrons from a metal, whereas removing successive electrons from one atom requires ever-increasing energies?

4. Why do alloys have lower electrical conductivities than the pure metals from which they are made?

5. Why does doping of pure silicon with either phosphorus or aluminum greatly increase the conductivity of silicon?

6. What is a transistor? Explain how one functions.

7. Explain the phenomenon of ferroelectricity.

14

Optical & Thermal
Properties of Materials

14.1 Optical Properties

Optical properties concern the interaction of photons, especially of visible or near-visible light, with materials. Light may be emitted from materials, or it may be absorbed, transmitted, reflected, or refracted. Phosphorescence and chemiluminescence are additional properties of interest. Lasers are also of growing importance, as are photovoltaic devices which convert light into electrical power.

Photons are small forms of energy having both a precise wavelength and energy, as well as an associated mass. Photons have no rest mass: when a photon is absorbed by a material, the photon usually is transformed into heat energy. The photon then ceases to exist. *Visible light* is comprised of photons of a given wavelength or photons having many different wavelengths. White light is comprised of photons of many wavelengths, whereas other colors consist of one wavelength of light or several wavelengths of light, usually having approximately the same wavelength.

The *visible spectrum* is made of light having wavelengths ranging from relatively short (violet) to relatively long (red). The spectrum consists of photons having wavelengths ranging from 400 to 700 nm. A typical spectrum is formed when white light passes through a prism. Then the various wavelengths are separated, with red being on one end of the spectrum and violet on the other. Intermediate colors are orange, yellow, green, blue, and indigo. The spectrum may easily be memorized by remembering the following: "*R*ead *o*nly *y*our *g*ood *b*ooks *in v*acation." The beginning of each word is a color of the spectrum. Another mnemonic device is Roy G. Biv.

The energy of photons is given by the following equation: $E = hc/\lambda$, where h is Planck's constant ($h = 6.62 \times 10^{-34}$ J·s), c is the velocity of light ($c = 2.998 \times 10^8$ m/s), and λ is the wavelength of the photon in meters.

Example 14.1A. Calculate (1) the energy of a mole of red photons having a wavelength of 700 nm and (2) the energy of a mole of ultraviolet light having a wavelength of 250 nm.

Solution: For red light, $E = [(6.62 \times 10^{-34}) \times (2.998 \times 10^8)/(700 \times 10^{-9})] \times (6.022 \times 10^{23}) = 1.71 \times 10^5$ J/mol (4.09×10^4 cal/mol). Recall that Avogadro's number is 6.022×10^{23}.

For ultraviolet light, E = $[(6.62 \times 10^{-34}) \times (2.998 \times 10^8)/(250 \times 10^{-9})] \times (6.022 \times 10^{23})$ = 4.78×10^5 J/mol (1.14×10^5 cal/mol).

Note: Compare these energies with the energies required to break a mole of chemical bonds. For instance, to break a mole of C—C bonds requires 3.48×10^5 J/mol (8.32×10^4 cal/mol). It is apparent that ultraviolet light possesses enough energy to break carbon-carbon bonds in polymers, but red light is not energetic enough to break these bonds.

It is no wonder, then, that light, and ultraviolet light in particular, has enough energy to degrade materials. Of course, solar energy may also be used to produce an electric current or to produce thermal energy. *Photovoltaic* devices convert sunlight (or light in general) to an electric current. These are usually solid-state materials. Already most persons have seen applications of these devices in the form of calculators that need no other source of power. The future is indeed bright for these materials. Inexpensive photovoltaic devices could revolutionize life.

Electrons whiz around nuclei with various fixed energies. Electrons may be promoted to higher energy levels by absorbing energy from a variety of sources: light, x-rays, or heat (thermal energy). When electrons fall from a higher energy level to a lower level, the energy difference is given off in the form of a photon, such as an x-ray or as visible light. However, the energy difference between two levels is exact, and therefore the energy given off is exactly the same for a given transition.

When *any* solid is heated to elevated temperatures, photons are emitted. *Black-body radiation* is energy given off from solid or liquid materials heated to various high temperatures. This radiation is the red glow from metals when heated in the range of 600 to about 1000 °C. When temperatures exceed this range, the radiation tends to become whiter. All materials emit black-body radiation upon heating because electrons are in energy bands, and the differences between the excited energy and the *ground state,* which is the state of lowest energy, is variable. This variability gives rise to photons having a range of wavelengths. For instance, when a tungsten filament in an incandescent light bulb is heated to a high temperature by an electric current, electrons are thermally excited to higher energy levels. Because many energy levels are involved, light of many wavelengths is emitted when the electrons decay back to lower energy levels.

Materials in the form of individual atoms also emit a characteristic line spectrum of various colors when heated to very high temperatures. These specific colors are emitted due to electron transitions from a high energy state to a lower energy state. Advantage may be taken of this property in the following ways:

- Materials may be analyzed by heating them to very high temperatures and measuring the wavelengths of various lines in the spectrum and the intensities of the lines (this is *emission spectroscopy*).
- Materials may be analyzed by heating them to high temperatures and shining a light of a given wavelength through the material; analysis by measuring the amount of absorption of light of certain wavelengths is called *atomic absorption*.
- Other uses include the light produced in various lighting fixtures (e.g., sodium vapor lamps).

When electrons are excited to higher energy levels than the ground state, sometimes photons are not emitted immediately: it takes time for the electron to fall from one energy level to a lower one. *Luminescence* is the glowing (visible light) of a material which is exposed to ultraviolet light (invisible). Luminesence may either be *fluorescence,* which is the glowing of the material only when it is being irradiated with ultraviolet light, or *phosphorescence,* which is the glowing of a material after the irradiation with ultraviolet light stops. Remember that ultraviolet light has much more energy than visible light, and when an electron falls from one energy level to another, it may take some intermediate steps; the emitted light, therefore, often has a different wavelength from the ultraviolet light that was absorbed. Heat, in the form of infrared radiation, may also be emitted. Energy is conserved overall: the total energy absorbed equals the total energy emitted. An important application of luminescence is in television picture tubes.

Chemiluminescence is the use of energy from chemical reactions to "pump" the electron to a higher energy level. Then the electron may drop back down by one or more steps in the process of luminescence. Obviously, a firefly does not produce light by achieving very high temperatures or using high-voltage electric currents! The process is chemiluminescence, the same as used in some children's toys or in temporary flashlights without batteries. *Electroluminescence* is supplying electrical energy to boost electrons to higher energy levels prior to luminescence.

A most useful application of luminescence phenomena is that of lasers. A *laser* is light amplification by stimulated emission of radiation (look at the first letter of each word in the definition). For example, a ruby laser comprises a cylindrical rod of a single crystal of Al_2O_3 containing a small amount of Cr^{+3} ions as an impurity. The rod has two parallel, flat ends, one of which is heavily silvered to reflect light. The other end is lightly silvered for partial reflection of light. A lamp containing xenon gas is flashed, emitting light of a characteristic wavelength (about 560 nm). When the photons from the xenon tube enter the Al_2O_3 (alumina) crystal, many electrons are promoted to a higher energy level, which is metastable (they do not fall to a lower energy level immediately). When some of the first electrons fall back to the ground state, they emit light having a precise wavelength. This light stimulates the emission of other photons, and a burst of light is obtained. Furthermore, the photons all have their wavelengths in phase (the light is said to be *coherent*). Most of the emitted photons are emitted in phase in a given direction. Consequently, the beam remains compact and does not tend to diverge as light from a flashlight does. There are many applications of lasers, ranging from scanning codes in grocery stores, to eye surgery, to cutting steel.

Absorption of light involves the interaction of a photon with electrons in an atom in such a way that the photon is converted to longer wavelength radiation. This may be heat (infrared) or light having a longer wavelength than that of the photon absorbed. *Color* is how we perceive visible radiation of a narrow range of wavelength or a blend of narrow ranges of wavelengths. Color may arise from the emission of radiation of a given wavelength such as by heating or through use of an electric current. Color may also arise by the selective absorption of some wavelengths of light, which usually leaves light of another wavelength, which is perceived as a certain color. For instance, a potassium permanganate solution has a deep purple color because it heavily absorbs light in the red-orange part of the spectrum. The deep purple photons

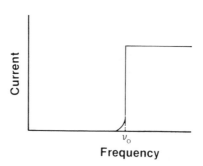

Fig. 14.1 Photoelectric current as a function of frequency (energy) of incident light, with light intensity being constant

pass through the potassium permanganate solution unaffected, whereas the red-orange photons are strongly absorbed.

When electrons absorb energy, usually they are promoted to higher energy levels. If the energy is great enough, the electron may leave the material. In a single atom, this is called the *ionization energy;* in a solid material, it is called the *work function* (see below also). The work function for alkali metals is relatively low when compared with that of other metals due to the relative ease of knocking an electron out of the atom. Because energy levels in a band in a solid are extremely close together, the work function remains constant for the emission of many electrons.

The *photoelectric effect* is the emission of electrons from a metal surface resulting from exposure to electromagnetic radiation, including visible light. Electrons are not emitted from a metal unless the frequency, or energy, of the light exceeds a specific value, v_o, characteristic of the metal. All the energy of one photon is transferred to a single electron. If $v < v_o$, then the light is merely scattered, and the electron remains in the metal. If $v > v_o$, then the excess energy is retained by the electron after it escapes from the surface (see Fig. 14.1). The energy, hv_o (h is Planck's constant), is called the work function of the metal. As the frequency (energy) of light increases above v, then each electron ejected from the metal possesses more energy (see Fig. 14.2). Photoelectric cells rely on this principle, and their applications include exposure meters in photography, and sensors for doors, gates, signals, or alarms. An even more important application is solar cells, which convert sunlight to useful current.

Single crystals of compounds containing transition metal ions frequently are colored due to the presence of unpaired electrons in the transition metal ions. For instance, crystals with the following ions usually have the corresponding characteristic colors: Cu^{+2} (blue; the ion has an unfilled d shell), Fe^{+3} (orange, as in rust), Fe^{+2} (green), Ni^{+2} (green), Co^{+2} (pink). Zn^{+2} is colorless because it has a filled d shell of 10 electrons.

Reflection is the interaction of photons striking the surface of a material in which the photons are not absorbed or transmitted through the material. Photons interact with electrons around atoms and are emitted such that the angle of incidence equals the angle of reflection. Because silver is highly reflective over the entire visible range of light, its surface color is white metallic. Copper and gold exhibit red-orange and yellow colors, respectively, because incident

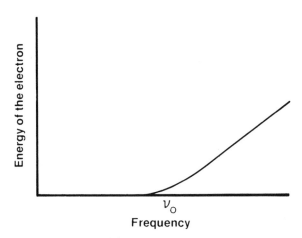

Fig. 14.2 Energy of photoelectrons as a function of frequency (energy) of incident light

light of short wavelengths excites electrons in filled d bands to empty levels in the s bands. The electrons decay by a different path, resulting in absorption of green, blue, and violet light, whereas yellow, orange, and red light are reflected. Many metals such as nickel and iron absorb light of various wavelengths and therefore have grayish or dull colors, or luster, due to relatively low reflectivity.

Transmission is the passage of photons through a material. This occurs only through extremely thin sections of metals. Because metals have overlapping bands of electron energy levels, photons are easily absorbed by metals: it is easy for electrons to be promoted to higher energy levels. Also, it is easy for these electrons to fall back to the ground state, emitting a photon in the process. The typical luster of metal surfaces is due to the absorption and re-emission of electrons in various directions, depending on the roughness of the surface. If the surface is polished smooth, then reflection is, of course, mirrorlike. If the metal surface is matte (slightly roughened), then the photons come off in various directions, and a reflected mirror image is not obtained.

In ionic and covalent solids, the electrons are bound strongly to atoms. A photon would have to have a great amount of energy before it could promote an electron to a higher energy level (recall that there is an energy gap which must be "jumped"). Consequently, when photons enter a single crystal of an ionic or covalent material, they pass through largely unaffected. If impurity transition ions or other transition ions are present, then certain photons are absorbed. This, of course, leads to the perception of color, as described above. Most single crystals, particularly of ionic materials, are transparent.

Opaque materials do not transmit photons. Either they absorb photons or reflect them. Semiconductors, such as germanium or silicon, are opaque to visible light because light is able to promote some electrons to higher energy bands. Therefore, photons cannot in general pass through these materials. However, photons of infrared radiation do not have sufficient energy to promote electrons to higher energy levels. Consequently, these semiconductors are transparent to infrared radiation but opaque to visible light.

Photoconductivity is the use of light to promote electrons to the conducting band of energy levels. Semiconductors may be altered so that the energy gap between filled bands of electrons and empty conducting bands is the same as the energy of photons of a given wavelength (color). Photons having longer wavelengths do not have enough energy to excite electrons to the conducting band. Likewise, photons having much shorter wavelengths are strongly absorbed so that they do not penetrate into the material. This technique is used to detect certain infrared radiation in military applications.

Aluminum oxide, Al_2O_3, forms a colorless crystal. When impurity atoms, such as chromium, are dissolved in the crystal, the material becomes colored: the famous red color of rubies. Often the impurity atom will create vacancies, and then electrons are stabilized in the vacancies. However, the energy needed to promote the electron to a higher energy level is modest; photons are able to do this, and the result is that the material absorbs photons of a certain wavelength. The remaining photons continue through the transparent material and are perceived as having a certain strong color.

Photochromic glass darkens on exposure to light due to the presence of silver ions. Photons may knock electrons free from some atoms, and these electrons reduce Ag^+ to Ag^0, which causes a darkening effect in the glass. In the absence of light, the reverse reaction occurs, and the glass lightens in color.

Refraction is the bending of light when it passes from one medium to another. Refraction is due to the different velocity of light in the two materials. The *index of refraction* is n = c/v, where n is the index of refraction. c is the velocity of light in a vacuum, and v is the velocity of light in the material. Liquids or solid single crystals having heavy elements (high atomic number) or high densities usually have high indices of refraction: the velocity of light in these materials is lower than in a vacuum. In noncubic crystals, the index of refraction may be anisotropic, or larger in one crystallographic direction than in another. Some crystals, such as calcite ($CaCO_3$) or quartz (SiO_2), split light into two polarized beams. This is called *birefringence*.

Polarized light is light having its electromagnetic vibration in one plane. Light may be polarized by passing through a Nicol prism or through polaroid films in which many small, doubly refracting crystals are oriented in a transparent plastic. When some materials are placed in a beam of polarized light, the plane of polarization is rotated either to the right or to the left. The extent of the optical rotation may be measured in a polarimeter. Optical activity occurs in compounds which contain a carbon atom attached to four different atoms or groups of atoms. Such a carbon atom is called an *asymmetric* carbon atom.

14.2 Thermal Properties

Thermal properties include melting point, boiling point, coefficient of thermal expansion, thermal conductivity, and thermodynamic properties. The following are brief definitions of these terms:

Melting point: the temperature at which the solid and liquid phases of a pure material are in equilibrium.

Boiling point: the temperature at which the vapor pressure of a liquid equals the pressure of the surroundings; normally, this pressure is taken to be 760 torr or 1 atm.

Thermal expansion: the fractional increase in dimensions of a material when heated resulting from increased thermal vibrations. When a material is at 0 °K, its atoms have an irreducible amount of vibrational energy. As the temperature is increased, the atoms begin to vibrate much more; the vibration is nonuniform because the atoms cannot compress each other's electron clouds very much. However, it is easier for the atoms to vibrate away from each other. The net effect is that atoms on the average become farther apart as temperature increases, due to the nonlinear vibration of atoms (essentially what happens is that the atoms force each other apart as they vibrate more and more).

In polymers, the bonding between chains of atoms is not very strong, and the coefficient of thermal expansion is relatively high. In metals, the coefficient of thermal expansion is roughly proportional to the strength of the metallic bonding (which is reflected in the boiling points). Tungsten, which has an extremely high boiling point and very strong metallic bonding, has a low coefficient of thermal expansion, particularly when compared with the coefficient for metals such as aluminum (see Table 3.6). Ceramics have very strong bonding, and their coefficients of thermal expansion are very low. Quartz, which softens only at very high temperatures, has a very low coefficient of thermal expansion. Pyrex (borosilicate glass) is not as strongly bonded as silica, but its coefficient of thermal expansion is still small; this accounts for many applications of Pyrex in the kitchen and in the laboratory. Soda-lime glass has a much lower softening temperature, and its coefficient of thermal expansion is rather high, making it unsuitable for many laboratory or kitchen applications.

When composites or different materials are joined together, the difference between coefficients of thermal expansion is important if the material experiences temperature cycling. In some instances, this difference may be used to advantage, as in bimetallic strips in temperature-control relays. The strips will bend when the temperature is changed; the degree of bending depends on the metals and their relative coefficients of thermal expansion. Many composite materials fail, however, due to differences in the coefficients of thermal expansion between the materials. In the design of composites subjected to thermal cycling, it is important to match closely the coefficients of thermal expansion of the various materials.

Thermal conductivity: the rate of thermal energy transport in a material. Thermal conductivity occurs by two processes: (1) the rapid movement throughout the material of loosely bonded valence electrons, and (2) the motion of *phonons,* which are quantized elastic waves. Elastic waves—for example, sound or heat waves—may move through solids by successive compression of planes of atoms as the wave front moves through the solid, and these elastic waves are capable of transmitting heat energy. In metals, a large fraction of thermal energy is transferred by electrons, and thermal conductivity is therefore proportional to electrical conductivity. This also explains why metals are much better conductors of heat than other materials such as polymers or ceramics.

Thermodynamic properties: these include heat capacity, internal energy, enthalpy, entropy, and free energy (refer to standard chemistry texts for details).

Heat capacity is the number of calories or joules that a gram (or mole) of a material absorbs when its temperature increases by 1 °C. The energy absorbed increases atomic vibrations and sometimes molecular or chain-segment rotation in materials.

Internal energy is the sum of all types of energy in a material: energy associated with mass, vibrational energy, rotational energy, and energy of translation. Internal energy cannot be measured, unfortunately, but changes in internal energy can be measured.

Enthalpy (sometimes called heat content) equals the internal energy plus the pressure times the volume ($H = E + PV$). Similar to internal energy, enthalpy cannot be measured; only the *change* in enthalpy or internal energy can be measured ("Δ" indicates a change in a given parameter): $\Delta H = \Delta E + P\Delta V$. Because most materials systems under study comprise either solids or liquids, or a mixture of solids and liquids, the difference between ΔH, the enthalpy change, and ΔE, the internal energy change, is insignificant: the quantitative difference is $\Delta H - \Delta E = P\Delta V$. Because the volume change is very small in systems involving only solids and liquids, the $P\Delta V$ term is insignificant (that is, ΔV is close to zero).

Entropy, S, is a measure of the disorder in a system. It may be measured experimentally by determining the heat capacities of materials as a function of temperature. According to the second law of thermodynamics, entropy always increases in a spontaneous reaction. Even when reactions are not spontaneous, the overall entropy increases; to carry out a nonspontaneous reaction, it is necessary to supply energy, which comes from another reaction that is spontaneous. Further details may be found in standard chemistry or thermodynamics texts.

The Gibbs free energy is the energy that is available to do work. Only changes in G can be measured experimentally. If the change in G, ΔG, is negative, then the reaction under consideration is spontaneous. If ΔG is positive, then the reaction is not spontaneous. ΔG is a most useful quantity because it not only enables one to determine whether a given reaction will occur, but it also permits one to calculate the extent of reaction (the equilibrium constant). In terms of the above parameters, $\Delta G = \Delta H - T\Delta S$. Whether a reaction occurs spontaneously (ΔG is negative) depends on two terms: (1) an energy term and (2) an entropy or probability term.

14.3 Problems

1. The energy required to break a mole of C—C bonds is 3.48×10^5 J/mol. Calculate the wavelength of light that has exactly this energy.

2. What is the difference between emission spectroscopy and atomic absorption spectroscopy?

3. What are the characteristics of a laser beam compared with a beam of ordinary light?

4. What causes color in various ions? Why are certain ions, such as sodium, chloride, calcium, and aluminum, colorless?

5. Why do materials expand when heated? In general, which class of materials has the highest coefficients of thermal expansion? Why is it important to control the coefficients of thermal expansion in many composites?

15

Failure of Materials

Most applications of materials rely on the integrity of the material: fracture usually represents failure. However, some uses of materials involve easy fracture or failure, such as the pull wire on a fire extinguisher or the easy tearing of aluminum foil wrap. Fracture of materials often occurs suddenly, as in the shattering of glass. However, fracture may also occur after the cyclic application of stress at relatively low levels, such as a gear tooth that has been subjected to millions of stress cycles (absence of stress followed by a state of stress as the gear tooth turns and is engaged momentarily). This type of failure by fracture is called *fatigue*. A second time-dependent type of fracture often occurs after excessive creep: a material that is subjected to stress at a level lower than the yield stress will elongate over a period, usually at elevated temperatures. When creep has occurred to a point where the elongated material can no longer sustain the applied stress, fracture will occur; this is called *stress rupture*.

15.1 Brittle Fracture

When fracture occurs in a material in which essentially no plastic deformation has occurred prior to fracture, then the failure is called *brittle fracture*. Brittle fracture occurs as a result of (1) crack formation or crack initiation and (2) crack growth or crack propagation. Fracture occurs when a crack propagates across a material, separating it into two or more pieces.

Because the bond strengths between atoms in various materials have been measured, one may calculate the fracture stress of the materials as equal to the number of bonds broken per square centimeter times the force required to break one bond. However, such calculations of fracture strength lead to gross errors; results are usually higher than experimental values by a factor of between 10 and 1000. The reason for the large error in brittle materials is due to stress concentrations formed at the tips or leading edges of cracks, as pointed out by A.A. Griffith.

The reason for stress concentrations in brittle solids may be seen in Fig. 15.1, in which the crack extends across half the cross section. If the tensile force is 1000 lb and the entire cross section is 1 in.^{2}, then obviously the intact ½ in.^{2} must bear the full load of 1000 lb. However, because the tensile force is applied evenly over the ends of the bar, it is impossible for the intact material to be stressed uniformly at 2000 psi. Actually, the material at the edge of the crack is stressed far in excess of 2000 psi.

Griffith derived the following equation to calculate the stress concentration at the leading edge of a crack: $\sigma_c = 2\sigma\sqrt{c/r}$, where σ_c is the stress concentration at the leading edge of the

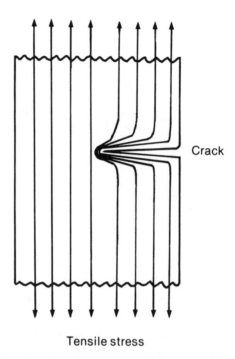

Tensile stress

Fig. 15.1   Lines of force showing the stress concentration at the tip of a crack in a brittle material

crack; σ is the applied tensile stress; c is half the length of an interior crack, or c is the "depth" of a surface crack (see Fig. 15.2); and r is the radius of curvature of the leading edge of the crack.

Example 15.1A. Calculate the radius of curvature of the leading edge of a crack in glass when the stress concentration is 1,500,000 psi, the applied stress is 8000 psi, and the crack depth is 2 μm (0.0002 cm).

Solution: $\sigma_c = 2\sigma\sqrt{c/r}$, or $c/r = \sigma_c^2/4\sigma^2$ and $r = (4\sigma^2 \times c)/\sigma_c^2 = (4 \times 8000^2 \text{ psi}^2 \times 0.0002 \text{ cm})/1,500,000^2 \text{ psi}^2 = 2.28 \times 10^{-8}$ cm, or r = 0.228 nm (10^7 nm = 1 cm).

Note: The tensile strength of glass is frequently about 8000 psi, but the strength of crack-free fibers has been measured in the vicinity of 1.5×10^6 psi, which is close to the stress concentration in the above example. Usually the radius of curvature of a crack tip in glass is approximately equal to atomic dimensions, indicating essentially no plastic deformation prior to failure.

For the spontaneous growth of a crack in a brittle material, it has been shown that $\sigma = [2\gamma E/\pi c]^{1/2}$, where σ is the applied stress necessary for spontaneous crack growth, or fracture stress; γ is the surface energy of the material in ergs per cm^2; E is the Young's modulus of elasticity, expressed in dynes per cm^2 (1 psi = 7×10^4 dyne/cm^2); and c is the crack depth, or radius of an interior crack.

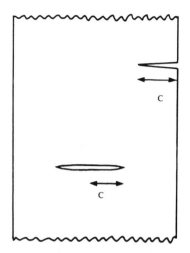

Fig. 15.2   Crack radii in a brittle material

Example 15.1B. Calculate the applied stress needed to propagate a crack in glass having a surface energy of 600 erg/cm^2, E $= 1.0 \times 10^7$ psi; the surface crack is 2 μm deep.

Solution: $\sigma = [2\gamma E/\pi c]^{1/2} = [2 \times 600$ erg/cm$^2 \times 1.0 \times 10^7$ psi $\times 7 \times 10^4$ dyne/cm^2/1 psi]$^{1/2} = [1.338 \times 10^{18}]^{1/2}$ dyne/cm^2, or $\sigma = 1.157 \times 10^9$ dyne/cm^2, or 1.157×10^9 dyne/cm$^2 \times 1$ psi/7 $\times 10^4$ dyne/cm$^2 = 16,500$ psi.

If a brittle material is placed under compression, note that the microcracks simply close up and transmit the compressive stress, as shown in Fig. 15.3. Even if the crack extends completely through the material, it may still have very high compressive strength. A dry wall made of bricks without any mortar is strong in compression, but, of course, has zero tensile strength. Brittle materials such as glass or some metals or polymers almost always have surface flaws or microcracks and hence are usually only moderately strong in tension, although they may be very strong in compression.

Because brittle fracture depends on the length of the largest microcrack in the specimen, and the lengths of microcracks are highly variable, brittle fracture strength is also highly variable for a given material. The tensile strength of a given brittle material often varies by at least ±30% when measured in several specimens. The consistency of ductile fracture strengths is much better—frequently about ±5% in several specimens.

Cleavage is the splitting of a crystal along planes of low Miller indices or relatively high interplanar spacing. For instance, many face-centered cubic crystals, such as the mineral fluorite (CaF$_2$), cleave easily along {111} planes. The net result is a crystal that is a perfect octahedron (a double pyramid with a square base and hence eight sides to the crystal). BCC materials, such as NaCl, cleave on {100} planes, resulting in a cube, or an elongated cube (inspect grains of salt closely to see this).

Brittle fracture may occur along grain boundaries in metals, particularly when films of a hard, brittle second phase occur there. An example is bismuth in copper, where a very low bismuth concentration can seriously embrittle the copper.

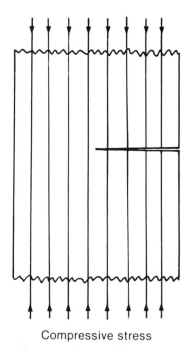

Compressive stress

Fig. 15.3 Lines of force showing no stress concentration for a brittle material in compression
Crack is closed by compressive forces.

15.2 Ductile Fracture

Because *toughness* is the ability of a material to absorb relatively large amounts of energy upon impact, brittle materials are not very tough. All ceramics, some metals, and some polymers are brittle. Brittle metals include some elements, such as chromium, and many intermetallic metals, such as calcium aluminide ($CaAl_2$). Brittle polymers include many thermosets, such as phenol formaldehyde (i.e., Bakelite), and some thermoplastics, such as PVC when it has not been plasticized.

Ductile fracture is fracture that occurs after a material has been plastically deformed. Good toughness often results from plastic deformation prior to failure. Work hardening in metals causes dislocation buildup and strengthening due to cold work: sometimes dislocations move through each other with great difficulty, thereby strengthening the metal. When metals are strengthened, they also are toughened unless strengthening results in excessive brittleness.

Strengthening of polymers by plastic deformation is due to orientation of molecules, resulting in greater crystallinity and, therefore, increased numbers of van der Waals' forces of attraction between polymer chains. Good toughness may also result from elastic deformation, such as in elastomers: although elastomers are not very strong, they are tough because of their extremely high elastic deformation, which enables a relatively large amount of impact energy to be absorbed. Plastic deformation often provides an "early warning" that failure may soon result from higher levels of stress.

Ductility in materials causes the blunting of leading edges of cracks: the greater the ductility, the more blunting will occur. Blunting of crack tips means that r, the radius of curvature of crack tips in the Griffith equation, is made large. Because r is relatively large in ductile materials, the stress concentrations at the tips of cracks are small, and fracture is much less likely to occur.

In most metals, particularly at relatively low temperatures compared with their melting points, strain hardening occurs due to the buildup of dislocations and the increased difficulty in dislocation movement. This strain hardening decreases the plasticity or ductility of the metal, meaning that less blunting occurs at crack tips. Therefore, higher stress concentrations result and crack propagation occurs more easily; fracture will occur when the material can no longer sustain the applied stress (see Fig. 15.4).

Necking occurs when a material strain hardens markedly: the cross-sectional area of the plastically deformed material is visually much smaller than that of the nondeformed material. Because the necked-down material is much stronger, further plastic deformation occurs at the end of the neck. Eventually a flaw such as a notch or crack may cause the material to fail in the necked-down region. Both metals and polymers experience this phenomenon.

When metals are at relatively high temperatures compared with their melting points, strain hardening is not appreciable because dislocations can anneal out and disappear. Also, dislocations can move much more easily through other dislocations. Consequently, many metals have high ductility and low stress concentrations at the leading edges of cracks at elevated temperatures. The metal can plastically deform so extensively that fracture does not occur due to the growth of cracks, but rather simply due to necking down to essentially zero cross-sectional area, at which time fracture occurs. This type of ductile fracture is called *chisel-edge rupture* and is shown in Fig. 15.5.

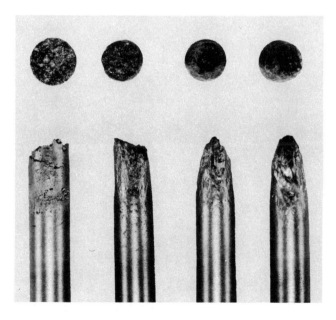

Fig. 15.4   Ductile fracture
Strain hardening causes necking; 6, 37, 63, 83% reduction in area, respectively

Fig. 15.5   Chisel edge fracture at elevated temperatures
99% reduction in area

In ductile polymers, plastic deformation is due to chain straightening and chain slippage: van der Waals' forces between chains are broken and reformed as slippage occurs. Eventually, the polymer chains are straightened as much as possible, and slippage finally causes rupture when the stress exceeds the strength of the polymer. Some polymers are highly ductile and undergo plastic deformation of several hundred percent.

In metals, ductile fracture surfaces have a dirty-gray appearance due to the rough, irregular contour of the fracture surface. Much of the surface of the fractured metal is inclined sharply to the average plane of fracture. Most ductile-fracture specimens exhibit a "cup-and-cone" type of fracture, as shown in Fig. 15.6. Failure begins with the formation of a crack in the center of the necked region in a plane parallel with the cross section (perpendicular to the tensile stress). As the metal is deformed, the crack grows in the above plane toward the outside of the specimen. The final zone of fracture is at an angle of about 45° with the plane of the crack; failure occurs very rapidly in this zone. The resulting appearance causes one portion of the tensile specimen to appear similar to a cup and the other surface as the mating cone.

Fig. 15.6   Cup-and-cone fracture (3×)
Rupture began at an interior pinhole, which is visible in the photograph.

Nonmetallic inclusions, such as carbides, nitrides, sulfides, or oxides, which are found in most commercial metals, act as stress concentrators leading to the formation and growth of cracks. These inclusions probably nucleate voids or pores in the center of tensile specimens. These voids ultimately join to form a crack in the cross-sectional plane. Extremely pure metals, having very few inclusions, are much more ductile than metals of slightly lower purity. Ultra-pure metals often neck down to a point before failure.

15.3   Ductile-Brittle Transitions

Many materials may fail in either a ductile *or* brittle manner, depending on temperature, strain rate, level of stress concentration, and grain size. Low temperatures, high strain rates, notches or cracks, and coarse grains favor brittle fracture in metals, while high temperatures, low strain rates, low stress concentrations (absence of notches, cracks, or inclusions), and fine grains favor failure by ductile fracture. Asphalt shatters when struck with a hammer, but plastically deforms at low strain rates. Most polymers, such as polyethylene or polyvinylchloride, plastically deform prior to failure at room temperature, but exhibit brittle fracture at liquid-nitrogen temperatures (below their glass transition temperatures). Metal parts should be designed to avoid notches, because they often lead to brittle fracture.

If the yield strength is higher than the fracture strength, a material will fail due to brittle fracture, as shown in Fig. 15.7. However, if the yield strength is lower than the fracture strength, the material will plastically deform prior to failure.

The *ductile-brittle transition temperature* is the approximate temperature at which the yield strength and the fracture strength are equal: at lower temperatures, brittle fracture occurs, and at higher temperatures, ductile fracture results. The transition temperature is *not* a specific temperature for any given material: it is represented by a short range of temperature. This fact is readily apparent from impact tests, as shown in Fig. 15.8.

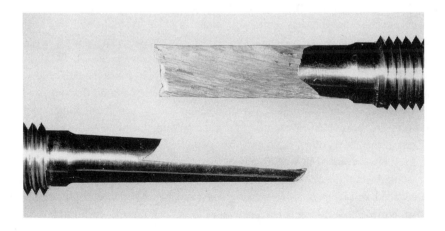

Fig. 15.7   Brittle fracture (3 ×)

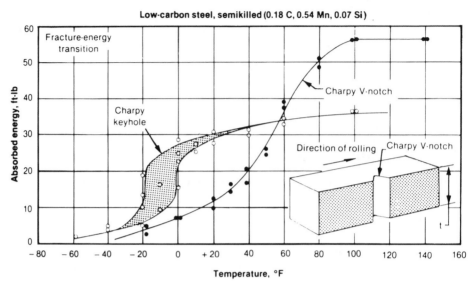

Fig. 15.8   Impact strength as a function of temperature, showing the ductile-brittle transition

Impact strength is the energy absorbed when a test specimen is fractured. A common impact test is the Charpy V-notch test: a metal specimen 1 cm × 1 cm × 5.5 cm has a V-notch machined across it (the notch is a maximum of 2 mm deep). The specimen is held with the notch facing out, and a weighted pendulum hammer is allowed to fall, breaking the notched specimen. The height that the weighted pendulum swings after breaking the specimen indicates the energy absorbed. Figure 15.8 is a representation of typical data for a material undergoing a ductile-brittle transition as a function of temperature. Failure by brittle or ductile fracture may be determined by inspecting the fractured surfaces: ductile fracture imparts a rough or fibrous texture to the surface, while brittle fracture gives rise to a surface having an irregular array of small, bright facets. Each facet is a cleavage plane in an individual grain that has failed by cleavage (brittle fracture).

Catastrophic failures have occurred due to the shifting from ductile fracture to brittle fracture in steel at relatively low temperatures (near freezing). Because the hull of a welded ship is one large piece of steel, a crack may rapidly pass completely around the ship, causing it to break into two pieces. This has, in fact, occurred several times, usually in winter on rough seas (high stress levels), particularly to some of the Liberty ships hastily constructed during World War II. Even in the 1970s other ships have been known to crack in two. Cracks as long as a kilometer have also occurred in welded gas pipelines.

The ductile-brittle transition temperature increases sharply with increasing carbon concentration. Quenched and tempered steel usually has a much lower ductile-brittle transition temperature than normalized and tempered steel with the same composition. Phosphorus adversely affects the transition temperature; this is a good reason why phosphorus is undesirable in carbon steels.

15.4 Fatigue Failure

Fatigue is the fracture of materials resulting from cyclic stresses. The maximum stress is below the yield stress, and it may be a tensile, shear, or compressive stress, but in commercial materials stresses are usually a combination of the three. The stress cycle often involves alternating tensile and compressive stresses, as in a loaded, rotating shaft. Teeth in gears alternate between a condition of no stress and one of high shear and tensile stresses, as well as compressive stresses. Many metal parts are designed for moving, rotating, flexing, or vibrating functions, and these are all subject to fatigue failure.

Fatigue is characteristic of ductile metals. It is a function of the number of stress cycles and the level of stress: at high stress levels, the number of cycles to failure is relatively low, while lower stress levels enable a given metal to withstand a much larger number of cycles. Many metals, particularly steels, have *endurance limits,* or a level of stress below which fatigue does not occur regardless of the number of stress cycles (see Fig. 15.9). It is common practice to obtain data from fatigue tests and to plot them as in Fig. 15.9, in which stress is plotted versus the logarithm of the number of stress cycles. These graphs are referred to as *S-N curves.*

Most fatigue failures start at the surface of a metal part, usually as a microcrack at some stress concentrator (screw hole, notch, tool mark from grinding, etc.). As each cycle of stress is applied, the crack grows very slightly until the remaining intact metal is finally unable to support the stress. The remainder of the metal then fails rapidly. The fatigued portion of the metal surface appears burnished or polished due to the rubbing of one surface over the other during cyclic stressing. Of course, the rubbing of the surfaces is over very short distances. During the last stage, in which fracture propagates rapidly, the surfaces are rough and irregular, similar to brittle fracture surfaces (see Fig. 15.10).

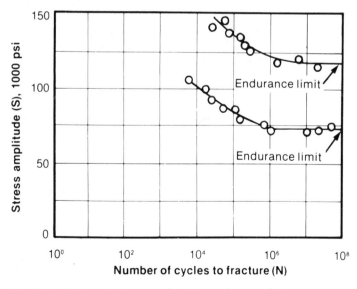

Fig. 15.9   Fatigue S-N curves showing endurance limits

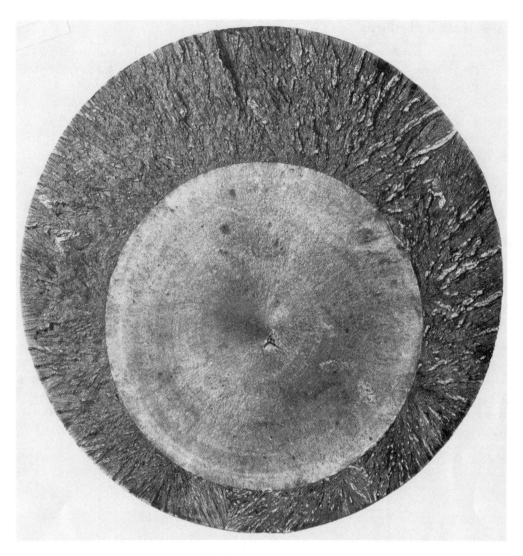

Fig. 15.10   Fatigue failure surface of a piston rod
Failure began at a forging flake near the center. Fatigue propagated outward slowly; outer circular region is final failure by brittle fracture.

Because stress concentrators usually initiate microcracks at the surface of metals, the condition of the surface has an important effect on the fatigue resistance of metals. Smooth surfaces, or surfaces that have been shot-peened, cause surface compressive stresses which result in higher endurance limits for metals. Fatigue life may also be shortened by even mildly corrosive environments: the endurance limit is lower for many metals when they are exposed to moist air compared with their limit in dry air. Case hardening with carbon or nitrogen can also lead to improved fatigue resistance in steels.

Fatigue tests may involve alternate (1) stretching in tension and relaxing, (2) stretching and compressing, (3) twisting in one direction and then in the reverse, or (4) reverse bending by rotating a bent bar. Of course, there are many other fatigue tests simulating actual conditions of use. A very important fact is that the *rate* of applying stress does not affect the fatigue strength at normal temperatures: therefore, accelerated tests may be run in the laboratory, where stress can be applied and relieved very rapidly—up to several hundred thousand cycles per minute. This is important because fatigue failure may occur only after 1 million or even 10 million, or more, cycles.

The endurance limit of many steels is about 40 to 50% of the tensile strength. This is only a rough rule of thumb, however, upon which one should not rely. The endurance limit of tempered martensite is about 60% of its tensile strength, compared with about 40% for pearlite. Machinery parts are frequently annealed, quenched to form martensite, and tempered. It is important to obtain as much martensite as possible to obtain maximum endurance limits. Most nonferrous metals do not have an endurance limit; however, the S-N curve is not linear: as the stress level decreases, the number of cycles to failure increases sharply. Many metal parts are subjected to a maximum of 10^7 to 10^8 stress cycles during their expected lifetime, so in effect one can design for stress levels low enough to avoid fatigue in the expected life of the metal.

The mechanism of fatigue failure involves (1) plastic deformation on a microscale via slip or twinning, (2) crack initiation at a stress concentrator on the surface, (3) crack propagation due to cyclic stress, and finally (4) rapid brittle fracture of the remaining intact metal. Slip lines form in many metals after many cycles of stress, but long before failure occurs. Strain hardening can occur due to dislocation movement along limited slip planes. In some metals, this strain hardening eventually causes small extrusions of ribbons of metals (see Fig. 15.11) to form due to slip in one direction along certain slip planes and then slip in the reverse direction along parallel slip planes. Imagine a deck of cards: if half the deck is slipped to the right on a given card and then the top quarter is slipped to the left on a different card, an ''extrusion'' of cards remains. Likewise, intrusions may be formed by a similar mechanism in which crevices into the

Fig. 15.11   Fatigue extrusions
Electron microscope replica after cyclic straining. (From A.H. Cottrell and D. Hull, *Proc. Royal Soc.*, Vol A242, 1957, p. 211)

metal may arise. Intrusions may act as stress concentrators to initiate cracks and ultimately cause fatigue.

Crack growth can occur during most of the life of the specimen. Growth rate is proportional to the square root of the crack length and to the applied stress. Of course, nonmetallic inclusions increase the growth rate of cracks, or they may nucleate cracks.

In some applications, the maximum number of stress cycles is known fairly precisely to be a relatively small number such as 10^4 or 10^5, e.g., a starter spring in an automobile. In this case, fatigue data may be used to design the part most economically—there is no sense in designing a part that will last a thousand times the expected life.

15.5 Creep and Stress Rupture

When stress is applied to metals, dislocations form in numbers proportional to the stress. As long as the yield stress is not exceeded, dislocation movement is minimal, particularly if the given stress is maintained for a time. For instance, a steel beam in a building is under constant stress; the beam contains a certain concentration of dislocations, and these dislocations are essentially stationary. If they moved appreciably, then the beam would plastically deform.

If steel is stressed at elevated temperatures, say in a gas turbine engine, dislocations are formed due to the stress, but the relatively high temperature enables the dislocations to move slowly. Therefore, plastic deformation may occur over time, and this is called *creep:* time-dependent plastic strain. Creep is appreciable only at temperatures above approximately 40 to 50% of the melting point of the metal in degrees Kelvin. Any structural application of a metal at elevated temperatures may involve creep. If extensive creep occurs, failure will occur when the ability of the metal to support the stress is exceeded. Failure due to creep is known as *stress rupture,* demonstrated in Fig. 15.12.

Creep of metals is usually tested by loading a metal at a given elevated temperature and measuring the strain (in./in. or cm/cm) as a function of time. Strain is then plotted against time, as shown in Fig. 15.13(b). Because creep involves plastic deformation and hence a continually changing cross-sectional area, it would be difficult to conduct a test employing constant stress.

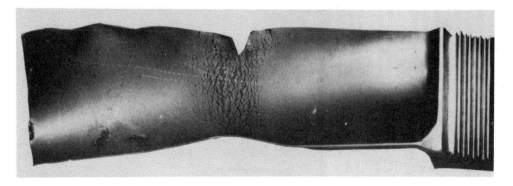

Fig. 15.12   Failure due to creep
Stress rupture of a jet engine turbine blade

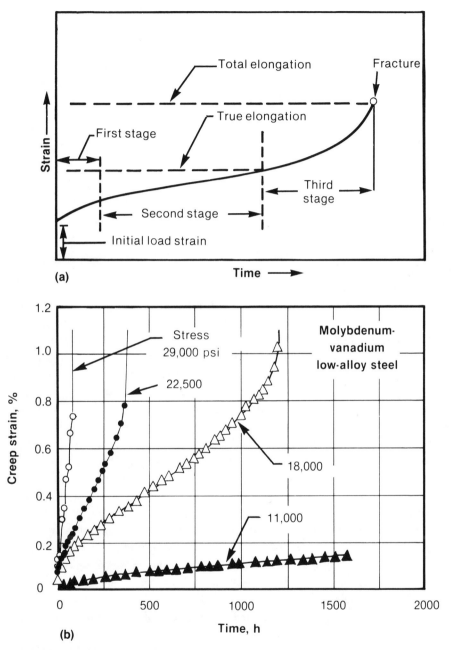

Fig. 15.13 (a) Schematic tension-creep curve, showing the three stages of creep. (b) Creep curves for a low-alloy steel at four stress levels at 600 °C

Figure 15.13(a) shows the three common stages often found in creep of a material. The first stage represents an immediate strain that is due to the elastic deformation of the material under stress followed by a sharp increase in strain hardening. The elastic deformation is usually very small and scarcely measurable except by special instruments.

The second stage of the creep curve in Fig. 15.13(a) is nearly a straight line; this is due to an approximate balance between strain hardening, which tends to decrease the rate of creep, and the decreasing cross-sectional area, which tends to increase the stress and therefore the creep rate. The linear second stage of creep represents fairly steady creep: the strain rate is more or less constant. *Strain rate* is elongation per unit of time, e.g., centimeter per centimeter per hour (cm/cm × h).

If either the stress or the temperature is relatively low, then the second stage is pronounced. However, if either the stress or the temperature, or both, is rather high, the second stage may be very small or absent.

The third stage in the typical creep curve occurs just prior to failure (stress rupture): it is a highly accelerated strain rate. As the material necks down to a smaller cross-sectional area, the constant load produces an increasing level of stress. The increased stress will therefore accelerate the creep rate. Acceleration continues until failure occurs because the yield stress is finally exceeded.

Stage three is always present whenever stress rupture occurs. However, many creep tests are limited by the time available and are only carried out to somewhere in the second stage. If the second stage is extrapolated to determine the useful life of the metal part, one must be careful to remember that the third stage of accelerated creep could begin almost at any time, thereby nullifying the extrapolation. In other words, it is dangerous to extrapolate unless good information is available to indicate that stage three is very unlikely to occur during the extrapolated period. At times, one may be interested in tolerances rather than failure so that stage-two data are very important and extrapolation is useful. For instance, in a gas turbine engine, rotating parts could not extend very far due to creep without trouble occurring.

The effects of temperature and stress are similar in the phenomenon of creep: increased temperatures have the same effect as increased stresses (see Fig. 15.14). Strain rate is exponentially related to temperature. In nearly all creep tests and in actual practice, the true stress is a variable that increases with time because the cross-sectional area continually decreases.

Because creep in metals involves the slow motion of dislocations at relatively elevated temperatures, the mechanisms of creep depend on the passage of dislocations around obstacles. Obviously, it takes time for the dislocations to pass by an obstacle because creep generally occurs over relatively long periods. Obstacles to dislocation movement include (1) other dislocations, (2) grain boundaries, (3) solute atoms, (4) nonmetallic inclusions or other second-phase particles, (5) incipient precipitates as found in age hardening, and (6) vacancies.

Even at elevated temperatures, thermal energy is capable of moving only a few atoms in a dislocation at one time; hence, only segments of dislocations can move at one time due to thermal energy. This is a primary reason why time is involved in the passage of a dislocation around an obstacle.

Polymers may undergo creep when stressed over periods of time. However, the mechanism for creep is different from that of metals. Chain straightening and slippage occur. Polymers that depend only on van der Waals' bonding among chains are most vulnerable. Engineering plastics, such as nylon or acetal, have many hydrogen or polar bonds between chains, and these

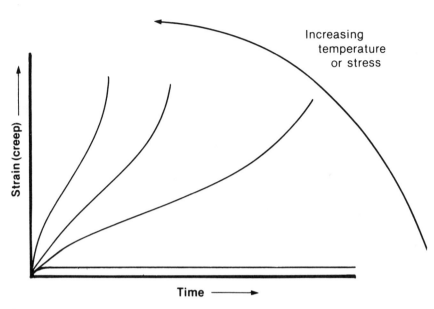

Fig. 15.14 Equivalency of temperature and stress on creep

bonds reduce or prevent creep from occurring. Accordingly, these polymers are highly dimensionally stable, which is another way of saying that they are not very susceptible to creep. Engineering plastics are used where good creep resistance is necessary, such as in plastic gears. Most plastics are used at or just above room temperature, where creep resistance is feasible for many polymers. At even 200 to 300 °C, most polymers are either unstable or highly susceptible to creep (remember that thermoplastics above their melting temperatures are simply highly viscous liquids).

Surprisingly, even ceramics are slightly susceptible to creep. Glass is amorphous and a supercooled, rigid liquid. However, it can flow ever so slightly over a period of years. When heavy stones are placed on one another in monuments, such as in ancient buildings, it is possible for creep to occur over a period of years. This partly explains the amazing fit that large stones have in ancient monuments (the other half of the explanation is that ancient artisans were very skilled in fitting large stones together). In ceramics, slip occurs only with great difficulty; consequently, creep is extremely low for most ceramics.

Much work has been and is being devoted to developing improved creep-resistant materials. Creep-resistant polymers are good alternatives to metals in a variety of applications. Creep-resistant alloys show promise in saving energy due to improved efficiency of operating motors at higher temperatures. In order to have outstanding creep resistance, an alloy should have (1) good high-temperature strength with as few grains as possible, (2) good high-temperature strength in grain boundaries, and (3) slow grain growth. The recrystallization temperature should also be high. Most of the usual hardening mechanisms increase the creep resistance of alloys: solid-solution hardening (alloying), age hardening, and hardening by cold work. If age hardening is used, care must be taken so that the alloy does not overage. Additionally, recovery and particularly recrystallization can remove the hardening obtained as a result of cold work. Second-phase precipitates are quite useful for increasing creep resistance. Excessive strain hardening—more than about 20%—should be avoided because this can lead to recrystallization at somewhat lower temperatures.

Not only is the creep resistance of alloys important in the practical application of metals at high temperatures, but often the oxidation resistance is of critical importance, as are the effects of oxygen and nitrogen from the atmosphere. Some of the most creep-resistant metals are the refractory metal alloys of niobium, tungsten, and molybdenum, but these alloys have relatively poor oxidation resistance, and frequently nitrogen and oxygen diffuse into the metals, causing an intolerable weakening due to embrittlement. Diffusion coatings are available to increase oxidation resistance, but nearly all of these coatings are brittle and susceptible to cracking, either due to thermal cycling or to impact by some airborne particle. Cracking of coatings usually leads to excessive oxidation of the underlying metal. One can easily understand the difficult materials problem in the design of gas turbine engines.

15.6   Corrosion

In general, ceramics have excellent corrosion resistance. The primary problem is one of solubility in the environment; an example is the effect of acid rain on limestone statues. Many statues are being corroded at an alarming rate after having been around for several hundred years. Another example is the usage of bricks in lining furnaces. If the slag that contacts the bricks is basic, then the bricks must be basic to prevent an acid-base reaction. Bricks containing an excess of silica (SiO_2) are acidic, whereas bricks made from materials such as magnesia (MgO) are basic. If the proper choice is made for a ceramic, then corrosion may be avoided. Ceramics often are very stable compounds because they are oxides or silicates. If the materials are insoluble in water, they are also chemically inert in most applications.

Because polymers are large organic molecules, they are almost always insoluble in water, and many common polymers, such as polyethylene, have outstanding resistance to chemicals— both acids and bases. The primary problem with polymers is their resistance to various solvents. Organic solvents often attack or dissolve polymers, or cause them to swell. Because of the great variety in the chemical nature of polymers, it is usually straightforward to find one or more polymers that are chemically resistant to a given solvent. Polymers are often resistant to aqueous systems because they are quite different chemically from water. Some polymers potentially could react with acids or bases because certain groups are present; however, if the polymer is incompatible with water, then the extent of reaction is negligible.

Corrosion of metals is chemical oxidation by the environment, usually by the atmosphere, seawater or solutions, which often are acidic. *Chemical oxidation* is the *loss* of electrons: for example, Fe − 2 electrons → Fe^{+2}. This is a simplified representation of the corrosion of steel. Because corrosion involves the loss of electrons by a metal, it is electrochemical by nature, and a solution is usually required to transfer electrons externally. Corrosion costs many billions of dollars annually, and it contributes to the depletion of our natural resources because metal corrosion products usually are widely dispersed in the environment, making it uneconomical to recover them. Corrosion is nature's way of transforming metals into "ores."

Corrosion is important because it may cause one or more of the following: (1) a poor or unacceptable appearance, (2) failure in a structural metal, (3) contamination of products such as food, (4) a health problem, or (5) loss of valuable materials such as nickel or chromium.

Because corrosion involves the loss of electrons by a metal, the electrons must be accepted by some other chemical: it is impossible to carry out an isolated chemical half reaction. Typical half reactions include the following:

$$Zn - 2e^- \rightarrow Zn^{+2} \quad (e^- \text{ means an electron})$$
$$Al - 3e^- \rightarrow Al^{+3}$$
$$M - 2e^- \rightarrow M^{+2} \quad \text{(corrosion of a metal)}$$
$$M - 3e^- \rightarrow M^{+3} \quad \text{(corrosion of a metal)}$$

The loss of electrons is chemical oxidation, so the above "half-cell" reactions all occur at the anode.

$$M^{+3} + 1e^- \rightarrow M^{+2} \quad \text{(metal ion reduction)}$$
$$M^{+2} + 2e^- \rightarrow M \quad \text{(electroplating)}$$
$$2H^+ + 2e^- \rightarrow H_2$$
$$O_2 + 4H^+ + 4e^- \rightarrow 2H_2O \quad \text{(acidic solution)}$$
$$O_2 + 2H_2O + 4e^- \rightarrow 4OH^- \quad \text{(basic or neutral solution)}$$

The gaining of electrons is chemical reduction, so the above half-cell reactions all occur at the cathode.

For corrosion to occur, two half reactions must occur together, one involving the *loss* of one or more electrons, and the other involving the *gain* of the same number of electrons. Electrons are simply transferred from one ion or element to another. An example is the corrosion of iron by dilute hydrochloric acid:

$$Fe + 2HCl \rightarrow FeCl_2 + H_2$$

This net reaction is the sum of two half reactions (often called "half-cell" reactions):

$$Fe - 2e^- \rightarrow Fe^{+2}$$
$$2H^+ + 2e^- \rightarrow H_2$$

When these two half reactions are added, the electrons cancel; the chloride ions are not included because they do not enter the reactions.

When corrosion occurs, a galvanic cell exists. A *galvanic cell* is a system in which an electrochemical reaction occurs spontaneously, with chemical oxidation (loss of electrons) occurring at the anode and chemical reduction (the gain of electrons) occurring at the cathode. The two electrodes, anode and cathode, may be very close together or they may be separated by several centimeters or more.

Because the reaction occurs spontaneously in a galvanic cell, it is theoretically possible to obtain work from the system. For example, a battery is a galvanic cell. In analyzing the galvanic cell in which the reaction $Fe + 2HCl \rightarrow FeCl_2 + H_2$ occurs, the reaction may be understood by referring to Fig. 15.15. Here, the iron dissolves at the anode, where it is oxidized chemically, and hydrogen forms nearby at a cathode. (The *anode* is the electrode where electrons are lost, or where chemical oxidation occurs. The *cathode* is the electrode where electrons are gained, or where chemical reduction occurs.) Note that in any galvanic reaction, the total number of electrons lost at the anode is equal to the total number of electrons gained at the cathode. Corrosion may occur uniformly on surfaces of iron in HCl, because what is a cathode at one time may later become an anode and dissolve.

Chemically active metals displace less chemically active metal ions from solution:

$$Zn + Cu^{+2} \rightarrow Zn^{+2} + Cu$$
$$Cu + 2Ag^+ \rightarrow 2Ag + Cu^{+2}$$
$$3Ag + Au^{+3} \rightarrow Au + 3Ag^+$$

When metals are arranged in order of their ability to displace others from solution, the arrangement is called the *electromotive series* (or galvanic series) (see Table 15.1). Electrochemists have developed this concept quantitatively, and the result is a powerful method for predicting what reactions occur. For details of the method, consult standard chemistry textbooks.

In the field of corrosion, "a little knowledge is a dangerous thing." Corrosion often occurs in a way that is opposite to one's intuitive reasoning, or it occurs in a bizarre manner that, at first glance, defies reason. If one confronts a problem in corrosion, it is wise either to obtain expert help or to study the area thoroughly before "solving" the problem. At times, solutions to corrosion problems have created worse problems than the initial one. Various types of corrosion are discussed below.

Uniform corrosion is an attack of the metal when an electrochemical reaction, or solution reaction, proceeds uniformly over the entire surface of a metal. It is the most common type of corrosion and is the least insidious form of corrosion because it is completely predictable. Unprotected steel exposed to moisture and air undergoes uniform corrosion rapidly to form rust.

Galvanic corrosion occurs when two different metals or alloys are in electrical contact in the presence of a corrosive environment. The more chemically stable metal acts as a cathode, and the more chemically active metal is the anode, where corrosion occurs. Galvanic corrosion

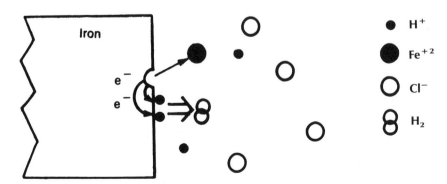

Fig. 15.15   Galvanic cell: corrosion of iron by hydrochloric acid

Table 15.1   The electromotive series

Noble metals (cathodic)	Platinum
	Gold
	Titanium
	Silver
	Austenitic stainless steel
	Ferritic stainless steel
	Nickel
	Monel
	Cupronickel
	Bronze
	Copper
	Brass
	Tin
	Lead
	Tin-lead solder
	Cast iron
	Steel
	Cadmium
	Aluminum
	Zinc
Base metals (anodic)	Magnesium

may be much more rapid than uniform corrosion of a metal in a given environment. For instance, if an iron pipe is attached to a copper pipe, the iron will corrode much more quickly in a given environment. If the relative area of copper is large compared with the iron, then, surprisingly, corrosion is accelerated even more. Galvanic corrosion is used to advantage in galvanized steel (steel coated with zinc), where the zinc corrodes preferentially to iron, even if the coating is scratched through to the iron substrate. The zinc behaves sacrificially: it corrodes instead of the iron. However, zinc forms a highly protective oxide coating on itself, and therefore the zinc corrodes very slowly. Galvanized steel coatings may last for many years.

Crevice corrosion is due to small volumes of stagnant solution collecting in crevices under bolts, under surface deposits such as dirt, corrosion products, or loose paint, and in holes under gaskets. Intense localized corrosion occurs when the crevice is wide enough to permit liquid entry but sufficiently narrow to create a stagnant solution. Although this type of corrosion occurs in openings of about 10 to 100 μm, it rarely occurs in wide grooves or slots (0.1 cm or larger). Fibrous gaskets are notorious for creating stagnant solutions on flanges. A surprising example of crevice corrosion is the fact that 304 stainless steel (Cr-Ni-Fe) may slowly be cut in two by placing a stretched rubber band around the stainless steel and immersing it in a chloride solution such as seawater.

Pitting is a form of severe localized attack, and it results in holes in a metal, usually of small diameter. Pits frequently go undetected until the metal has been severely corroded, because pits may be covered with corrosion products or by a protective coating having only a pinhole opening. Pitting is difficult to predict by laboratory tests, so it is another particularly insidious form of corrosion. An example is the formation of corrosion pits in auto bumper substrates beneath the chromium-nickel electroplated surface.

Intergranular corrosion occurs when grain boundaries are corroded far more easily than the grains themselves. Intergranular corrosion is caused by impurities collected at grain boundaries or by enrichment or depletion of elements in and near grain boundaries. An example is the sensitization of austenitic stainless steels (e.g., type 304) during welding. Parallel to the weld, a precipitate of chromium carbide, $Cr_{23}C_6$, forms in grain boundaries. This results in the depletion of chromium in a narrow zone parallel with the grain boundaries. The depleted zone then corrodes much more easily and rapidly than the bulk alloy.

Stress corrosion is an insidious cracking of a metal due to the combination of stress and an environment which is specific to the metal. Stress cracking may occur at relatively low levels of stress compared with stress needed for mechanical failure and at relatively low concentrations of chemicals. An example is the stress corrosion cracking of austenitic stainless steels (e.g., Fe-Cr-Ni) when stressed mildly and exposed to chloride solutions, particularly warm solutions. This has occurred frequently in chemical plants where the austenitic piping is carrying warm solutions, and the insulation around the pipe contains some chlorides. Rain or moisture then causes stress cracking.

Selective leaching is the preferential removal of one metal from a solid solution as a result of corrosion. The most common example is the *dezincification* of brass, in which zinc is preferentially removed from brasses by a process of corrosion of the brass followed by chemical reduction of copper ions back to copper metal. The reduced copper is usually porous and deposited on the brass.

Erosion corrosion is accelerated corrosion due to the relative motion of the metal and the environment, usually a liquid inside a pipe. The metal may corrode, and the corrosion products are swept away, particularly at bends of tubing.

Liquid metal corrosion is a nonelectrochemical form of corrosion in which metals are corroded when exposed to flowing liquid metals at elevated temperatures, particularly when a temperature gradient is present. The corrosion process often involves mass transfer due to solution of a given element in a liquid metal at a relatively high temperature and deposition of the element at another location in the system where the temperature is lower.

Frequently engineers are interested in combating corrosion, particularly in an economical way. Inexpensive parts may be replaced, or design can be altered to avoid certain types of corrosion mentioned above. Choosing a different metal, or replacing a metal with a polymeric material, may be the best answer. Coating the metal with another metal or with a polymer may prevent or reduce corrosion.

Some design principles include the following:

- Choose one metal only for a system or electrically insulate unlike metals from each other.
- Use a large surface area of anode compared with the surface area of cathode.
- Eliminate or reduce to a minimum crevices, lap joints, nuts and bolts, rivets, small recessed areas, grooves, or scratches.
- Provide for easy maintenance to prevent buildup of dirt, scale, rust, etc.
- Allow for uniform and moderate flow rates of fluids through systems having gentle bends.
- Use the proper material for a given environment.
- Reduce the stress in parts that may stress crack.
- Use corrosion inhibitors, corrosion-preventing films, or coatings where necessary.

Some common coating techniques for metals include the following:

- *Chemical conversion*: building up a relatively thick oxide film by anodizing (anodic oxidation, usually of aluminum) or immersing in a chromate or phosphate solution
- *Cladding*: hot rolling a sandwich of two metals to obtain metallurgical bonding between them
- *Diffusion*: high-temperature treatment in which one or more metals are diffused into the substrate metal to form an alloy diffusion coating
- *Electroplating* (or simply "*plating*"): electrodeposition of a metal such as chromium, nickel, copper, tin, zinc, cadmium, gold, silver, or platinum in a thin layer on a metal substrate
- *Enameling*: fusing a glassy composition on the surface of steel or fusing a polymeric composition on a metal substrate
- *Flame spraying* (or *metallizing*): blowing metal in finely divided form through a melting flame onto a metal substrate. The finely divided hot metal is obtained from a wire or powder.
- *Hot dipping*: immersing metal substrates in liquid metals to coat them with zinc (as in galvanizing), tin, lead, or aluminum
- *Painting*: applying a pigmented organic carrier that reacts with air or dries by evaporation

15.7   Oxidation

Corrosion of metals involves chemical oxidation, or the loss of electrons. *Oxidation* refers to the reaction of oxygen with metals to form oxides, usually found on the surface of metals. Very active metals, such as Group IA and Group IIA elements, generally react with oxygen in the atmosphere to form a partially protective oxide coating at the surface. Frequently, moisture in the air causes a hydroxide to form on the surface of these metals.

Most pure metal surfaces react rapidly with oxygen in air to form a very thin, protective oxide film. Some metals, such as aluminum, titanium, and stainless steel, form extremely protective oxide films, known as *passive oxides,* and these films form spontaneously and rapidly in air when the metal is scratched or otherwise damaged. Nearly all metals exposed to air form at least a monomolecular film of oxide on their surfaces. Only in a very high vacuum is it possible to have really clean metal surfaces. In fact, not only do oxides occur on metal surfaces, but many react with water in air to form hydroxides or hydrates at the interface between the oxide and air. These reactions result in part from the high surface energies of metals and the moderately high surface energies of oxides. Perfectly clean metal surfaces are highly reactive, and even metal-oxide surfaces are reactive enough to adsorb water. When the surface oxide has a smaller volume than the metal from which it forms, the oxide is nonprotective, and oxidation occurs at a linear rate because the oxide is unable to prevent further oxidation of the substrate metal.

Oxide films may cause several problems in the use of metals: (1) aluminum oxide films make soldering of aluminum most difficult; (2) thick oxide films, such as "scale" on steel, may cause excessive wear in processing because of the hardness of the oxide; (3) oxides are poor surfaces for painting or finishing because they are brittle and are easily chipped or cracked off the surface; (4) oxides may be embedded in metal surfaces during rolling, drawing, or stamping unless they are first removed.

Oxide films offer some outstanding properties in some cases; for example, passive films are responsible for the excellent corrosion resistance of many metals. Other oxide films are somewhat protective: even rust on steel decreases the subsequent rate of oxidation.

Like corrosion, the oxidation of metals is an electrochemical process: electrons are lost by the metal at the metal-oxide interface, and electrons are gained by oxygen at the oxide-air interface. New oxide sites are produced either at the metal-oxide interface or at the oxide-air interface. The oxide serves to conduct electrons from the metal-oxide interface, where they are given up, to the oxide-air interface, where the electrons are used to reduce oxygen. In order to balance charges, the metal ions must have a net relative motion away from the metal-oxide interface, and the oxide ions must have a net relative motion toward the metal-oxide interface (see Fig. 15.16).

At high temperatures, steel oxidizes in air to form a complicated structure: a relatively thick layer of FeO forms next to the metal substrate, a middle layer of Fe_3O_4 is formed, and a relatively thin Fe_2O_3 layer forms at the oxide-air interface. Oxidation proceeds by the diffusion of Fe^{+2} ions and movement of electrons from the metal-FeO interface toward the Fe_2O_3 interface. Oxygen diffuses in from the atmosphere and oxidizes Fe^{+2} ions to Fe^{+3} ions.

Most oxides formed on metals such as iron, nickel, copper, and chromium grow principally at the oxide-air interface by diffusion of cations to the surface. This results in a counterflow of vacancies toward the metal-oxide interface, where voids frequently appear due to condensation of vacancies.

Metals such as niobium, tantalum, and titanium form oxides in which a large volume increase occurs when the metal forms the oxide. This may cause cracks to form in the oxide

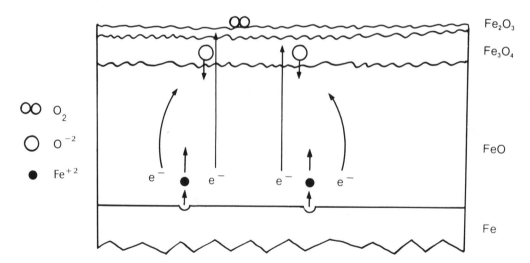

Fig. 15.16   Oxidation of iron at elevated temperatures
Note: Fe^{+2} diffuses more rapidly than O^{-2} because it is much smaller.

scale, and oxygen molecules to diffuse into the cracks and rapidly oxidize the metal because the effective scale thickness is greatly reduced. Hence, these oxides are not very protective at elevated temperatures and provide poor oxidation resistance for these pure metals.

Many metal oxides are *nonstoichiometric compounds,* which are compounds in which the atoms do not have a fixed, whole-number ratio. This is particularly true when the metal has more than one common oxidation state. The actual compositions of nonstoichiometric oxides are close to that of a stoichiometric compound. For instance, FeO is the stoichiometric formula for iron(II) oxide. In actuality, however, the compound is a mixture mainly of Fe^{+2} ions with a small concentration of Fe^{+3} ions as well. The actual formula is something like $Fe_{0.936}O$. For more details about nonstoichiometric oxides, refer to Chapter 9.

Cation vacancies or interstitial cations are most responsible for diffusion in oxides, and therefore the particular metal-oxygen system affects the concentrations of defects and hence the rate of diffusion. Diffusion-controlled oxidation, which, practically speaking, is the most important type, may be altered by decreasing the concentration of cation defects in the structure.

The oxidation rate of most diffusion-controlled oxidation processes is expressed by $W^2 = kt + C$, where W is weight gain, t is the time in seconds, and C is a constant determined by experiment. Most oxides which are free of cracks follow this equation. Two rules for decreasing oxidation rates are the following: (1) for oxides such as ZnO, which contains an excess of zinc ions and electrons in interstitial sites, introduce a higher valence cation into the lattice to decrease the number of interstitial cations while increasing the number of excess electrons; (2) for oxides such as FeO, which may have many cation vacancies, introduce lower valence cations to decrease the concentration of cation vacancies.

Spalling is the separation of an oxide scale from a metal due to differences in the coefficient of thermal expansion of the metal compared with that of the oxide and due partly to the mechanical properties of the oxide (ease of cracking, etc.). Certain elements, such as aluminum, generally reduce spalling of oxides on thermal cycling. Steel that has been protected by aluminum-iron alloy coatings may be heated to elevated temperatures, cooled to room temperature, and reheated to elevated temperatures many times without spalling. Metals are frequently alloyed with aluminum, silicon, chromium, nickel, and yttrium to increase their oxidation resistance.

Table 15.2 contains maximum operating temperatures at which oxide formation is negligible for various metals. In order to use metals at high temperatures, not only is oxidation resistance important, but also mechanical properties wherever metals undergo more than nominal stress.

Because most ceramics are either silicates or oxides, oxidation is normally not much of a problem even at very high temperatures. However, some ceramics are susceptible to oxidation; an example is graphite, which reacts with oxygen at elevated temperatures to form CO or CO_2. Graphite is an excellent material for high-temperature usage in the absence of oxygen.

Most polymers become unstable when heated to 200 to 300 °C. Carbon-carbon bonds break, and polymeric materials break down or otherwise react with oxygen. A few polymers may be used at elevated temperatures (but usually less than 500 °C).

Table 15.2   Maximum operating temperatures for various metals

Metal	Composition, wt. %	Maximum operating temperature, °C
1010 steel	99+ Fe, 0.10 C	480
Stainless steels		
410	88 Fe, 12 Cr	760
430	84 Fe, 16 Cr	840
446	73 Fe, 27 Cr	1040
302	74 Fe, 18 Cr, 8 Ni	900
309	64 Fe, 24 Cr, 12 Ni	1100
310	55 Fe, 25 Cr, 20 Ni	1150
316	58 Fe, 18 Cr, 12 Ni, 2 Mo	900
Hastelloy X	22 Cr, 9 Mo, 20 Fe, 49 Ni	1200
Hastelloy C	53 Ni, 6 Fe, 19 Mo, 17 Cr, 5 W	1150
Brass	70 Cu, 30 Zn	700

One of the greatest challenges facing materials scientists and engineers is the problem of materials for high-temperature applications. Undoubtedly, important advances will be made in the future, perhaps mostly with ceramic materials.

15.8   Problems

1. What is the stress concentration for a glass specimen having an applied stress of 9000 psi, a surface crack 1.50 μm deep, and a radius of curvature of 0.30 nm at the tip of the microcrack?

2. What is the length of an interior crack in a glass specimen in which the stress concentration was 200 times the stress, and the radius of curvature of the tip of the crack was 0.28 nm?

3. Calculate the surface energy of glass having a surface microcrack 2.8 μm deep; an applied stress of 9500 psi was sufficient to cause the crack to grow spontaneously.

4. Explain why table salt (FCC NaCl) is comprised of small cubes or elongated solids having 90° angles. What are the Miller indices for these planes?

5. What are the Miller indices for the surface planes of an octahedral crystal of fluorite (FCC CaF_2)?

6. Contrast the failure of a ductile metal with a ductile plastic on the atomic or molecular scale.

7. Do ceramics undergo ductile-brittle transitions? If so, under what conditions?

8. What sort of failure is likely responsible for the fracture of a gear tooth after many years of extensive use? After sudden stopping of the gear while in use?

9. What sorts of failures are most likely to occur in a metal in a high-temperature gas turbine engine?

10. What type of corrosion is responsible for each of the following failures:

(a) The surface coating only of a garbage can corrodes very slowly even though the coating is scratched through the substrate metal (iron) in places.

(b) An old automobile bumper rusts in many small spots.

(c) A keg of nails is left out in the rain, and they all begin to rust quickly.

(d) In a do-it-yourself plumbing job, a copper pipe and a steel pipe were joined, and the steel pipe rusted severely.

(e) A nickel-containing stainless steel pipe was bent into a U-shape and immersed in a warm solution containing sodium chloride, whereupon it cracked catastrophically.

(f) A nickel-containing stainless steel sheet was welded; corrosion developed parallel to the weld.

(g) A painted metal surface developed some small corrosion spots.

(h) A Roman brass mirror lost much zinc from its surface on being buried for 2000 years.

11. How could corrosion in parts b, c, d, e, f, and g in problem 10 have been reduced or eliminated?

12. Which of the following metal oxides could be nonstoichiometric (explain your choices): (a) CaO, (b) K_2O, (c) Al_2O_3, (d) Cr_2O_3, (e) MnO, (f) NiO?

13. How could *each* of the oxides in problem 12 be made to contain a relatively high concentration of vacancies?

Appendix

Values for Miscellaneous Constants

Avogadro's number: 6.022×10^{23} atoms/mol
Faraday constant: 96,487 coulombs/g-equivalent weight
              23,061 cal per volt/g-equivalent weight
Charge of electron: -1.602×10^{-19} coulomb
Molar Boltzmann constant (gas constant):   1.987 cal/mol°K
                        8.205×10^{-2} liter-atm/mol°K
Speed of light (in vacuum): 2.998×10^{10} cm/s
Mass of one ^{12}C atom: 1.9926×10^{-23} g
Electron rest mass: 9.1096×10^{-28} g
Proton rest mass: 1.6726×10^{-24} g
Planck constant: 6.6262×10^{-27} erg s
Boltzmann constant: 1.3806×10^{-16} erg/°K
              8.611×10^{-5} eV/°K
Acceleration of gravity: 9.781 m/s^2
Gas volume (STP):   22.4 liter/mol

Miscellaneous Conversion Factors

1 meter (m) = 39.37 inches (in.)

1 decimeter (dm) = 0.1 m

1 centimeter (cm) = 0.01 m

1 millimeter (mm) = 0.001 m

1 nanometer (nm) = 10^{-9} m = 10^{-7} cm

1 kilometer (km) = 1000 m

1 angstrom (Å) = 10^{-8} cm = 100 pm

1 micron (μm) = 10^{-6} m = 10^{-4} cm

1 yard = 0.9144 m

1 inch = 2.54 cm

1 mile = 1.609 km

1 liter (L) = 1000 cm^3 = 1 dm^3

1 milliliter (mL) = 0.001 L = 1 cm^3

1 liquid quart = 0.9463 L

1 cubic foot = 28.316 L

1 gram (g) = 0.001 kilogram (kg)

1 milligram (mg) = 0.001 g

1 microgram (μg) = 1×10^{-6} g

1 tonne (metric) = 1000 kg

1 ounce = 28.35 g

1 pound = 453.6 g

1 ton (short) = 2000 lb = 907.18 kg

1 ton (long) = 2240 lb = 1.016 tonnes

1 calorie (cal) = 4.1840×10^7 ergs
              = 4.1840 joules (J)

1 Btu = 1.055×10^3 J

1 kilocalorie (kcal) = 1000 cal

1 electron volt (eV) = 1.6021×10^{-12} erg

1 eV/atm = 23.061 kcal/mol

1 liter-atm = 24.217 cal

1 erg = 1 dyne cm = 10^{-7} J
          = 6.242×10^{11} eV

1 ampere (amp) = 1 coulomb/s

1 foot-pound = 1.355 J

1 g/cm^3 = 62.4 lb/ft^3

1 newton (N) = 0.2247 lbf = 10^5 dynes
       = 1 kg m/s^2

1 atmosphere (atm) = 1.013×10^5 N/m^2
             = 101.3 kPa
             = 760 torr (mm Hg)
             = 1.013×10^6 dyne/cm^2
             = 14.7 lb/in.^{2} (psi)

1 nutritional calorie = 1 kcal

1 kilowatt hour (kWh) = 3.600×10^6 J

1 Pascal (Pa) = 1.45×10^{-4} lb/in.^{2} (psi) = 1 N/m^2

1 watt (W) = 1 J/s

1 horsepower (hp) = 745.7 W

1 psi = 6.90×10^3 N/m^2 = 6.90 kPa
    = 6.90×10^4 dynes/cm^2

1 dyne/cm^2 = 1.4504×10^{-5} psi = 0.1 Pa

1 bar = 10^2 kPa

Chemical Compositions of Various Alloys

Carbon steels: See Table 6.2

Alloy steels: See Table 6.2

Stainless steels: See Table 6.3

Miscellaneous alloys: See Table 6.3

Cast irons and other ferrous metals:

   Gray cast iron, class 20: 3.6 C, 0.6 Mn, 2.5 Si, rem Fe

   Gray cast iron, class 40: 3.1 C, 0.6 Mn, 2.1 Si, rem Fe

   Gray cast iron, class 60: 2.9 C, 0.6 Mn, 2.0 Si, rem Fe

   Malleable cast iron, ferritic: 2.5 C, 0.55 Mn, 1.6 Si, rem Fe

   Malleable cast iron, pearlitic: 2.4 C, 0.4 Mn, 1.2 Si, rem Fe

   Nodular cast iron, 80-55-06: 3 C, 0.4 Mn, 2.55 Si, 0.5 Ni, rem Fe

   White cast iron: 3.2 C, 0.6 Mn, 0.9 Si, rem Fe

   Cast low-alloy steel: Fe-C, small alloy content

   Cast carbon steel: (Fe-C)

   Cast stainless steel: 0.08 C, 1.5 Mn, 2 Si, 19 Cr, 9 Ni

Index

About the Authors

Giles F. Carter received his Ph.D. in physical chemistry at the University of California at Berkeley. From 1952 to 1967 he was research chemist and staff scientist at E.I. du Pont de Nemours & Company. Since 1967, he has been affiliated with Eastern Michigan University. He received the University's Distinguished Faculty Award for Research in 1983, and recently retired from his position as professor of chemistry, which he held for nearly 24 years.

Dr. Carter has authored more than 50 publications and 17 patents. He is a member of ASM International, the American Chemical Society, the American Numismatic Society, the Royal Numismatic Society, and the Society for Archaeological Science. His research interests are chemical compositions of copper-based Roman coins and diffusion alloy coatings formed in liquid-metal systems.

Donald E. Paul holds a Ph.D. in physical chemistry from Washington University, St. Louis. Between 1955 and 1972, he held research and development positions with E.I. du Pont de Nemours & Company and Cabot Corporation (including group leader Central Research and project manager Corporate Development Department), and was senior consultant at Arthur D. Little, Inc. He is currently President of D.E. Paul Associates, where for the past 18 years he has served industrial clients in product development in energy, specialty chemicals, plastics, and high-performance materials.

Dr. Paul is a member of the American Chemical Society, the Materials Research Society, Sigma Xi, the Chemical Market Research Association, the Boston Computer Society, and the New York Academy of Sciences. He holds patents for flame processes for the manufacture of titanium dioxide and other pigments.

Atomic Weights of the Elements

Element	Symbol	Atomic weight	Element	Symbol	Atomic weight	Element	Symbol	Atomic weight
Actinium	Ac	(227)	Gold	Au	196.9665	Praseodymium	Pr	140.9077
Aluminum	Al	26.9815	Hafnium	Hf	178.49	Promethium	Pm	(147)
Americium	Am	(243)	Helium	He	4.00260	Protactinium	Pa	231.0359
Antimony	Sb	121.75	Holmium	Ho	164.9304	Radium	Ra	226.0254
Argon	Ar	39.948	Hydrogen	H	1.0079	Radon	Rn	(222)
Arsenic	As	74.9216	Indium	In	114.82	Rhenium	Re	186.207
Astatine	At	(210)	Iodine	I	126.9045	Rhodium	Rh	102.9055
Barium	Ba	137.33	Iridium	Ir	192.22	Rubidium	Rb	85.4678
Berkelium	Bk	(247)	Iron	Fe	55.847	Ruthenium	Ru	101.07
Beryllium	Be	9.01218	Krypton	Kr	83.80	Samarium	Sm	150.4
Bismuth	Bi	208.9804	Lanthanum	La	138.9055	Scandium	Sc	44.9559
Boron	B	10.81	Lawrencium	Lr	(256)	Selenium	Se	78.96
Bromine	Br	79.904	Lead	Pb	207.2	Silicon	Si	28.0855
Cadmium	Cd	112.41	Lithium	Li	6.941	Silver	Ag	107.868
Calcium	Ca	40.08	Lutetium	Lu	174.97	Sodium	Na	22.9898
Californium	Cf	(251)	Magnesium	Mg	24.305	Strontium	Sr	87.62
Carbon	C	12.011	Manganese	Mn	54.9380	Sulfur	S	32.06
Cerium	Ce	140.12	Mendelevium	Md	(258)	Tantalum	Ta	180.9479
Cesium	Cs	132.9054	Mercury	Hg	200.59	Technetium	Tc	98.9062
Chlorine	Cl	35.453	Molybdenum	Mo	95.94	Tellurium	Te	127.60
Chromium	Cr	51.996	Neodymium	Nd	144.24	Terbium	Tb	158.9254
Cobalt	Co	58.9332	Neon	Ne	20.179	Thallium	Tl	204.37
Copper	Cu	63.546	Neptunium	Np	237.0482	Thorium	Th	232.0381
Curium	Cm	(247)	Nickel	Ni	58.70	Thulium	Tm	168.9342
Dysprosium	Dy	162.50	Niobium	Nb	92.9064	Tin	Sn	118.69
Einsteinium	Es	(254)	Nitrogen	N	14.0067	Titanium	Ti	47.90
Erbium	Er	167.26	Nobelium	No	(255)	Tungsten	W	183.85
Europium	Eu	151.96	Osmium	Os	190.2	Uranium	U	238.029
Fermium	Fm	(257)	Oxygen	O	15.9994	Vanadium	V	50.9414
Fluorine	F	18.998403	Palladium	Pd	106.4	Xenon	Xe	131.30
Francium	Fr	(223)	Phosphorus	P	30.9738	Ytterbium	Yb	173.04
Gadolinium	Gd	157.25	Platinum	Pt	195.09	Yttrium	Y	88.9059
Gallium	Ga	69.72	Plutonium	Pu	(244)	Zinc	Zn	65.38
Germanium	Ge	72.59	Polonium	Po	(210)	Zirconium	Zr	91.22
			Potassium	K	39.0983			